GREEN BACKLASH

Global Subversion of the Environmental Movement

ANDREW ROWELL

LONDON AND NEW YORK

First published 1996
by Routledge
11 New Fetter Lane, London EC4P 4EE
Simultaneously published in the USA and Canada
by Routledge
29 West 35th Street, New York, NY 10001

Typeset in Perpetua by Keystroke
Printed and bound in Great Britain by Biddles Ltd, Guildford and King's Lynn

British Library Cataloguing in Publication Data
A catalogue record for this book is available from the British Library

Library of Congress Cataloguing in Publication Data
Rowell, Andrew.
Green backlash: global subversion of the environmental movement/Andrew Rowell.
p. cm.
Includes bibliographical references and index.
(cloth: alk. paper).
(paper: alk. paper)
1. Anti-environmentalism. I. Title.
GE195.R68 1996
363.7-dc20
95-53941
CIP
ISBN 0-415-12827-7 (hbk)
0-415-12828-5 (pbk)

GREEN BACKLASH

The tide is turning against the environmental movement worldwide. Environmental activists are increasingly being scapegoated by the triple engines of the political Right, corporations and the state. The backlash has one simple aim: to nullify environmentalists and environmentalism.

Green Backlash is the first controversial exposé of global anti-environmentalism movement. Drawing on interviews with leading environmental activists and researchers around the world, Rowell reveals the extreme violence and brutality suffered by environmental campaigners including the activists working against oil operations in Nigeria and those trying to save forests in Brazil. In Asia, environmental activists have suffered severely for daring to challenge the *status quo*. In Canada and Australia most of the activity has been in the forestry debate, whilst in Europe marine resource issues are the key rallying point. The anti-roads movement in the UK has also been subject to intimidation and violence.

Organisations such as Greenpeace now suffer a severe backlash for successful campaigns as the media reiterate the views of the people in power. Industry dollars are funding politicians, think-tanks and forming front organisations to con the public whilst companies are spending billions on perpetuating a green and caring image and public relations companies are spending millions on anti-green PR.

Green Backlash offers an important new understanding of the challenges and threats now facing the environmental movement worldwide. The backlash is set to worsen as the resource wars of the coming decades intensify, as more people fight over less. Conflicts over water, wood, whales, metals, minerals, energy, cars and even consumerism will all happen and all inevitably create backlash. Stressing the urgent need for the environmental movement to re-evaluate and change itself, Rowell points to opportunities to halt the backlash before it is too late.

Andrew Rowell is a freelance environmental consultant who has undertaken pioneering research and written extensively on contemporary environmental issues.

To Ken Saro-Wiwa

CONTENTS

LIST OF PLATES

PREFACE

In November 1995 the world watched in horror as an illegal military dictatorship in Nigeria, insulated from world opinion by oil money, executed the environmental campaigner, Ken Saro-Wiwa, along with eight other Ogoni. The events leading up to their execution are told graphically in this book.

I first met Ken Saro-Wiwa back in 1992, whilst he was trying to whip up international support for his people, the Ogoni, whom no-one had ever heard of. Ogoni would still be an unknown place in a faraway land if it had not been for the raw courage and sheer persistence of Saro-Wiwa. The struggle cost him his life. Ironically, it was his death that forced the world to listen to the tragedy of the Niger Delta, but memories fade quickly as the media moves on.

Many people have argued that Saro-Wiwa was no environmentalist, that he was a human rights activist and his was a political struggle – ecology was the weapon he used to gain international attention. But there can be no greater human right than access to clean water, clean land and clean air. The oil industry has stripped the inhabitants of the Niger Delta of this basic fundamental right; they have devastated the Delta, dehumanised an impoverished people. The Ogoni fight is an ecological one, but it is more: it involves the issues of social justice, equity and self-determination, as do many ecological conflicts around the globe.

There is no doubt that globally environmentalists cover an enormously broad spectrum, and some people would not immediately be thought of as 'environmentalists'. In the book when I talk about the environmental movement or environmentalists, I mean someone who is fighting for ecological protection. This may be a professional environmentalist in America or an Amazonian inhabitant attempting to conserve his eco-system. Chico Mendes, for example, was a person who died fighting for the preservation of the Amazon, a cause which he adopted late, after years of fighting for the rights of rubber tappers.

In the USA, differences can be made. When I talk about the mainstream environmental movement, I generally refer to what is commonly known as the Group of Ten. These mainstream groups mainly focus their agenda on changing legislation in Washington, although they may have a supporter base throughout the country. Although the membership of this group has changed, it is widely perceived to be the Defenders of Wildlife, Environmental Defense Fund, National Audubon Society, National Wildlife Federation, Natural Resources Defense Council, Friends of the Earth, Izaak Walton League, Sierra Club, Wilderness Society and the World Wide Fund for Nature. Greenpeace has always been on the fringe of what is considered mainstream environmentalism, and has traditionally worked with grassroots groups such as the Citizens' Clearinghouse for Hazardous Waste. There is increasing friction between the mainstream groups and their grassroots counterparts, as is briefly outlined in chapter 1.

To define an anti-environmentalist is a difficult and dangerous task, because often they either see themselves, or at least portray themselves, as the true environmentalists. For this purpose, I shall use the simple analysis that they are actively working *against* someone who is working *for* ecological protection. In many cases, such as in the USA, anti-environmentalists are working to repeal legislation that enhances ecological protection. In many cases this includes industry. I first met Saro-Wiwa whilst researching and writing about the oil industry. Like any major industry, it pursues a deregulatory agenda, and one which hopes to pre-empt international accountability of its operations. It was also when I came face to face with the full force of industry public relations. The sheer capacity to misrepresent the truth startled me. The astounding difference between the companies' rhetoric and the reality on the ground was as deep as some of the wells that they drill. Many of these companies, whilst funding environmental groups or ecological projects, were also spending twice as much to undermine them.

When I was approached by Greenpeace to research the anti-environmental movement, it was the bitter taste that the oil industry had left that made me accept the offer. The critics will argue that Greenpeace has a vested interest in exposing the forces that are attempting to undermine it, and that 'they would say that anyway'. Others will dismiss the book as a cynical fund-raising effort. I reiterate that Greenpeace had no editorial control over the book, and as readers will hopefully see, there are some extremely serious reasons for people to be worried. The Green Backlash is a disturbing story to tell. Activists working with Greenpeace have been killed, one of Greenpeace's campaigners has had her house burnt down, countless others have been intimidated and become victims of violence, as have hundreds of other ecological activists around the world.

If you had seen a colleague murdered, you yourself had been beaten up, or someone had burnt your house down, you would want someone to investigate it. The book is the result of my investigations and, if nothing else, I hope it will help highlight and stop the violence.

Sadly, though, it is not just environmentalists who have been scapegoated, demonised and attacked. Every movement for social change has created its own backlash, something that always goes under-reported. This book hopes to ensure that this particular backlash is at least documented to some degree, so that lessons can be learnt for other activists who dare to challenge the *status quo*.

As anyone who has ever written a book of this nature will know, it can really test your endurance for a period that has stretched over two and a half years. Just a couple of months into the research, Ken Saro-Wiwa was framed and imprisoned. I would regularly get reports of his torture and appalling conditions of incarceration. During the long hours sitting in front of my computer, of all the many atrocities and tragedies outlined in this book, the one thought that outraged me more than anything else was Ken's predicament and the apathy of people in power to intervene and help him. Ken's death only intensified that anger.

For Ken, the media sound-bites came too late, but one piece of poetry still haunts me. The following poem was written by Ben Okri, before Ken was sentenced to death:

FOR KEN SARO-WIWA

That he should be jailed
And tortured
For loving the land
The earth
And crying out at its
Defilement
Is monstrously unfitting
We live in unnatural times;
And we must make it
Natural again
With our wailing.

For unnatural times also become
Natural by tradition
And by silence
That is why the lands
Today ring with injustice
With lies
With prejudice
Made natural.

The earth deserves our love.
Only the unnaturals
Can live so at ease
While they poison the lands
Rape her for profit
Bleed her for oil
And not even attempt
To heal the wounds.

Only the unnaturals
Jail those whose loves are bold
Who are weepers of the earth's agony
Handmaidens of her quiet vengeance
Sybils of her future rage.

Only the unnaturals
Rule our lands today
So deaf to the wailing
Of our skies, of the hungry,
Of the strange new disease
And of that dying earth

Bleeding and wounded
And breeding only deserts
Where once there were
The proud trees of Africa
Cleaning their rich hair
In the bright winds of heaven.

That he should be jailed
For loving the land
And tortured
For protecting his people
And crying out
As the ancient town-criers did
At the earth's defilement
Is monstrously unfitting.
And we live in unnatural times.
And we must make it
Natural again
With our singing
And our intelligent rage.

ACKNOWLEDGEMENTS

For any project of this size, there will be countless people to thank for their assistance, advice and support. Regrettably some of the activists who have been interviewed for the book have had to remain anonymous because of the severe repercussions of speaking out. Others who have spoken publicly may have increased the threat to their personal safety by doing so. It goes without saying that I am extremely indebted to their courage.

With a project that stretches over two years, as this has done, it is difficult to remember everyone who has assisted in some way to help the book to fruition. If you feel you should have been mentioned below and fancy being put in for the reprint, I'm afraid the anti-green PR industry may see to it that you don't get the chance.

It should be added, in a sentence that will be repeated often by the critics of this book, that financial and logistical support for this project was provided by Greenpeace. This said, Greenpeace had *no* editorial control over the content. Without the support of Greenpeace and the book's publishers, Routledge, it would never have been written and many activists would continue to suffer in silence.

Certain people who still work for, or used to work for Greenpeace have all assisted in some way of significance: Patrick Anderson, Martin Baker, Cindy Baxter, Aled Davies, Blair Palese, Duncan Currie, Arni Finnsson, Jonathan Hall, Eilis Gallagher, Andrea Goodall, Desley Matther, Goeren Olenburg, Michael Nielsen, Dephra Rephan, Ali Ross and Richard Titchen. Jim Sweet's tea-making capability is recommended. Josselien Janssens should be singled out for her long-suffering assistance and I am extremely indebted to her.

Many other people have contributed too. In the USA, some of the key researchers tracking the anti-environmental movement and political Right had done much of the ground work. They offered answers to the many ignorant questions that were asked as a foreigner attempted to grapple with the internal politics and make-up of modern America. A special thank-you has to be given to Sheila O'Donnell, a Californian licensed private investigator who has studied many of the cases of violence against environmentalists. Sheila worked on the book, interviewing key activists. She often had her morning cup of coffee shattered by intrusive phone calls from England as I sounded out my latest line of thinking with her. I owe Sheila an enormous debt. British Telecom probably owe us for record profits.

Dan Barry from CLEAR, the Clearinghouse on Environmental Advocacy and Research in Washington; Chip Berlet from the Political Research Associates in Boston and Tarso Ramos from the Western States Center in Oregon also offered invaluable advice, information and assistance. Other people who gave time generously were the leading political analysts: Sara Diamond, Dan Junas, Russ Bellant and Paul de Armond. Dave Helvarg allowed a sneak preview of his key book *The War Against the Greens* which remains the definitive study on the Wise Use movement and anti-environmental violence in the USA.

John Stauber, editor of *PR Watch*, and co-author of *Toxic Sludge is Good for You*, also provided invaluable insights into the PR industry. The staff at Global Response also contributed key information on atrocities committed outside of the USA. Two environmentalists should also be singled out for assisting, despite personal tragedies that are documented in the book. Judi Bari provided key information on the bombing of her vehicle. Pat Costner still investigates the industry many suspect of burning her house down.

Moving up to Canada, journalists Joyce Nelson, Kim Golberg, and Colleen McCrory and the staff at the Valhalla Society all offered valuable information, as did Mike Mason from the University of North London, who had studied the Share Movement. Bob Lyons and the Greenpeace staff in Vancouver also provided more mega-bytes of information than a normal mind or computer can deal with. Tamara Stark from Greenpeace Canada added her experience too. For information on Latin and Central America, I thank Sue Brandford from the BBC World Service, Madeleine Adriance of the Rio Maria Committee, Judy Kimerling, and Helena Paul of the Gaia Foundation.

The research for the chapter on Australia and New Zealand (apart from the bombing of the *Rainbow Warrior*) was provided by Bob Burton of the Wilderness Society. I owe Bob a special thanks for his work and his unrelenting energy in following up all my endless questions, when I know he had far more pressing things to attend to. It is Bob's turn now to write a more detailed analysis of the Green Backlash down under.

Meanwhile, for moral support in the dark times covering the tragic murder of Ken Saro-Wiwa in Nigeria, I thank specifically Shelley Braithwaite, Glen Ellis, Kay Bishop, Steve Kretzmann and Cindy Baxter. Several Ogoni who will have to remain anonymous, Professor Claude Ake, Sister Majella McCarron, Nick Ashton-Jones, and Oronto Douglas all have to be thanked for invaluable information too. Vandana Shiva, Martin Miriori from the Bougainville Resistance Movement, Alex Wilkes of the *Ecologist*, Simon Counsell at Friends of the Earth and other activists who also have to remain anonymous helped with the chapter on Asia and the Pacific.

Mike Swartz from Bindman and Partners, Rebecca Lush and the activists at Road Alert!, Jason Torrance and Paul Morotzo, Thomas Harding and Small World Productions, Barbara Dinham and the staff at the Pesticides Trust, all helped with the chapter on Britain. Other people in Europe or England who helped are Susan George, Claudia Peter, the staff at Article 19, and Mark Campanale of National Provident Institution (NPI).

Helena Paul, Shelley Braithwaite, Ruth McCoy, Bob Burton and Colleen McCrory all read drafts of relevant chapters. Thanks to Nigel Dudley and Sheila O'Donnell for reading the monster draft manuscript. Nigel should also be thanked for his general assistance and guidance. Thanks also for help with photographs from Daphne Christelis from the Environmental Picture Library and Liz Sommerville at Greenpeace Communications, Jay Townsend at the Greenpeace USA Photo Desk, Dan Barry at CLEAR, and Paul de Armond. Other people to thank for the photographs themselves are: Dylan Garcia, Julia Guest, Rod Harbinson, Alois Indrich, Steve Johnson, Ulrich Jurgens, Tim Lamson, Jeff Libman, Michael McKinnon, *The Mercury*, Hobart, Miller, Steve Morgan, Sheila O'Donnell, Ros Reeve, Andrew Testa and the *Walawala Chieftan*. Photographic inspiration came from Johnie Novis. I also thank Ben Okri for allowing me to reproduce his poem about Ken Saro-Wiwa.

Lastly thanks once again to Joss and Sheila. Couldn't have done it without you.

LIST OF ABBREVIATIONS

ACSH	American Council on Science and Health
ACTU	Australian Council of Trade Unions
AECO	Costa Rican Ecological Association
AEWC	Alaska Eskimo Whaling Commission
AFA	Alliance for America
AFC	American Freedom Coalition
AIAM	Association of International Automobile Manufacturers
ALC	Association of Liquid Paper Carton Manufacturers
ANWR	Arctic National Wildlife Refuge
AONB	Area of Outstanding Natural Beauty
APPM	Associated Pulp and Paper Mills
ARE	Asian Rare Earth
ASI	Adam Smith Institute
B-M	Burson-Marsteller
BCSD	Business Council for Sustainable Development
BCSEF	Business Council for a Sustainable Energy Future
BCTV	British Columbian Television
BGH	bovine growth hormone
BLM	Bureau of Land Management
BNFL	British Nuclear Fuels Ltd
CAP	Consumer Association of Penang
CAUSA	Confederation of Associations for the Unification of the Societies of America
CCRKBA	Citizens Committee for the Right to Keep and Bear Arms
CDFE	Center for the Defense of Free Enterprise
CEC	Citizens Electoral Councils

CEI	Competitive Enterprise Institute
CIR	Center for Investigative Reporting
CIS	Centre for Independent Studies
CISPES	Committee in Solidarity with the People of El Salvador
CITES	Convention on International Trade in Endangered Species
CJA	Criminal Justice and Public Order Act 1994
CLEAR	Clearinghouse on Environmental Advocacy and Research
CNP	Council for National Policy
COFI	Council of Forest Industries
COINTELPRO	counter-intelligence programme
CONAP	National Commission of Protected Areas
CORE	Conserve Our Residential Environment (Australia)
CORE	Commission on Land Use and Environment (Canada)
CPPA	Canadian Pulp and Paper Association
CSE	Centre for Science and Environment
CVC	Coalition for Vehicle Choice
CWIT	Canadian Women in Timber
DGSE	Direction Générale des Services Externes
DoT	Department of Transport
ECDC	European Chlorine Derivatives Council
ECO	Environmental Conservation Organisation
ECSA	European Chlorinated Solvent Association
EDA	Earth Day Alternatives
EDF	Environmental Defense Fund
EEEG	extremist environmentalist elitist groups
EF!	Earth First!
EIA	Environmental Impact Assessment
EIR	*Executive Intelligence Review*
ELP	European Labour Party
EPA	Environmental Protection Agency
EPSM	Environmental Protection Society of Malaysia
EPZ	export processing zone
ESA	Endangered Species Act
FAFPIC	Forests and Forest Products Industry Council
FAIR	Fairness and Accuracy in Reporting
FAO	Food and Agricultural Organisation
FCF	Free Congress Foundation
FDA	Fertiliser and Pesticide Agency

FIAT	Forest Industries Association of Tasmania
FICA	Forestry Industry Campaign Association
FPS	Forest Protection Society
FTZ	free trade zone
GATT	Global Agreement on Tariffs and Trade
GBN	Green Business Network
GCC	Global Climate Coalition
GGT	Global Guardian Trust
GMO	genetically manipulated organisms
GOP	Grand Old Party
H and K	Hill and Knowlton
HNA	High North Alliance
HOPE	Help Our Polluted Environment
IBAMA	Brazilian Institute for the Environment and for Renewable Natural Resources
ICE	Information Council on the Environment
ICLC	International Caucus of Labor Committees
ICPP	International Climate Change Partnerships
IEA	Institute of Economic Affairs
ILMA	Interior Lumber Manufacturers Association
IPA	Institute of Public Affairs
IPCC	Intergovernmental Panel on Climate Change
ISA	Internal Security Act
ISO	International Standards Organisation
ITTO	International Tropical Timber Organisation
IWA	International Woodworkers of America
IWC	International Whaling Commission
IWMC	International Wildlife Management Consortium
JBS	John Birch Society
L-P	Louisiana-Pacific
LWD	Liquid Waste Disposal
MBD	Mongoven, Biscoe and Duchin
MICC	Moresby Island Concerned Citizens
MOM	Militia of Montana
MOP	Mothers Against Pollution
MOSIEND	Movement for the Survival of the Izon (Ijaw) Ethnic Nationality in the Niger Delta
MOSOP	Movement for the Survival of Ogoni People

MPF	Mobile Police Force
MSLF	Mountain States Legal Foundation
MTIDC	Malaysian Timber Industry Development Council
MVMA	Motor Vehicle Manufacturers of America
NAFI	National Association of Forest Industries
NAMMCO	North Atlantic Marine Mammal Commission
NASA	National Aeronautics and Space Administration
NBA	Narmada Bachao Andolan
NCLC	National Caucus of Labor Committees
NDES	Nigerian Delta Environment Survey
NEFA	North East Forest Alliance
NFLC	National Federal Lands Conference
NHTSA	National Highway Traffic Safety Administration
NIREX	Nuclear Industry Radioactive Waste Executive
NNPC	Nigerian National Petroleum Corporation
NPCIS	National Poisons Control and Information Service
NPI	National Provident Institution
NRA	National Rifle Association
NRDC	National Resources Defense Council
NVDA	non-violent direct action
NWC	National Wetlands Coalition
NYCOP	National Youth Council of Ogoni People
NZFIA	New Zealand Fishing Industry Association
O and M	Ogilvy and Mather
OED	Operations Evaluation Department
OLC	Oregon Lands Coalition
OPEC	Organisation of Petroleum Exporting Countries
Opic	Overseas Private Investment Corporation
PACs	Political Action Committees
PAN	Pesticides Action Network
PAN–AP	Pesticide Action Network Asia and the Pacific
PARC	Perak Anti-Radioactive Committee
PERC	Political Economy Research Center
PFW!	People for the West!
PILF	Public Interest Law Firm
PLF	Pacific Legal Foundation
PNG	Papua New Guinea
PPF	Putting People First

PR	public relations
QCC	Queensland Conservation Council
RECAP	Romulous Environmentalists Care About People
SACTRA	Standing Advisory Committee on Trunk Road Assessment
SAF	Second Amendment Foundation
SAM	Sahabat Alam Malaysia
SEAC	Student Environment Action Committee
SEPP	Science and Environmental Policy Project
SLAPP	Strategic Lawsuits Against Public Participation
SNOCO PRA	Snohomish County Property Rights Alliance
SPDC	Shell Petroleum Development Company
SSD	Sardar Sarovar Dam
SSSI	Site of Special Scientific Interest
STOP	Supporters to Oppose Pollution
TFM	Tree Farm Licence
TFP	Society for Defense of Tradition, Family and Property
TIC	Transnational Information Centre
TNC	transnational corporation
TTRLUF	Tasmanian Traditional Recreational Land Users Federation
TWS	The Wilderness Society
UARS	Upper Atmosphere Research Satellite
UDR	União Democrática Ruralista
UN	United Nations
UNCTC	United Nations Centre for Transnational Corporations
UNEP	United Nations Environment Programme
WACL	World Anti-Communist League
WBCSD	World Business Council for Sustainable Development
WCWC	Western Canadian Wilderness Committee
WHO	World Health Organization
WICE	World Industry Council for the Environment
WIRF	Wilderness Impact Research Foundation
ZOG	Zionist Occupation Government

INTRODUCTION

On 10 November 1995, Ken Saro-Wiwa, the leader of the Movement for the Survival of Ogoni People (MOSOP), and eight other Ogoni were hanged by an illegal military junta in Nigeria after a trial condemned by the British government as 'judicial murder'. Saro-Wiwa's real crime was to highlight to the world the ecological devastation caused by the multinational oil company Shell and to campaign for a greater share of the oil wealth that had been drilled from under Ogoniland. The Ogoni paid the ultimate price for fighting for environmental and social justice. Over 1,800 have been killed, 30,000 made homeless, and countless others imprisoned, raped, tortured and beaten up.

The Ogoni join a growing list of activists around the world who have been killed, tortured and harassed for speaking out. All over the globe, environmental activists are currently facing a growing backlash, which is designed to intimidate them into inactivity and silence. It has one simple aim: to nullify environmentalists and environmentalism.

'I think one has to know that if you are being effective, there will be backlash,' says Vandana Shiva, a leading environmental activist and government advisor, from the Research Centre for Science and Ecology, in India. 'In fact, that backlash is occurring is a tribute to the environmental movement, because it shows that the environmental movement is making a difference. If someone does not make a difference, there is no backlash.'[1]

In India, for example, Shiva's fellow environmental activists have been beaten up, vilified and shot for campaigning against the building of dams and the relocation of multinational corporations on their home soil. Elsewhere in Asia and the Pacific, a region not normally known for its tolerance of dissent, activists have suffered severely for daring to challenge the *status quo*. Thousands of miles away in Latin America, environmentalists opposing oil companies exploiting the Amazon in Ecuador have also been threatened in their work. The violence against people trying to save the forests in Brazil has also been brutally bloody.

For many people living in North America and Europe, the green backlash is not just some phenomenon happening in a faraway land. It is happening at home, too. In America, anti-green sentiment has actually formed itself into a movement of people who are out to destroy environmentalism itself. Here the anti-environmental movement has also helped elect Republican politicians, who have an openly anti-green, pro-industry and free enterprise agenda. The success of the environmental movement has contributed to its growth, but so have its failures. Mainstream environmentalism has been blamed for neglecting social concerns and focusing too much on lobbying in Washington, rather than grassroots activism.

The anti-green movement in America has links to the political Right, but also the far Right and the militia, the fastest growing movement of anti-government feeling ever seen in the USA. Commentators on the Right are actively making scapegoats of the environmental movement and polarising the environment debate, away from solutions into direct confrontation. The logical extension of such vitriolic rhetoric is violence, which is on the increase.

Corporations are also funding anti-environmental groups and right-wing think-tanks that are promoting an assault on environmental legislation. This means that some of America's landmark environmental laws are likely to be gutted. Whilst funding this deregulatory push, industry is spending vast amounts of money on public relations in an attempt to demonise the environmental movement, on the one hand, and co-opt the environmental debate, on the other.

Just as the green backlash has occurred around contentious resource issues like forestry in the USA, so the forestry debates in Canada and Australia have spawned their own green backlashes. In Europe, meanwhile, many of the key anti-environmentalists are beginning to network with their counterparts in the USA on marine resource issues. Also in Europe, the fastest growing grassroots movement, the anti-roads movement in Britain has created its own backlash.

Because the green backlash is so advanced in the USA, that is where the story begins. The first half of the book draws mainly examples from the USA, although, where relevant, other examples are given. In the second half of the book, the green backlash is documented in other countries and regions around the world. So the story starts in the USA, where the growth of the anti-environmental movement is detailed and how the anti-greens are linked to the Right is then documented in chapter 2. Chapter 3 examines how far corporations will go to achieve a deregulatory agenda whilst public relations companies and their key role in the green backlash are documented in chapter 4.

In chapter 5, it is argued that there is a concerted effort to force a paradigm shift on the environmental movement in order to try and marginalise it, so that it loses popular support. This is mainly being done through the use of language and by debunking environmental science. The logical extension of this paradigm shift, that of harassment and violence, is then revealed in the following chapter. The backlash that occurred in the forestry debate in Canada, especially the use of corporate front organisations, is elaborated on in chapter 7. Meanwhile, chapter 8 looks at the backlash in Central and Latin America, but concentrates on Ecuador and Brazil.

The story of the green backlash in Australia is examined in chapter 9. Links between key Australian anti-greens and their North American counterparts are also documented. The backlash in Asia and the Pacific focusing on India, the Philippines, Malaysia, Indonesia and the island of Bougainville is then documented in chapter 10. The horrendous backlash against the Ogoni in Nigeria is detailed in chapter 11.

Moving on to Europe, the backlash against the anti-roads movement in Britain is the main theme of chapter 12, although the controversy over Shell's decision to dump the Brent Spar is also explored. The last chapter looks at the backlash against Greenpeace and other organisations, centred around marine resource issues. Finally, in the Conclusion some ideas are put forward for environmentalists to beat the green backlash.

1

ROLL BACK THE MILLENNIUM

America leads the way

They have all the money, all the paid staff, what they don't have is science, truth, or the interests of humanity, or even the environment at heart. That's why they know we're going to whip their butts.[1]

Ron Arnold

There are, at the time of writing, four years left until the year 2000. Soon we will begin the preparations for the biggest party of all time, as we celebrate a new millennium. Those celebrations will not only reflect the past but look to the future and central to this debate will be how we want to live in relation to the earth and each other. Discussions will take place of what kind of society we want to live in for the next thousand years, or even if we believe that the planet can survive that long. America, one of the strongest economic nations, is going to lead the way in this global re-evaluation.

Part of the American millennium celebrations will be Earth Day 2000. This festival will be an environmental extravaganza, funded by corporations who want to show that they are the true stewards to steer us into the coming century. Will America's environmental movement be able to challenge these new corporate environmentalists? Will America still believe in environmentalism, or was it just a passing fad of the twentieth century? Has environmentalism succeeded or are there still challenges ahead?

There are some real reasons to celebrate. The achievements of the environmental movement, over the twenty-five years since the first Earth Day in 1970, should have been a cause for celebration. In the regulatory and legislative arena, Congress has passed twenty-eight major environmental laws, including the Clean Air and Clean Water Acts.[2] On the whole, the air, water and land are cleaner. America no longer dumps its sewage at sea. Tens of millions of acres have been added to the nation's protected wilderness areas. A few species have been hauled back from the abyss of extinction. Environmentalism has gone from a fringe idea to being embedded in the psyche of most Americans. The press have woken up to cover environmental issues – back in 1970 most sent their architecture reporters to cover Earth Day, as there was no-one else interested.

However, there is also real cause for consternation and concern. Numerous pressing problems remain. It appears that the very future of the environmental movement itself is threatened: the messenger may be slain before the message has been broadcast. The environmental movement is under attack both from outside and from within. A growing backlash movement, under the umbrella of the Wise Use movement, has mobilised thousands of people on an anti-environmental ticket. In some instances, they have spectacularly beaten environmentalists at their own game – mobilising the grassroots.

Meanwhile grassroots environmental activists complain about the lack of contact between them and the large environmental groups. They criticise the mainstream organisations for selling out to the authorities and to companies eager to add a green veneer to their polluting activities. The big environmental groups, activists allege, have been co-opted, compromised and corrupted. Furthermore, they have lost touch with the grassroots and forgotten how to work with activists who are increasingly facing a vitriolic campaign of scape-goating, intimidation and violence, all trade marks of the green backlash.

With the election of President Clinton, and 'Green' Al Gore, having spent years attempting to gain political influence in Washington, many mainstream environmental groups thought that their time of political affluence had come. 'After twelve years out in the cold, when this President snapped his fingers the big environmental groups wagged their tails like lost puppies happy to find a home, any home,' wrote Peter Montague in *Rachel's Environment & Health Weekly*.[3] After the honeymoon period ended, these groups realised that a marriage between those in power and those who traditionally oppose people in power, was never going to be easy. In fact, in this case, it was never going to work. 'What started out like a love affair turned out to be date rape,' said Jay Hair from the National Wildlife Federation.[4] Many of the most effective staff, co-opted into Clinton's administration, soon realised that they had become another piece of bureaucratic machinery.

Another painful realisation was that Clinton and Gore were not working for them. By the end of the 103rd Congress, the Clinton administration had a worse environmental record than either the Bush or Reagan administration.[5] A poll by the League of Conservation Voters reported that the 1994 session of Congress fared the worst in its twenty-five years of environmental ratings, having rejected virtually every major piece of environmental legislation brought before it.[6] After the November mid-term elections when the Republicans took control of both houses of Congress, the situation went from bad to worse.

The mainstream environmental groups found themselves slowly swinging in the cold breeze of political ineptitude. Meanwhile in the warm corridors of power, the Republicans, with the 'Contract With America' as their political carving knife, were about to gut many landmark environmental statutes that had spearheaded the new environmental consensus a generation before. The Clean Air Act, the Endangered Species Act, and Superfund were all up for reauthorisation before the Republican anti-environmental chopping block.

Like its successful legislation, environmentalism was in serious danger of being slain as the prodigal advocator of government regulation whose time had passed. In the new right-wing, pro-industry revolution gripping America, environmental regulation is now seen as government oppression not protection, and therefore environmentalists are seen as oppressors not guardians.

'Environmentalists are on the run', reported *Fortune Magazine*,[7] in a headline that could have been repeated across newsstands of the nation. One influential writer, Mark Dowie, even condemned the environmental movement 'as dangerously courting irrelevance'.[8] By the mid-1990s, the critics were already writing the green obituaries. The political Right and a new anti-environmental movement seemed to be growing in political and grassroots strength day by day. Environmental leaders faced a barrage of criticism for failing to respond to the Republican rout,[9] for failing to heed the warnings of the grassroots and underestimating the serious nature of the green backlash that is gripping America. If America votes in a Republican president this year, the environmental movement's influence and effectiveness could be in terminal decline.

'The only option is to go back to the people and talk to them, to try to find out why these issues don't seem to resonate,' said Interior Secretary Bruce Babbitt, who had been head of the League of Conservation Voters, before being co-opted by the Clinton administration. 'To try to find the embers and fan them into life once again.'[10] To understand what is happening in the USA, we have to go back some time to see the spark that kindled the anti-environmental movement.

SIGNS OF TROUBLE

One of the first signs of trouble to appear for the environmental movement stretches back over a quarter of a century to 1971. The event was a speech given to the United States Chamber of Commerce by Lewis Powell, who was to become the Supreme Court Justice in the late 1970s. Powell, a corporate lawyer at the time, warned business that public policy for that decade, including the environmental agenda, would be fought in the court-rooms of America.[11] Some landmark legislation on ecological issues had been passed in the previous decade such as the Wilderness Act of 1964, the Federal Wild and Scenic Rivers Act of 1968 and the National Environmental Policy Act of 1969. Policy-makers were considering further fundamental legal reform on environmental issues, that would see such laws as the Endangered Species Act of 1973 entering the American statute books.

Industry was ill prepared to act within this new legislative arena, Powell advised. The answer was simple – to learn from and mirror the opposition. Set up your own independent legal firms, he told industry, that are pro-business, but label them as being in the public interest, and call them 'public interest law firms' (PILFs).[12]

The year before Powell's speech, on 22 April 1970, hundreds of thousands of people had celebrated the first Earth Day, an event which had helped propel environmentalism into the consciousness of America. In 1969, the year before the Earth Day celebrations, over eighty members of the House of Representatives had proclaimed the 1970s as the decade of the environment. With Powell's speech, the seeds of the corporate counter-attack against the environmental movement had already been sown.

Within two years, the Pacific Legal Foundation (PLF) had been set up in Sacramento by the California Chamber of Commerce. 'For nearly twenty years, it has come to the defense of chemical manufacturers, oil producers, mining and timber companies, real estate developers, the nuclear power industry, and electric utilities, to name a few,' wrote Mark Megalli and Andy Friedman in *Masks of Deception: Corporate Front Groups in America* in 1991.[13] The PLF was quickly followed by other 'public interest law firms' and by 1975, a coordinating body, the National Legal Center for the Public Interest, had been set up in Washington.[14] 'What we cannot accept,' the National Legal Center's founding president, Leonard Theberge, said, 'are mindless proposals that would sacrifice the people of the US on the altar of nature.'[15] By talking in such religious undertones, Theberge sparked off a line of attack that would be used time and time again against the green movement – environmentalists were engaged in a holy war whereby nature was considered more important than humans.

Joe Coors sat on the first board of the National Legal Center.[16] The ultra-conservative Coors family, the beer barons, who have bank-rolled the growth of the New Right in America, would take pride in being named as champion crushers of the counter-culture. Russ Bellant, a journalist who has studied the Coors family's activities, believes it to be the 'most right wing corporation in the United States.'[17] Coors have consistently funded key activists in the fight against the environmental movement. Even the conservative *Readers Digest* magazine has labelled Joe Coors, 'one of the country's leading anti-environmentalists'.[18]

The National Legal Center, in turn, helped create seven regional PILFs, one of which has particular historical and current relevance to the continuing green backlash that is sweeping America – the Mountain States Legal Foundation (MSLF) in Denver. The MSLF, a non-profit organisation, is 'dedicated to the values and concepts of individual freedom, our right to private property, and the private enterprise system'.[19] The National Legal Foundation donated $50,000, and Coors $25,000 to set up MSLF. Other early funding came from big oil companies – Amoco, Chevron, Marathon, Phillips and Shell.[20] Jo Coors, who sat on MSLF's Board for three years, appointed a friend of his, James Watt, as the Foundation's first president. During Watt's tenure in office, the MSLF

> tackled such enemies of western business interests as the Environmental Protection Agency, the Sierra Club, the Environmental Defense Fund – and the Department of the Interior. In the process, the organisation picked up the reputation of being anti-consumer, anti-feminist, anti-government and above all, anti-environmentalist.[21]

WATT A STORM OF PROTEST

It is as a vehement anti-environmentalist that James Watt is a name that still sends shivers down the most ardent environmentalist's spine. During Reagan's early years as President, Watt's tenure as Secretary of State of the Interior resulted in over 100 million letters of complaint being mailed by environmentalists.[22] Described during his tenure at Interior as 'public enemy No. 1' by the

Sierra Club,[23] Watt advocated a policy of a 'free for all' for mining, ranching and drilling companies on public lands in the USA, wanting to open up vast protected areas for exploitation, as well as opening up the offshore continental shelf for oil exploration. Watt also wanted to reduce the number of national parks and roll back the restrictions preventing industrial usage of such protected areas. In a sense, Watt typified the Reagan years, which were openly hostile to the environmental community and where a right-wing agenda of unrestricted exploitation of natural resources was openly advocated.

'James Watt: The apostle of pillage' raged the New York *Village Voice* in disgust.[24] The Sierra Club declared war on 'Watt-ism' and the 'entire Reagan anti-environmental assault'. Senate Congressional Records from 1981 show what the Sierra Club meant by Watt-ism, 'Watt-ism views our public lands, forests and other resources not as a legacy for the future, but as a bank balance to be drawn down as quickly as possible in the name of immediate development and a fast buck.'[25]

Watt was also firing back at his critics, calling environmentalists 'extremists' and 'preservationists', who were 'a left-wing cult which seeks to bring down the type of government I believe in'.[26] Due to his hardline policies Watt was controversial. Part of this unpopularity was due to Watt's belief of dominion theology, whereby proponents use sections of Genesis in the Bible which state that God gave 'man' dominion over the earth.[27] If this is the case, they argue, then 'man' can basically do what he wants to the animals, plants and resources of the earth, with God's express permission. This effectively means that there are no limits to exploiting the earth.

'You can't really hurt the planet because God wouldn't allow that. God wouldn't have given man chain-saws if he didn't think they were benign,' says Chip Berlet, a leading expert on the political Right, explaining the thinking behind dominion theology. This belief is also held by many in the Christian Right, in business, and the anti-environmental movement in the USA, who also argue that because 'man' was given dominion by God, it does not matter if 'man' makes species extinct through industrial activity.

Watt was also one of arch advocates of the Sagebrush Rebellion, a campaign to transfer lands under federal control in the Western USA to the states or to private ownership. Attempts to gain state control over federal lands in the west had periodically occurred throughout the century, but without success. In 1979, however, the Nevada Legislature demanded 50 million acres of federal lands be given over to state control and other states followed suit.[28] Described as 'naive and disorganised', the Sagebrush Rebellion had fizzled away by the early 1980s.[29] This was due, in part, to the realisation that much of the land might be privatised and sold to the highest bidder, if it was given over to state control. This would not suit the average rancher who had originally backed the Rebellion.[30] Resentment over federal land ownership and regulatory control issues that had constituted part of the melting pot of the Rebellion would boil over later in the decade.[31]

Watt was becoming unpopular and in 1982 the Free Congress Foundation (FCF), commissioned an unknown writer, Ron Arnold, to write a biography on him. The FCF is headed by the New Right guru Paul Weyrich, a leading archconservative and ideological and business associate of Joe Coors. It was Weyrich's idea to produce a glowing biography of the controversial Secretary of the Interior.[32] Not only did the book exonerate Jim Watt to near sainthood, but Arnold, like Watt, launched a broadside attack against the environmental movement. 'The implicit goals of environmentalism to drastically reduce or dismantle industrial civilisation and to impose a fundamentally coercive form of government on America are real, even though they tend to be hidden in the complex structure of the movement,' wrote Arnold in his book, *Eye of the Storm*.[33]

Arnold, a self-professed former environmentalist and member of the Sierra Club, claimed that environmentalism stood for a new religion, which was anti-humanity, anti-civilisation, anti-technology as well as pro-alarmism and terrorism, in his 'coherent critic' of the environmental movement.[34] *Eye of the Storm* gave Arnold the political break he had been seeking. However, he later conceded that the book was more ideological than objective and written from a conservative viewpoint.[35] Gaylord Nelson, the Chairman of the Wilderness Society, was unimpressed. 'People like Watt and Mr Arnold really don't see that there is much value to anything unless you can put a dollar sign on it,' he said. 'They'll deny it and say they want "balance". But if you look at that "balance", it's all in favour of unrestrained exploitation of natural resources.'[36]

Arnold and Watt were not the only people talking in divine undertones. 'It's a holy war between fundamentally different religions,' Charles Cushman of the National Inholders Association was raging.[37] 'Environmentalism is a new paganism. It worships trees and sacrifices people' is a favourite Cushman saying.[38] Charles 'Chuck' Cushman, calls himself the 'tank commander' of the green backlash and has spent much of the last fifteen years warning rural audiences about the US National Parks service and the environmental movement, or preservationists, as he calls them. Cushman, who had originally formed the National Inholders Association to represent 'inholders' or people who live inside National Parks, received national notoriety when he was appointed to the National Park System Advisory Board by Jim Watt. Already a friend of Ron Arnold's and a highly controversial figure, Cushman enthusiastically defended Jim Watt, rallying support for the ailing Secretary of the Interior. 'The Secretary is doing a hell of a job,' said Cushman.[39]

Neither Arnold's book nor Cushman's support could prevent Watt's early demise from office. Watt, who had become renowned for public gaffes, described the make-up of a newly formed commission: 'We have every kind of mixture you can have. I have a black, I have a woman, two Jews and a cripple.' One mistake too many, Watt resigned as Secretary of the Interior in the autumn of 1983.[40]

However, much of the rhetoric that is used against the environmental movement now, labelling activists as 'extremists', 'preservationists', 'religious fanatics' and 'communists' began in earnest in the Reagan era with people such as Watt, Arnold and Cushman. In turn they seem to have drawn much of their terminology from political extremist and perennial presidential candidate, Lyndon LaRouche. The process of the marginalisation of the environmental movement had begun.

GUNNING OVER THE MOON

The same year as Watt resigned and headed back west, another archconservative was heading south, jetsetting off to an all-expenses paid conference in Jamaica. Alan Gottlieb, one of the foremost right-wing fundraisers in the USA, who describes himself as the 'premier anti-communist, free-enterprise, *laissez-faire* capitalist',[41] was attending a conference arranged by an organisation called CAUSA – the Confederation of Associations for the Unification of the Societies of America. This mouthful is a front group for the Reverend Sun Myung Moon. The Reverend Moon, whose popular image is one of the head of a religious cult, in fact heads a vast smokescreen multinational political and business empire whose tentacles span the globe. He owns 280 corporations in America alone. Moon's stated goal is a global automatic theocracy. 'I will conquer and subjugate the world,' says Moon, who sees himself as the son of God.[42]

CAUSA, the organisers of the conference in Jamaica, was set up in 1980, as Moon's transnational political front. 'The primary mission of CAUSA,' recalls leading Moon watcher, Dan Junas, 'was to support the Reagan foreign policy doctrine,'[43] which in turn 'sought to roll back the Soviet empire, and support such anti-communist "freedom fighters" as the Nicaraguan contras and UNITA in Angola.'[44] Besides attending the CAUSA conference, Gottlieb has had other dealings with Moon front organisations.

Gottlieb and Moon also share another thing in common: both are convicted tax felons. Gottlieb, like many on the political Right, considers his conviction 'a badge of honour' and views himself as an anti-tax hero.[45] Others consider Gottlieb a hero too. 'Those who love freedom are in his debt,' wrote William Perry Pendley, the current president of the Mountain States Legal Foundation, and another prominent anti-environmentalist.[46] Pendley had been a close personal friend of James Watt's since 1975 and had been appointed by Watt to the position of Deputy Assistant Secretary for Energy and Minerals in the Interior Department, whilst he was in office.[47] Pendley has since also become close friends with Arnold, Cushman and Gottlieb.[48]

Alan Gottlieb is the founder and Director of the Center for the Defense of Free Enterprise (CDFE), a right-wing think tank situated in the aptly named Liberty Park in the leafy suburbs of Seattle, on the west coast of the United States. He is also the President of the Citizens Committee for the Right to Keep and Bear Arms and the Second Amendment Foundation, two of the most powerful pro-gun advocate groups alongside the National Rifle Association. (His wife is editor of the magazine *Woman and Guns*.) Gottlieb is also a ferociously effective fund-raiser who has made tens of millions of dollars for various pro-gun and rightist causes and politicians, including Ronald Reagan. Furthermore, he owns a publishing house and a radio station that churns out right-wing rhetoric. A local newspaper depicted a picture of the bespectacled, balding Alan Gottlieb, and asked 'Is this the most dangerous man in America?'[49]

Just months after his trip to Jamaica, Gottlieb spent time in a minimum security jail for his tax conviction. Before being locked up, he met Ron Arnold, Watt's biographer, who was looking for a financial backer and a partner.[50] Arnold, who had spent the previous couple of years honing his anti-environmental message into a more coherent format, found in Gottlieb the perfect answer to his prayers. In the intervening years, Arnold had been hired by a variety of timber and chemical interests, including firms such as Weyerhaeuser, to produce training videos, and he had helped defeat the expansion of the Redwoods National Park.[51]

It was also in 1984 that Chuck Cushman warned a captivated audience that, after the local area had been designated 'wild and scenic' the National Park Service were 'going to come in and strangle you'. In the weeks after the talk, violence erupted in the community, with swastikas painted on the Park Service vehicles.[52]

Together, the right-wing triumvirate of the 'fund-raiser' Alan Gottlieb, the 'tank commander' Chuck Cushman and 'philosopher' Ron Arnold would become one of the most potent and profitable driving forces in the green backlash. Between them they set out to exploit the ripples of discontent against the environmental movement and turn them into waves of spite and wads of dollars.

A MOVEMENT AGAINST A MOVEMENT

'The most effective response to a movement is another movement,' concluded Luther Gerlach and Virginia Hine in their theoretical analysis of social movements, entitled *People, Power, Change* in 1970.[53] This little known but authoritative text, with case studies about the Pentecostal Movement and Black Power Movements, has been used as the basis to attempt to destabilise the green movement in North America. 'It takes a movement to fight a movement,' Ron Arnold said in a 1984 speech, a statement he has repeated time and time again over the last fifteen years.[54]

'American industry cannot save itself by itself in an activist society,' Arnold wrote in his aptly named book *Ecology Wars*, 'and an activist movement, can only be defeated by an activist movement.'[55] Industry, contended Arnold, cannot win purely with a public relations drive and therefore needs to initiate a pro-industry activist movement not only to win the public's hearts and minds but to fight the environmental movement. The end result is people fighting for industry, but with all the hallmarks of fighting for themselves. 'What we need is pro-industry, pro-free enterprise citizen action,' Arnold advocated.[56]

Arnold told the Ontario Forest Industries Association in 1988:

> The public is completely convinced that when you speak as an industry you are speaking out of nothing but self-interest. The public will never love big business. The pro-industry citizen activist group is the answer to these problems. It can speak as public-spirited people who support the communities and the families affected by the local issue. It can speak as a group of people who live close to nature and have more natural wisdom than city people . . . It can form coalitions to build real political clout. It can be an effective and convincing advocate for your industry. It can evoke powerful archetypes such as the sanctity of the family, the virtue of the close-knit community, the natural wisdom of the rural dweller

> . . . it can use the tactic of intelligent attack against environmentalists and take the battle to them . . . and it can turn the public against your enemies.[57]

And it is battle. 'This is a war zone,' says Arnold. 'Our goal is to destroy, to eradicate the environmental movement,' he told the *Toronto Star* in 1991. 'We're mad as hell. We're not going to take it any more. We're dead serious – we're going to destroy them.'[58] The most effective way of defeating the environmental movement is 'by taking their money and their members' with another movement.[59]

WISE USE

Ron Arnold, who has become the Executive Director of the Center for the Defense of Free Enterprise, along with Alan Gottlieb, are now two of the key puppet-masters of the anti-environmental movement in the USA: a growing coalition of ranchers, miners, loggers, farmers, fishermen, trappers, hunters, off-road vehicle users, property rights advocates, industry associations, corporate front groups and right-wing activists who are rising up against the environmental movement across the USA. They are advocates of multiple use of land, and call themselves the Wise Use movement.

In order to rally and unite the many disparate groups that are part of the backlash, Arnold wanted to find a unifying umbrella, specific enough for people to join, but vague enough to not really know what they were fighting for. He found it with Wise Use. Coined from a phrase by Gifford Pinchot, the first chief of the US forest service, who defined conservation as the wise and efficient use of resources,[60] 'Wise use' was the simple utopian term that suited Ron Arnold, with which to describe his fledgling creation. Serious thought had gone into his new definition. '"Wise Use" was catchy, and it took up only nine spaces in a newspaper headline, just about as short as "ecology",' he told *Outside Magazine*. 'It was also a marvellously ambiguous expression. Symbols register most powerfully in the subconscious when they're not perfectly clear.'[61] 'The Wise Use movement stands for man and nature living together in productive harmony,' he adds,[62] evoking further dreams of a paradoxical paradise that symbolises 'Wise Use'. 'Wise Use is just a generalised name that covers a multitude of sins and nonsense,' Arnold has also admitted.[63] 'We don't even care what version of Wise Use people believe in as long as it protects private property, free markets and limits government.'[64]

The Wise Use movement is where the green backlash has coalesced into a movement, manipulated by a few key right-wing activists, such as Arnold, Gottlieb and Cushman. 'We provide the Jello mould,' concedes Arnold.[65] Wise Use leaders say that environmentalists are anti-family, anti-Christian, anti-American, anti-people, anti-human, only interested in their own power and money and their ultimate agenda. This agenda, they contend, is to destroy your job and ultimately your life. The backlash aims to marginalise the environmental movement, to fight it to the death.

But would the concept of 'Wise Use' work, and would industry be interested in Arnold's concept? 'I think it is important for people to realise that Wise Use is not some sudden spontaneous thing,' says Paul de Armond, a researcher who lives near Arnold in Seattle, and who has spent considerable time studying Wise Use. He continues:

> Arnold started thinking and working on the concept of using citizens' groups to front for industry positions in the late-mid 1970s . . . brooded over it for years before he linked up with Gottlieb in 1984, spent another four years preparing for the launch of Wise Use in 1988, and has been working very hard at it ever since.[66]

By 1982, Arnold claims he had found nearly 800 groups that were defending industry in one form or other,[67] but it would take another six years before Arnold and Gottlieb could get people to attend a conference on Wise Use, which would officially herald the birth of the movement. So why did it take so long to get things off the ground?

RESENTMENT GROWS IN THE WEST

It was not until the late 1980s that a disparate collection of different economic, social and political forces amassed in such a way that was right for a movement such as Wise Use. Two political scientists, Ralph Maughan and Douglas Nilson, who have studied Wise Use, believe it was the unexpected continuing economic decline in the Western USA which resulted in a condition of deprivation ripe for the politics of resentment and another reactionary social movement. Furthermore, when the environmental movement began to make serious dents in the harvest of old growth timber in Oregon and Washington, the peoples of the West felt increasingly threatened. Their fears were exacerbated by the seemingly growing strength of the environmental movement. 'By the late

1980s,' conclude Maughan and Nilson, 'all the conditions were ready for the Wise Use Movement. Only leaders were needed to set it in motion.'[68]

Industry was not ready to join in the support for the movement until the late 1980s either. Tarso Ramos, from the Western States Center, in Portland, Oregon, who monitors anti-environmental activity in nine of the western states, believes the most important reason was that there was no need for an environmental backlash movement in the early and mid-1980s, not least because of Reagan's continuing occupation of the White House. Industry, particularly the natural resource industry, already had a friend in power.[69]

Following the election of President Bush, there was dissatisfaction brewing within industry, who saw the new President as much less sympathetic. At the same time, Ramos believes there were a number of impending crises for the resource industries which made them consider other strategies for achieving deregulation of their industries. This included the impending timber supply crisis in the North West and the forthcoming reform of certain industry-related laws, including the reform of the 1872 Mining Act, which provided incredibly favourable conditions for American and international mining companies.[70] Other laws were targeted for revision too: mainly the Endangered Species Act, the Clean Water Act, the Superfund Act. Industry wanted to find ways to weaken these regulations or to make them unworkable. Furthermore, there was a growing inclination within industry that, due to environmental successes, there needed to be a greater public relations effort made,[71] covert action, and greenwashing. Violence against environmentalists would also occur.

WISE USE IS OFFICIALLY 'BORN'

In August 1988, the first 'Multiple-Use Strategy Conference', sponsored by Gottlieb's Center for the Defense of Free Enterprise, took place in Reno, Nevada. This is the conference that is credited with heralding the birth of the Wise Use movement, although a similar conference had taken place three months previously, organised by the Wilderness Impact Research Foundation (WIRF). The baby from the conference was to be *The Wise Use Agenda*, which contained twenty-five goals of the movement, many which read like a wish-list for industry. There were just under 300 participants. Arnold wrote in *The Wise Use Agenda*:

Thus, for the first time, industry had a core of citizen supporters with a clear-eyed critical understanding of the problems of industry. They were a social force that had been growing for more than a decade without even being aware of its own existence. They were the founders of the Wise Use Movement.[72]

So who were they and who was there? Timber giants were in attendance with the likes of Macmillan Bloedel, Georgia Pacific, Louisiana-Pacific and the Pacific Lumber Company; pro-timber associations such as Alaska Women in Timber, Cariboo Lumber Manufacturers Association, and Californian Women in Timber were out in force. Big oil was represented by Exxon. The chemical colossus Du Pont was there. Mining associations were present. Cattlemen and ranchers were represented too, along with hunters and fishing enthusiasts. So were numerous snowmobile and mechanised off-road and powerboat advocates. The Right were in residence, with old favourites such as Pacific Legal Foundation, MSLF, the National Rifles Association and numerous Farm Bureau Associations, as well as other important organisations such as the Competitive Enterprise Institute and Accuracy in Academia.[73] Apart from funding from the CDFE, money and in-kind services were provided for the conference by among others, Chuck Cushman's National Inholders Association and the American Freedom Coalition. The National Inholders Association, the Farm Bureau, and the Citizens Equal Rights Alliance were three organisations at the conference with known anti-Indian connections too.[74]

The Wise Use Agenda became a battle-call for Wise Users, as they became known. It was also an agenda that attempted to ally right-wing activists bent on a deregulatory agenda with resource companies who sought to undermine environmental regulations at any political opportunity. The back cover of *The Wise Use Agenda* book depicted a smiling Gottlieb with an equally happy George Bush shaking hands, giving the impression that Wise Use had the then President's blessing, although the picture had been taken three years previously. 'Wise Use,' wrote Alan Gottlieb 'will *be* the environmentalism of the 21st Century' in the 'Call to Action' at the beginning of the book.[75] So what were the most notable goals of this new variety of environmentalist:[76]

- Immediate development of the petroleum resources of the Arctic National Wildlife Refuge (ANWR); [currently banned by Congress, but under review by the Republicans].
- The Global Warming Prevention Act. To convert 'all decaying and oxygen-using forest growth on the National Forests into young stands of

oxygen-producing, CO_2 absorbing trees to help ameliorate the rate of global warming and prevent the greenhouse effect. [A mandate to clear-cut old growth forest.]

- To allow 3 million acres of the Tongass National Forest in Alaska to be logged.
- Creation of a National Mining System whereby 'all public lands including wilderness and national parks shall be open to mineral and energy production'.
- To reform the National Parks Reform Act (advocating a twenty-year construction programme in all forty-eight National Parks).
- To change the patents on pest control and chemical products.
- To amend the Endangered Species Act to 'specifically identify endangered species as relic species before the appearance of man . . . and endemic species lacking the biological vigour to spread in range'. Projects intended to protect species would be scrutinised on a strict cost benefit analysis procedure.
- Global Resources Wise Use Act:

'Congress should enact a policy measure that explicitly recognises the shrinking relative size of the total goods sector of our world's economy and take steps to insure raw material supplies for global community industries on a permanent basis. Should include free-trade measures and incentives for developing nations that favour free enterprise.'

- Perfect the Wilderness Act. To change the definition of wilderness.
- Standing to sue in defence of industry: 'Pro-industry advocates should win standing to sue on behalf of industries threatened or harmed by environmentalists.'
- Federal gasoline taxes should be used to create off-road multiple use trails for vehicles.

THE 'MOONIE' CONNECTION

One of the funders of the conference had been the American Freedom Coalition (AFC), which had recommended opening up ANWR to drilling in its submission to the *The Wise Use Agenda*. The AFC called itself a conservative network which defended traditional American values, whilst promoting moral and ethical solutions to the political, economic and social struggles that threatened America's future.[77] In reality it is a Moon front group.[78] Robert Grant, the AFC's first chair, writing in the Moon-backed *Washington Post*, admitted in 1989 that in the two years since its inception, the Unification Church had given $5,250,000 to the AFC.[79] The AFC's Environmental Task Force, in wording implicitly similar to Arnold's and Gottlieb's, had also become dismissive of the environmental movement. 'What began as a wholesome movement to protect wildlife, clean air and water, is no longer wholesome. The time is at hand for a "Neo-Environmental" movement' and for a 'new breed of environmentalist' they declared.[80] In fact, both Arnold and Gottlieb were working with the AFC at the time. Arnold was the President of the Washington State Chapter of the AFC from 1989 to 1991 and Gottlieb served on the Board of Directors in 1989 and 1990.[81] Furthermore, Arnold was a speaker for CAUSA, the organisation that had organised the conference in Jamaica, which Gottlieb had attended.[82]

Moon-watcher Dan Junas contends that:

> If one looks in the *American Freedom Journal*, the publication of the American Freedom Coalition, very early in the beginnings of the Wise Use movement they [Moon] were sponsoring Wise Use conferences in a variety of places but specifically in the Pacific North West, in Oregon, Washington and Idaho. So here's a large-scale foreign-funded organisation in the United States that played an important role in getting the Wise Use movement going.[83]

Tarso Ramos agrees with Junas: 'Whatever the ultimate motivation for the AFC's involvement in pursuing what became the Wise Use Agenda, there is no question that their activity was very important in getting Arnold and Gottlieb and the rest of them onto this path.'[84]

'Resource-Use Conference had links to the Moonie Cult,' declared *The Vancouver Sun*[85] the following year, in the first of many articles that were to expose Arnold and Gottlieb's association with the Moonies, as well as question some of the alliances and allegiances of principal right-wing activists. The initial allegations were extremely damaging to Arnold and Gottlieb as people from industry refused to be associated with them. Moreover, in recent times, although the two worked with Moon earlier on, it seems that they have been associated less with Moon and more with another group of right-wingers centred around Paul Weyrich (see next chapter). Historically there has been no love lost between Moon's associates and those of Weyrich's.[86]

Many of the original articles in the press dismissed Wise Use as nothing more than a corporate scam. The corporate front group cliché was written time and again. Although some of the Wise Use groups do receive corporate money, many do not. Because of Arnold's links to a Moon-affiliated organisation, business was wary of jumping into bed with such a potentially politically damaging person. This did not mean that certain elements of his strategy were not adopted. According to Tarso Ramos:

> To a certain degree Arnold has had more success in promoting his strategy than in being the broker or being in control of the implementation of that strategy. Some of the natural resource trade associations, for instance, have adopted components of Arnold's strategy without necessary putting Arnold on the business.[87]

THE FIRST WISE USE CAMPAIGN: THE 'OREGON PROJECT'

What can be called the first official Wise Use campaign was instigated by the timber industry, using parts of Arnold's strategy, the citizen front group. Called the 'Oregon Project', in late 1988, it was initiated by Republican Senator Mark Hatfield working with timber interests in the region. Bill Grannell and his wife from the Western States Public Lands Coalition (renamed the National Coalition for Public Lands and Natural Resources) were asked to run the campaign. Its objective was to muster a broad-based coalition to support unrestricted access to timber and other resources on public lands.[88] After their pro-timber stance the Grannells would then move on to set up the pro-mining Wise Use group, People for the West!, another corporate front group (see next chapter).

The previous decade had seen the timber industry reap short-term profits over long-term stability. It had cut investment and reduced labour and wages wherever possible. 'The short-term economies of extraction, waste and plunder that have characterised the human use of American land,' argues T. H. Watkins of the Wilderness Society, 'have brought more ruin to individual lives than anything any environmentalist has promoted.'[89] However, the Grannells were to be given a publicity coup the following year when the spotted owl was listed as an endangered species. 'It provided a scapegoat in no small part for irresponsible industry practices such as overcutting, automation, and log exportation,' argues Ramos.[90] It meant that the timber industry, which was still in the greatest recession since the time of the Great Depression, had someone else to blame. A decade of low timber demand, of low prices, of lay-offs and closures could now be fairly and squarely blamed on environmentalists. It also provided a symbol for the Wise Use movement to argue that environmentalists cared more about owls than people. The Grannells, and other Wise Use activists, like many in the timber industry, quickly polarised the debate into 'Jobs versus Owls'.[91]

It would be the first of many attempts by the Wise Use movement to exploit peoples' natural and real fears over job security or private property rights. For example, Chuck Cushman's frenzied speeches often talk about 'cultural genocide'. Arnold warns audiences that the environmental movement are out to 'destroy industrial civilisation'.[92] Simply polarising the issue as your job versus the environment, your job versus the owl, good versus evil, ignores the complex issues at stake and makes any compromise difficult. As David Hupp wrote in the *Western States Center Newsletter* in the summer of 1992:

> The Wise Use Movement is built on a base of ordinary citizens with a respectable goal: trying to keep their jobs, but this alliance is led by right-wing strategists and corporate funders who see big money at stake and have direct ties to the White House. The leadership is exploiting a broad, angry and fearful base of working families, rural citizens, and employees of natural resource corporations.[93]

The Oregon Lands Coalition, a Wise Use group that evolved from the Oregon Project, showed the way for future resource use fights. Its principal impact on the debate was to polarise it, according to Tarso Ramos.[94] This process of polarisation and confrontation is no coincidence. People are told that the environmental movement is out to close down their industry, take away their property rights and their livelihood and life. The polarisation process does two important things. First it drives a wedge between workers and environmentalists; second, it instills fear into the community which means that workers are increasingly vulnerable to corporate and right-wing manipulation. 'The only thing new about the Wise Use movement,' wrote Robert Barry for the Montana Alliance for Progressive Policy in 1992, 'is that its leaders have become much more sophisticated about the techniques of using the fears and uncertainties of grassroots people to manufacture support for their pro-industry agenda.'[95]

As the recession hit rural areas of North West America in the early 1990s, the anti-green message was a successful and purposeful one to tell. Citizens groups began to mobilise around the region and yellow ribbons, the official symbol of pro-industry solidarity, were seen with increasing frequency. One group even named themselves the Yellow Ribbon Coalition. In April 1990,

10,000 timber industry supporters attended a rally in Portland, Oregon, organised by the Yellow Ribbon Coalition. If it suddenly seemed that the timber industry had found support from nowhere, over 300 companies had provided the day off for their employees, others even laid on free transport to the rally.[96] Industry had its grassroots support that had been lacking for so long.

Furthermore, it actually seemed that the popular tide was beginning to turn against the environmental movement. A poll commissioned by the *New York Times* and CBS in 1989 had found that 80 per cent of respondents felt that 'protecting the environment is so important that standards cannot be too high, and continuing environmental improvements must be made regardless of cost'.[97] In stark contrast though, three years later a similar poll by *Time* and CNN showed that 51 per cent of Americans thought environmentalists had 'gone too far', whereas the previous year the figure had stood at only 17 per cent. Results like these were to add fire and meaning to the spreading backlash.[98] A backlash that had growing support and increasing antagonism against an enemy that was seen to be too powerful.

THE ENVIRONMENTAL MOVEMENT'S ROLE IN THE BACKLASH

'Anytime anyone gets as much power as the environmental movement has achieved,' an Oregon logger told *Time* in 1992, 'a backlash can be expected.'[99] In short, both its accomplishments and its failures led to the growing environmental backlash over the next few years. 'It should be acknowledged that, in many ways, the growing professionalism of the environmental movement has also contributed to the backlash syndrome,' argues Antony Ladd, a Loyolo University Professor.[100]

'I think it is a truism that the more successful a social change movement is, the more aggressive the backlash movement is. That's predictable,' says Chip Berlet, from the Political Research Associates, a leading think-tank that tracks the American Right and anti-environmental activity.[101] Berlet and other activists have studied backlashes that have happened to other movements for social change by the political Right (see next chapter). Tarso Ramos agrees with Berlet:

> Certainly, I think, as with other backlash movements against progressive social movements in this country, to a certain degree the environmental movement has been a victim of its

own success. It had enough impact and became enough of a problem for certain established powers and forces in this country to raise a coordinated strategy to dismantle or reduce the effectiveness of the movement.[102]

But the blame for the backlash must also be laid squarely at the door of the mainstream groups in Washington. 'The environmental community has only itself to blame for the rise of Wise Use sentiments,' argues Philip Brick, an assistant professor of Political Science at Whitman University. Brick believes that 'the environmental movement has shown astonishing insensitivity to the plight of those who live near the areas they hope to save'.[103] 'Historically, mainstream environmentalists have not addressed the working man's issues,' says Lois Gibbs, the Executive Director of Citizens' Clearinghouse for Hazardous Waste, 'Their main focus has been to protect forests and concentrate on conservation issues.'[104] There is no doubt that if the environmental movement had listened and responded to the concerns of workers and rural communities more than they have done, then the backlash they are suffering today would not be nearly as intense or justified.

Tarso Ramos also argues that the failure of the environmental movement to address social and economic issues as well as frame environmental issues as community issues based on social justice, has made the movement vulnerable to this kind of backlash. Ramos contends that 'following the election of Clinton in 92, the attempts by environmental organisations on the national level to portray themselves, in effect, to become key players inside the beltway, and play insider politics, have hurt the environmental movement in a couple of ways'. First, Ramos believes that power did not manifest itself and 'secondly, it has reinforced the characterisation that Wise Use has given the environmental movement of being a powerful, DC-centred, rich, elitist movement without much popular base of support'.[105]

This Washington-based litigation strategy has been heavily criticised. 'By clinging to its legislative agenda and litigation strategies at the expense of more pro-active and aggressive tactics, the movement became increasingly outmaneuvered by the superior resources of the industrial lobby and its political action groups,' argues Professor Ladd. Furthermore, and maybe most damaging for the large environmental groups is the animosity it seems to have evoked from the grassroots. Ladd adds:

At the same time the shift also brought the national organisations under attack by the activist grassroots wing of the movement, charging them with being arrogant, racist, insensitive to local issues (especially toxins in minority communities) and more concerned with wildlife issues than human life. Indeed, one could conclude that the movement today is experiencing a backlash both from without and within.[106]

John Stauber, the editor of the journal *PR Watch*, also agrees that the environmental movement's strategy is also to blame, in part, for the backlash. 'Two things that are obvious to me are the very inherent weaknesses of our environmental organisations and the failures of environmental organisations to carry the fight,' says Stauber, adding that, 'so-called third-wave environmentalism has played into the hands of industry and their PR strategies who are attempting to divide and conquer and co-opt environmentalists.'[107] Third-wave environmentalism is where conflict between the environmental movement and industry is replaced by compromise and co-option. Mark Dowie, in his criticism of the modern-day American environmental movement, has also taken issue with this third wave of environmentalism, labelling it 'essentially anti-democratic' and the 'institutionalisation of compromise'.[108]

Dowie also argues that the mainstream environmental groups have 'also made two other, near-fatal blunders: one was to alienate the grassroots of their own movement; the other was to misread and underestimate the fury of the antagonists'.[109] He also takes issue with the strategy of many environmental groups, one of 'legislation backed up with litigation'. He calls it a folly that this kind of strategy will actually protect the 'environmental health of the nation'.[110] It is also this preoccupation with legislation in Washington at the expense of grassroots organising that has cost environmentalism *per se* a heavy price.

WISE USE SUCCESSES

The environmental community, or more precisely the major environmental organisations, were slow to respond to the Wise Use threat. When they did react, it was without much insight or imagination, and so it comes as no surprise that the Wise Users soon started achieving some of their goals.

By 1992 the annual Wise Use leadership conference at Reno, once again organised by the Center for the Defense of Free Enterprise, was receiving considerable attention from the press as well as from environmentalists, who were beginning to wake up to the threat. Under Arnold's old banner 'Man and Nature Living Together in Productive Harmony', the Programme listed many of the old Wise Use war-dogs as co-sponsors, such as the MSLF, Cushman's National Inholders Association, the corporate-backed Blue Ribbon Coalition and People for the West!, as well as newer organisations such as Putting People First, an anti-animal rights organisation.[111]

Harry Merlo of the titanic timber company, Louisiana-Pacific, received a 'Lifetime Industrial Achievement Award', for battling against green activists. Also honoured was Steve Symms, a Republican Congressman from Idaho and Wise Use supporter, with a 'Lifetime Policy Achievement Award' for securing the Wise Use movement's first legislative victory, the passing of the National Recreation Trails Fund Act through Congress. The Act diverts $30 million in federal gasoline taxes for the development of off-road bike, vehicle and snow-mobile trails on federal lands. Clark Collins, from the off-road advocate the Blue Ribbon Coalition, whose efforts really secured the bill, also received a Lifetime Award. Item Number 24 of *The Wise Use Agenda*, which had urged the creation of the 'National Recreation Trails Trust Fund', had been won.[112] The passing of the 'Steve Symms Act', as it became known, was a clear sign of victory, and was achieved 'against everything the environmental movement could throw at us', contends Arnold. 'We know how to lobby better than they do and we've got coalitions that can overwhelm them. That's never happened to them before. It frightened them big time,' he crowed to *Z Magazine* in 1992.[113]

ALLIANCE FOR AMERICA

A month after the Symms Bill became law, the Alliance for America was initiated, in a further show of growing grassroots support for the anti-green crusade. Representing a coalition of 125 groups, it professed to 'put people back in the environmental equation'.[114] The Alliance for America was formed by the Oregon Lands Coalition (OLC), the Wise Use group that had been formed by the Grannell's original Oregon Project. The 'Alliance for America is a dream come true for many hard-working Americans. The spawning of the Alliance evolved from a nationwide lobby effort in Washington DC, September 21–26 1992,' proclaimed the *Alliance News*[115] speaking for the 'Ranchers, farmers, loggers, private property owners, public property users, miners, etc. the messengers that lobbied Congress', on the First 'Fly-in for Freedom'. Since its inception, the Alliance has held 'Fly-ins for Freedom' in Washington where delegates are flown in to lobby on behalf of various corporate interests.

One trait that the Wise Use movement began to show early on in its existence was copying the successful techniques that the environmental movement had used to such success. The OLC had already learnt some of the environmentalists' tricks and had started putting 'a human face on the issue', says Jackie Lang, Legislative Director for the Coalition. Bumper stickers had appeared with pictures of children on, underneath the wording 'Oregon's Real Endangered Species'.[116]

The Alliance has copied other techniques. Fax and phone-ins are used. Lobbying of local and federal government has now become commonplace. The media is inundated with opinion pieces, letters, demands and denials. If people feel that the media have not been representative in their coverage of an issue then demonstrations outside media outlets are organised, such as the one outside CBS in 1993 for perceived biased reporting on the logging issue. Pro-industry anti-environmental demonstrations have spread. For example, the proponents of a WTI incinerator protested outside Greenpeace USA's office in December 1992.

The impression to be given to the press, to Washington and the world, from these Fly-Ins and demonstrations, is that average workers, not some other species, are at risk of extinction. Participants are told to 'Tell your personal story. Paint word pictures for the reporter that provide compelling and emotional images. Convey that hard-working American families like yours are being seriously hurt by environmental extremism.'[117] People 'Flying-In' to Washington, many for the first time in their lives, are ordered to look authentic. 'Wear work clothes to the rally, with special attention to gloves, boots, hard hats, bandannas, hearing protection, protective eye-wear and other work-related accessories,' attendants are told.[118]

Despite this, the goals of the Alliance for America, are basically a boiled down Wise Use Agenda:[119]

- To balance the needs of people and the environment.
- To advocate multiple use of public lands and natural resources.
- To restore and protect constitutional private property rights.

Many of the same players, such as the American Freedom Coalition and the National Inholders Association, Alaska Women in Timber, Alaska Miners Association, Blue Ribbon Coalition, California Farm Bureau Federation, California Women In Timber, Communities for a Great Northwest, National Federal Lands Conference, National Rifle Association, Multiple Use Association,

Plate 1.1 WTI incinerator supporters outside Greenpeace's offices.
Source: Jay Townsend/Greenpeace

Pacific Legal Foundation, and Western Forest Industries Association who had signed up at Reno, are now pledging their allegiance to the new Alliance.[120] In all, 125 groups originally signed up to campaign for 'sensible environmental policies'.[121] Although they say they are more moderate than Arnold's version of anti-environmentalism, much of the rhetoric is the same. Harry McIntosh of the Alliance for America warned that environmental activists 'are working to bring their brand of socialism to the forefront' of American politics.[122]

Many of the objectives are the same, too. Overturning wetlands policy, reform or abolition of the Endangered Species Act, and private property and 'takings' legislation. The basic stance of both Wise Use and Multiple Use advocates is one of paramount private property rights and unlimited unregulated resource extraction. Also important is the impression of numbers: to be seen as a broad-based moderate anti-environmental movement.

'The formation of the Alliance is the beginning of the grassroots revolution.'[123] That is the message. 'We are the true environmentalists. We are the stewards of the earth,' wrote William Perry Pendley in *It Takes A Hero: The Grassroots Battle Against Environmental Oppression* in 1994, adding that:

> We're the farmers and ranchers, the hunters and trappers, the fishermen and watermen – people who have cared for the land and the waters for generations. We're the men and women who have clothed and fed the nation and the world. We're the miners and the loggers and the energy producers who have provided this nation with the building blocks of a modern civilisation. We're the workers, the builders, the doers. Together, we hard-working Americans built this country, made it strong and prosperous, delivering to generation after generation not just a better and more abundant life, but also the hope for such a life.[124]

The book, a project of the Mountain States Legal Foundation, was produced by Alan Gottlieb's Free Enterprise Press and distributed by his own Merril Press.[125] It was clear evidence of Pendley, Gottlieb and Ron Arnold, who received special praise, of working closely together. Inside were profiles of fifty-three people 'who risked everything to stand up for the truth'. In addition, the book contained '1,000 leading grassroots fighters against environmental oppression: The Hero Network'.[126] An emotional tome, the book was a self-congratulatory pat on the back for many of the key players in the green backlash, and contained many of the old green adversaries who believe that 'environmental extremists and their allies in the US Government are good neither for people or for the environment'.[127] The Wise Use leaders obviously believe that the Wise Use movement is becoming a potent force.

HOW IMPORTANT IS WISE USE?

Is Wise Use a real movement, a coherent force for radical social change, with a broad base of support? Will it become the new environmentalism, as its leaders claim? Is it a self-sustaining and malignant movement that is growing across America, killing its nemesis, the environmental movement? Or is it a benign bunch of people being manipulated by corporations and the Right? Without that orchestration would the movement fall apart, losing direction and resources? These are questions that need to be answered, to understand whether Wise Use has a future.

In William Perry Pendley's book he lists the 'Hero Network', which names 1,000 key activists and groups throughout the USA. The Alliance for America now lists around 500 groups in 50 states. There is no doubt that the green backlash movement, whether advocates of Wise Use, property rights, or multiple use, does enjoy considerable grassroots support from within certain sections of society, mainly people concerned about their property rights all over the country, off-road enthusiasts, people employed or dependent on the extraction industries and right-wing activists. But the anti-green leaders often state that their membership runs into millions of people. Arnold, for example, in *The Wise Use Agenda* commented that the organisations who attended the first Wise Use conference 'represented a total membership in excess of ten million people'. At the time that was a considerable potential pot of promised support for a fledgling movement. But by the 1994 conference at Reno, six years into the Wise Use Rebellion, there were serious signs that Arnold and Gottlieb were having trouble finding committed activists who would fit into the mould of their intended mutiny. Less than one hundred people attended.[128]

The theme of the 1994 Reno Conference was 'Going Mainstream', but the rhetoric and agenda sounded more extreme than ever as Arnold, Gottlieb and Cushman searched for further sections of society to align themselves with. 'Those coalition partners who are not industry front groups are right-wing anti-semites (LaRouche), anti-tax advocates and right-wing religious people,' wrote private investigator Sheila O'Donnell who attended the conference.[129] Even the more moderate 1994 Alliance for America Fly-In for Freedom only drew some 200 participants. Having said this, there is no doubt that Wise Use can mobilise a considerable number of people against specific legislation at critical times.

However, Tim Egan, the Seattle correspondent of the *New York Times*, told fellow journalist Dave Helvarg:

> In terms of the media chain, we've elevated Wise Use to this large alternative movement, which I really don't see on the ground . . . I probably travel more in the West than any other national reporter, sixty to seventy thousand miles a year to all these small towns and rural communities, and I just don't see any signs of a real environmental backlash . . . This great countermovement is just not there.[130]

Egan's sentiments are backed up by Tarso Ramos.

> The Wise Use movement is not, as some have argued, simply a constellation of corporate front organisations, although it does involve some corporate front groups. Nor is it an authentic expression of citizen resentment and frustration and opposition to public interest, especially in environmental regulation, although there is that component to it now as well. There is no question that both corporate right-wing-funded and sponsored Wise Use activity has generated a certain degree of grass-roots opposition to environmental regulations that is, at least to a considerable degree, independent of the initial corporate and right-wing sponsors. That is, it has a life of its own.

This said, Ramos concludes that, 'Were the corporate sponsorship, or the right-wing activists, to wither and blow away, the movement, as a movement, could not persist.'[131]

'You do not have movement against movement, you have the corporate sector against the environmental movement and I think this is a very important point,'[132] says leading political expert Sara Diamond. The political analyst Dan Junas adds:

> I think in some respects it is an artificial creation. However, at the same time the people who are involved in the Wise Use movement at the grass-roots, they have, in some cases, legitimate concerns, although I don't necessarily agree with their view-point. It is not just that they are being manipulated, but I do see it as being an artificial creation.[133]

The Wise Use movement has had a major political success, and with the Republicans in Congress, other anti-environmental measures will seriously undermine many of the environmental gains of the last twenty years. This may be the Wise Use movement's greatest legacy, that, through its grassroots organising it helped one of the most vehement anti-green Republican administrations in the history of American politics into power.

KILLING THE BIODIVERSITY TREATY

To date, maybe Wise Users' biggest victory happened in 1994. During the 103rd Session of Congress, the United States Senate did not ratify the United Nations Framework Convention on Biological Diversity, which had originally been agreed at the Earth Summit in Rio de Janeiro in June 1992 to protect the world's flora and fauna. Although it had been expected to be ratified, it was blocked by the Republican Summit. The Senate opposition only materialised after a campaign by Wise Use activists, who in turn had been misleadingly primed by a colleague of the political extremist, Lyndon LaRouche.[134]

Roger Maduro (also known as Rogelio), an associate editor for *21st Century Science and Technology*, a magazine associated with LaRouche, wrote a scathing conspiratorial attack on the treaty for the American Sheep Industry Association, which later was widely circulated and used as evidence to kill the treaty. Maduro, who has been active in LaRouche's political movement since the late

1970s, had also urged Wise Use activists at the 1994 Reno Conference to oppose the Treaty, saying that otherwise, people would be governed by the United Nations.[135] Activists from Reno as well as from the Alliance from America were to take Maduro's conspiratorial rhetoric and spin it into an unstoppable barrage of faxes, phone-ins and letters that inundated Congress in July, asking for the Treaty to be buried. The Cattlemen's Association and American Farm Bureau were soon to follow suit.[136]

John Doggett, the director of governmental relations for the American Farm Bureau, which opposed the Bill, conceded that 'unfortunately, what we've seen is that certain groups tried to create a crisis where one doesn't exist'.[137] Moreover, Doggett was to concede that Maduro's campaign was the 'key to triggering the masses'. A former Chief Counsel to the Senate Agriculture Committee also agreed that the campaign had had a serious effect, 'slowing the treaty down and eventually stopping it'.[138]

The stopping of the Biodiversity Treaty could pale into insignificance compared to other items on the anti-green agenda being promoted by the Wise Use movement and now the Republican administration in Congress. 'The most prevalent theme running throughout all the speeches was anti-federal government with private property rights a close second,' said Sheila O'Donnell, of the Reno conference in 1994. 'Cost benefit analysis, risk assessment and unfunded mandates were also discussed.'[139]

THE 'CONTRACT WITH AMERICA'

The issues of property rights/takings, cost benefit analysis/risk assessment and mandates have become the three-pronged trident that has been struck in the heart of the environmental debate, which has left the green movement staggering from the potentially fatal wound. Dubbed by environmentalists the 'unholy trinity', it is easy to see why these three critically important issues are the Wise Use devil's advocate. More worrying, the 'unholy trinity' is also at the heart of Gingrich's 'Contract With America', and the new Republican revolution.

The emphatic return to Republican control of House and Senate after a forty-year absence, with the election victories in November 1994, signalled a turning point in the political fortunes of the Wise Use movement. Not since the days of Reagan and Watt did the advocates of the green backlash have such friends on Capital Hill. There is no doubt that certain Republicans surfed back into power on a wave of Wise Use, grassroots support and an anti-green, anti-regulatory, anti-Clinton, and anti-federal government platform. The vehicle by which the Republicans achieved this change was the 'Contract With America'. At every opportunity, the Contract would be read by Newt Gingrich, who was to become the new Speaker of the House. Although it sounded good, Gingrich's plan was not only bad for the environment, it was bad for the nation's children, and it was bad for the nation's women.[140]

'Contract With America' actually says nothing specifically about environmental protection. But the small print tells a different story. Whilst 83 per cent of voters at the November polls considered themselves environmentalists, they voted in the vehemently anti-green Gingrich. The 'Contract With America' is nothing less than a contract to allow unrestricted, unregulated mining, logging, grazing, drilling and extraction. Environmental protection will be repealed and rolled back or simply gutted. 'The fate of our nation's environmental laws now hangs in the balance in Congress,' warned fifteen major American environmental groups in the autumn of 1994, as the threat became apparent.[141]

In February 1995, one of the groups, the NRDC warned that 'the Republican Contract With America threatens to undermine virtually every federal environmental law on the books, meaning dirtier air, dirtier water and more species pushed to the brink' in a report entitled *Breach of Faith*. The NRDC was joined in condemnation by two former Senators and two current Democrat Congressmen. The Congressional proposals 'would halt twenty-five years of accomplishment and turn the clock back to the days when the special interests made the rules and the people absorbed the risks', said former Senator Edmund Muskie, the lead author of the original clean air and water acts.[142] 'Under the cover of a politically popular promise to limit big government,' said Democratic Congressmen, George Miller and Henry Waxman, the proposals were 'actually attacking key environmental and health laws that annoy or threaten big business and industry'.[143]

Many items of the Contract had a direct impact on the ecological debate in the USA and deep ramifications for the environmental movement, none more so than the 'Job Creation and Wage Enhancement Act'. Journalist Antony Lewis, writing in *The New York Times*, remarked with irony on the implications of the Act. 'An insufficiently noticed item in the contract would make it virtually impossible for any governmental body, state or Federal, to protect the environment,' he wrote. 'The Gingrich item is labelled the "Job Creation and Wage Enhancement Act". George Orwell would have appreciated the beauty of that title for a proposal that might better be called the Death and Desertification Act.'[144] Others were just as cynical: 'The Job Creation and Wage Enhancement Act would take us back to nineteenth-century environmental protection,' wrote Jessica Mathews, from the Council on Foreign Relations, 'which is to say, none.'[145] The House passed the Bill in March.[146]

THE UNHOLY TRINITY

The Unholy Trinity were also at the heart of the Job Creation and Wage Enhancement Act and it is necessary to explain the issues of property rights/takings, cost benefit analysis/risk assessment and mandates. Armed with the Fifth Amendment of the United States Constitution, which declares that 'nor shall private property be taken for public use, without just compensation', property rights, Wise Use activists and now Republicans are interpreting the Constitution in a way that furthers their own agenda, so much so that the National Audubon Society sees the property rights movement as the 'single greatest threat to continued progress in protecting the environment'.[147] Its proponents are cleverly marketing the strategy, by calling it patriotic and in favour of the individual property owner, when it is just another business deregulatory agenda.

Aside from politicians, leading proponents of takings are the right-wing Pacific Legal Foundation, the National Farm Bureau, the Defenders of Property Rights, and an expanding grassroots network that includes the Property Rights Alliance and the Independent Landowners Association and many Wise Use groups.[148] 'It is important to recognise,' argues Neil Hamilton, a law professor from Drake University, 'that the property rights movement is laden with individuals and organisations whose larger goal is promoting a conservative agenda to limit the power of government.' Hamilton believes that expansion of takings laws could lead to 'environmental anarchy, with individuals free to act without regard for public health or welfare'.[149]

Property rights activists initially argued on physical 'takings', for example, if someone's house was bought by the federal government for an infrastructure development, such as the building of a road, the owner was dutifully compensated for the government 'taking' their property. However, the argument now being put forward by 'takings' advocates is that any reduction in the price of a property due to any regulation constitutes a 'taking' and therefore has to be compensated. But the taking argument goes further too. If someone is prevented from developing a wetland because it is protected, then this too is a 'taking' of private property. If an oil company cannot increase production and thence air pollution from a refinery as this would breach air quality standards, this too becomes a 'taking'.

The logical extension to this is that nearly all government regulations affect private property and business in one form or another. This argument has the potential to bulldoze new legislation into the bowels of history before it even reaches the statute book. Existing legislation becomes unworkable and therefore is more vulnerable to be repealed. This has the potential to grind the whole system to a halt, tangled in a bureaucratic nightmare or simply gone broke. It is no small wonder that Mark Dowie calls 'takings' the '"Mantra" of the free market environmentalists and the war cry of landholders large and small'.[150] Not only are environmental laws at risk, so too are laws that protect worker safety, public health and other public welfare protection.[151]

As with much of the anti-environmental agenda, 'takings' had their roots in the Reagan administration, and it is easy to see why Gingrich's new Republicans are returning to the traditional anti-green laws, whilst trying to write new 'takings' laws, whilst Republican-appointed judges interpret the existing ones. Many of the Reagan and Bush appointees to the Supreme Court are still in office and it is up to them how to implement their legal interpretation of the Fifth Amendment. In fact, seven out of nine sitting Justices on the Court were appointed by Republican presidents.[152]

In the most famous of the 'takings' cases, a developer, David Lucas, who had bought nearly $1,000,000 investment in two pieces of land was prohibited from developing them due to the South Carolina Beachfront Management Act which prohibited development close to sensitive coastal eco-systems. Lucas challenged this decision as a 'taking' of his property. The South Carolina Supreme Court ruled in favour of the Act and against Lucas, but South Carolina was forced to prove that the buildings constituted a public nuisance. This change in the legal interpretation is considered an important victory for the green backlash,[153] and it was only a matter of time before other victories followed. New 'takings' legislation is being adopted at state level as well as federal level. A ruling that allows a property owner to claim compensation if their land lost 50 per cent of its value due to new regulation has already been passed in ten states.[154]

The advocates of 'takings' have played a powerful ace in the environmental movement end-game. 'As an organising strategy, takings is a kind of deviant genius,' says Tarso Ramos, 'It automatically puts environmentalists in the position of defending the federal government and appeals to anyone who has ever had any kind of negative experience with the federal government, which is a hell of a lot of people.'[155] However, what 'takings' advocates do not address is that in many cases it could be lack of public protection that reduces the value of someone's property, not vice versa. For example, someone living downwind from an unregulated industry could find that their property decreases as the pollution increases. Moreover, they do not consider any of the benefits that existing regulation brings to their property.

Unfunded mandates, the second spike of the 'unholy trinity', is another issue that has far-reaching consequences for the whole of the US legislature. Local and state authorities, as well as the anti-regulatory movement within the USA, which includes many sections of people involved in anti-environmental activity, want the federal government to stop imposing regulations on them for which there is no federal funding. These are called 'unfunded mandates'. The Tenth Amendment, which states that, 'The powers not delegated to the United States by the Constitution, nor prohibited by it to the States, are reserved to the States, respectively, or to the people', is being used by a growing number of states trying to reduce federal power. Some states want to rid themselves of any federal control altogether. As part of this process, a few states have passed resolutions ordering the government to 'cease and desist' passing down 'unfunded mandates'.[156]

Many of these 'unfunded mandates' are fundamental environmental regulations such as the Clean Air and Clean Water Acts, so it is understandable for anti-environmentalists to want these regulations severely restricted. 'Costs of unchallenged pseudo-environmental federal and state government mandates,' wrote the Property Rights Foundation of America in March/April 1994, 'are becoming an outrageous burden.'[157] It is not just environmental legislation that would be affected by such drastic constitution changes. 'The growing drumbeat for reduced "unfunded mandates" may sound like a catchy and worthwhile idea,' warned the American Federation of State, County and Municipal Employees in a Congressional testimony in 1994, 'but behind it lurks an outright assault on fundamental federal protections which benefit all Americans.'[158]

Benchmark laws such as the Civil Rights Act, Americans with Disabilities Act, Family and Medical Leave Act, National Child Protection Act, Nursing Home Reform Act, Fair Labor Standards Act and Safe Drinking Water Act, would all be affected. Indeed, the National Conference on State Legislatures has listed over 200 federal laws that they believe contain 'unfunded mandates'. All these would be severely weakened or made effectively useless.[159] Later in

1995, the unfunded mandate provisions in the 'Contract With America' were passed into law.[160]

The third and final prong of the trident, that of cost benefit analysis and risk assessment, has the same effect on the legislature. Cost benefit analysis in simple terms means the cost of an action should not outweigh the benefit of that action. The proponents of risk assessment, in this context at least, advocate that the extent of the risk being regulated by an agency, for example the Environmental Protection Agency, must be evaluated and compared to other risks people face. In fact, anti-environmentalists tried to pass a Bill that would have forced the EPA to conduct a risk assessment on all new proposed regulations and compare it with six non-regulated risks. Once again these measures are designed to weaken environmental and health legislation and grind the federal government to an abrupt halt. 'Paralysis by analysis' is how environmentalists see risk assessment and cost benefit analysis taken to these limits.[161]

Using these three arguments the green backlash movement now sits on the verge of repealing all the major pieces of landmark environmental legislation. The first step has been to stall present legislative reform. The next step will be the Endangered Species protection revision, water pollution laws, National Park reforms – all caught up in the 'unholy trinity' debate.[162]

THE ENDANGERED SPECIES ACT

The Wise Use movement has been waging an all-out assault on the Endangered Species Act (ESA) for a while, using it as another scapegoat to blame environmentalists and government for caring more about animals than humans. The Alliance for America calls the ESA a 'law out of control'.[163] The facts say otherwise: Americans spend forty times as much on popcorn at the movies as on the total federal effort on Endangered Species.[164]

While there are examples of individual property owners who have suffered because of the Act, the national picture that the Wise Use movement attempts to portray of thousands of farmers, ranchers, loggers and property owners ruined by rats, owls, newts and toads, does not really stand up to scrutiny. As we have seen, the spotted owl became the scapegoat for a timber industry that had pushed short-term economic gain over long-term sustainable forestry. The Endangered Species Act has become the new scapegoat, despite the fact that the ESA has not stopped development to any significant degree.

Between 1988 and 1992, federal officials reviewed 34,600 proposed developments with Endangered Species ramifications and refused only twenty-three. The Nature Conservancy point out that during the same period twenty-nine light aircraft struck buildings, so that a developer had a greater chance of his building being hit by a plane than his development being stopped because of the ESA.[165] Despite this, an endangered species Bill introduced by Republicans in March 1995 would effectively repeal the Act, no longer making it necessary to protect endangered species.[166] Many Wise Use and right-wing organisations want to substantially change the Act and groups have even been set up to that effect: the industry front group, the Endangered Species Act Reform Coalition, Chuck Cushman's Grassroots ESA Coalition and the Californians for Sensible Environmental Reform.[167] Gingrich himself has said that individual plants and animals should not necessarily be protected.[168]

Despite the widespread criticism of the Act, an expert scientific panel, ordered by Congressional leaders of both parties, concluded in May 1995 that ESA was 'critically important' and a successfully used tool for preserving biological diversity. Altogether, some 500 species have become extinct since the Declaration of Independence in 1776. It recommended strengthening the law. 'In general, our committee finds that there has been a good match between science and the ESA,' said the committee.[169] In June the Supreme Court upheld the broad scope of the 1971 Endangered Species Act, and overturned a lower court ruling. It was a rare victory for environmentalists in 1995.[170]

THE REPUBLICAN ROLL-BACK

In Congress the strength of the anti-environmental, pro-industry sentiment was becoming all too apparent, as a whole litany of measures was drawn up, some of which were successful and some of which failed as the politicians realised just how unpopular they would be with the voters. Under the soothing umbrella of balancing the budget, the Republicans still want to nullify most of the gains of the environmental movement over the last thirty years.

Key committee chairmen, Representative Don Young and Senator Frank Murkowski of Alaska both promote oil exploration and tree felling on protected areas, and opening up the Arctic National Wildlife Refuge (ANWR) which Murkowski calls the Arctic Oil Reserve, and which has long been a bone of contention between environmentalists and industry. It is well known that the

House Resources Committee Chair, Don Young, wants to gut the ESA. Gingrich himself sees the Environmental Protection Agency as one of the two 'jobkillers' in the USA, along with the Food and Drug Administration. The Republicans have blocked Clinton's plan to promote the EPA to Cabinet level. In fact Clinton had to veto a Bill that would have slashed the EPA's funding by a third.[171]

As the Republicans get their way, wetlands are being opened up, so too are protected logging areas, recycling is to be attacked, and air pollution measures are being scrapped. Money for National Park protection and upkeep is being scrapped or reduced. So too are auto fuel economy standards. The House budget proposal would freeze listings of new endangered species until Congress moves 'to balance the right of landowners and species'. It also would abolish money for enforcing federal protection of coastal and agricultural wetlands. Republican proposals would ease requirements for industry to provide toxic pollution information to the public. The Republicans have also slowed down energy efficiency savings and renewable energy expenditure.[172]

By February 1995, the House had moved to weaken some of the landmark environmental legislation of the last quarter century including laws affecting clean air, endangered species and forest protection. House action on the environmental provisions of a Bill rescinding $17.1 billion in previously approved federal funding foreshadowed upcoming debates on those laws.[173] A House task force studying the deficit endorsed a proposal by the Heritage Foundation, a conservative think-tank, to sell federal grazing lands, some wilderness and refuge areas and parklands to help balance the budget. Heritage Foundation analysts estimate the government could save up to $3.6 billion in management costs alone by selling these lands.[174]

The EPA could be prohibited from imposing new dioxin regulations on the pulp and paper industry, and to regulate emissions from oil refineries. It would not be able to limit pollution from cement kilns, or limit sewage disposed of in rivers. A proposal for uniform federal water quality standards for the Great Lakes could be scrapped, as could new limits for radon and arsenic. New automobile emission testing to reduce smog could also be done away with. Pesticide regulations would also be weakened. 'This hit list has nothing to do with balancing the budget,' says David Driesen, an attorney for the Natural Resources Defense Council, a leading environmental group. 'It reads like a list of special interests that oppose environmental protection.'[175]

Jim Baca was ousted from his position as director of the Bureau of Land Management (BLM) because of attempts to reform grazing practices on federal lands.[176] In March, the Republicans pushed through a Bill to collect dying timber that would effectively double the timber harvested on federal lands. An attorney with the Sierra Club Legal Defense Fund, Kevin P. Kirchner, said, if enacted by Congress, the programme would be 'the most far reaching assault on national forests in the thirteen years I've been here'. President Clinton signed the Bill in July, despite having promised to veto it.[177]

Also in March, the GOP drafted a Bill to reform the antiquated mining law. Interior Secretary Bruce Babbitt said that the legislation 'continues the giveaway of valuable publicly owned hard-rock minerals like gold, silver and platinum for peanuts'.[178]

A Bill reluctantly signed by Clinton in July suspended the Endangered Species Act and other laws to accelerate salvage of timber harvests in an effort to reduce the threat of fire in federally owned forests nationwide. It also directed the Forest Service to log, free from the normal environmental constraints, some of the Northwest's oldest forests where the threatened northern spotted owls and marbled murrelets reside.[179] In July, too, Republican politicians were introducing legislation that would open up the Tongass National Forest, a plan which was condemned by Native Indians, fishermen and local environmentalists.[180] However, also that month some pro-environment Republicans refused to back the measure to curb the Environmental Protection Agency's enforcement of clean air and water standards. The vote was lost 212-to-206, and was considered a 'significant setback' for the anti-environmental agenda.[181]

Also in July, the European Union Environment Commissioner, Ritt Bjerregaard, severely criticised the Republican assault on environmental laws. 'The US has a responsibility to play its role in global leadership. And what is going on in the US Congress now is very discouraging,' said Bjerregaard, adding that, 'More and more often you hear claims that environmental legislation is bad for business or bad for competition. I do not agree with the thinking behind this and I do not think that the public – in the US or Europe – agrees with this thinking.'[182] In the autumn, the battle lines were drawn up again. The EPA once again stood to have its budget cut by 35 per cent. 'This is about shutting us down, there can be no mistake. This is a concerted effort,' says Carol Browner, the head of the EPA, 'it means our air, our food, our drinking water, the water we fish and swim in, will not be as safe.'[183] Unless other sources of revenue totalling $1.3 billion could be found, the Arctic National Wildlife Refuge would be opened up to oil and gas development.[184]

By February 1996, Carol Browner once again issued a dire warning that 'we cannot assure the American people their air is clean, their drinking water is safe, the health of their children is protected,' after announcing that EPA inspections were down 40 per cent since October 1995 as a result of spending cuts approved by Congress the previous year. Other Clinton officials charged that Congressional Republicans were using budget cuts in order to achieve changes in environmental laws they could not pass in 1995.[185]

If nothing else, the Republican proposals in Congress forced the major environmental groups to re-evaluate their thinking and leave Washington, highlighting the dangers in store for environmental protection. The effort included a $1.3 million television advertising campaign.[186] In October 1995, environmental activists handed in a petition with one million signatories, urging Republican leaders to maintain air, water and wildlife protection.[187] The backlash against Gingrich's reforms showed in the membership of the environmental organisations, some of which showed an increase for the first time in four years.[188] However, the major groups have a long way to go to regain the momentum that has been so successfully taken away from them by the Wise Use movement.

Not only are the Republicans following a key right-wing agenda, it is very much a corporate one too. The Republicans' commitment to 'get the government off our backs' is nothing less than a case where the politicians are in the pockets of industries. For example, senators who support oil and gas drilling in the Arctic National Wildlife Refuge in northern Alaska get much more money on average from oil and gas interests than those who oppose it.[189] What has also become public is the finance behind the new Republican assault on the environmental. It came from Project Relief, a coalition of 115 corporate and industrial lobby organisations that donated $10.3 million to Republican congressional campaigns. Project Relief has targeted key Congressmen such as David Mackintosh, Chairman of the House Regulatory Committee and committee chairmen Representatives Don Young and Senator Frank Murkowski of Alaska.[190]

It has been Project Relief's lawyers who have been drafting the new Republican laws and not the government's own experts. So water company experts have been drawing up legislation that affects their industry. It has been the car companies' lobbyist who drafted laws stopping the courts from imposing new clean air requirements on exhausts. So it is too, with the Chemical Industries Association who wrote legislation watering down pollution compliance laws. The lobbyist for the petrochemical industry even drafted a law that there should be no more federal regulations of any kind.[191]

Basically the Republican revolution, whilst pretending to benefit the average American, is simply a deregulated corporate agenda, which does not serve the health of the nation nor the state of the environment. In this sense, the Republican rout very much reinforces the Wise Use movement and *The Wise Use Agenda*. We now have two important parallel forces at work both pushing an anti-environmental agenda: Wise Use and the Republicans.

This gutting of environmental protection could intensify this year, if America votes in a Republican President. He (and it will be a he) is likely to strengthen ties between the Right and corporations. The green backlash and the Republican assault on the environmental movement are set to worsen, even if President Clinton clings to power. It is not just an isolated attack, though: it is part of something much bigger and broader that is gripping America. Roll on the millennium, so that regulations can be rolled back.

2

CULTURE WARS AND CONSPIRACY TALES

All that I have found are a baby's finger and an American flag.

Firefighter at Oklahoma, 19 April 1995[1]

THE CULTURE WARS

WELCOME TO THE WAR ZONE

Welcome to the war zone. However, there are no tanks, warships or heavy weapons here. There are no heroic war correspondents despatching stories of horror to an eager audience back home. The majority of people did not even know that the war had been declared until 19 April 1995. They were to be rudely awakened when the Alfred P. Murrah federal government building in Oklahoma was blown up, killing 168 people, many of whom were children.

The Wise Use movement and the anti-environmental assault in Washington are just another battle in this wider full-blown war that is waging across the country. It's a culture war, and it's being fought by sections of the political Right against the government and more progressive elements of society. The Right even call it a culture war, a battle for the soul of America. The green backlash is part of a greater right-wing backlash gripping the country that stretches from the Republicans in Washington to the breeding grounds of the militia in the Northwest.

After the Republican Party took control of Congress for the first time in thirty years in November 1994, a group of progressive researchers, academics and activists met to discuss the rise of the Right. The 'Blue Mountain Working Group', as the activists labelled themselves after the location of their meeting, issued the following warning: 'We see the current general right-wing backlash as one of the most significant political developments of the decade, combining well-funded national institutions with highly motivated grassroots activists,' said their release. They continued:

> While the political Right in the US can be bewildering in its complexity and shifting identities and allegiances, its players historically have assembled their core tenets and shared agendas from the same set of beliefs. They include conscious or unconscious support for white privilege; male supremacy; subservience of women and people of color; hierarchical religious and family structures; the protection of property rights over human rights; preservation of individual wealth; a rapacious form of unregulated free market capitalism; aggressive and unilateral military and foreign policies and authoritarian and punitive means of social control. They also include opposition to the feminist movement and abortion rights; democratic pluralism and cultural diversity; gay rights; government regulations concerning health, safety, and the environment; and minimum wage laws and union rights.[2]

Therefore the environmental movement and environmental regulation are not alone in being targeted by the Right. This is not to say that groups from the whole spectrum of the Right all oppose environmentalists, or work against environmental issues. Indeed, many on the Right would consider themselves the 'true environmentalists' as do many in the Wise Use movement. They also see themselves as pro-family and pro-morality, whereas environmentalists are portrayed as evil, totalitarians, socialists, communists, against liberty and freedom of the average individual.[3]

BACKLASH POLITICS

All right-wing movements are, according to political analyst Chip Berlet, who has been studying the Right for twenty years, 'anti-democratic in nature, promoting in various combinations and to varying degrees authoritarianism, xenophobia, conspiracy theories, nativism, racism, sexism, homophobia, demagoguery, and scapegoating'.[4] The Blue Mountain researchers believed that:

> [the] main goal of the anti-democratic Right is to craft a reactionary backlash movement to co-opt and reverse the gains of the progressive social movements of the 1960s and 1970s which sparked the ongoing civil rights, student rights, anti-war, feminist, ecology, and gay rights movements.[5]

It is widely perceived that the modern-day environmental movement grew out of the social change movements of the 1960s. Therefore the green movement, in a sociological context at least, is considered to be a social change movement, just like the other movements mentioned by the Blue Mountain Group. All these movements were asking people and society to re-evaluate certain relationships and attitudes between different groups or individuals. The environmental movement questioned the very relationship human society had with the planet, and asked people to re-evaluate the impact many everyday processes were having on the natural world. At its very heart was a belief that change was needed. A change in attitudes, a change in industry, a change in government, a change in society itself.

The political Right have resisted this rocking of the *status quo*. Furthermore, every single social change movement that has fought for equality, for justice, for basic human rights, has experienced, at some stage or other, a severe backlash for speaking out.[6] So not only are the Right opposed to the environmental movement, but they are actively creating a backlash against it. The green backlash follows the backlash that other social change movements, such as the civil rights movement, the anti-war movement, the Indian Rights movement, the women's movement and the gay rights movement, have already experienced and continue to endure. Some of the key figures within the Right who have been instrumental in targeting other scapegoated sections of society are now part of the growing green backlash.

Whilst championing the evils of these social change activists, the Right has lined their pockets, making millions by fund-raising against these activists by portraying them as threats to society, tradition and jobs. These new scapegoats are portrayed as bastions of evil, who threaten the bedrock of conservative values. The New Right made millions with the politics of resentment over abortion and women's rights.[7] They are now making millions off the greens, but in order for that to happen, greens had to become the new scapegoats for American society.

When the international threat of communism receded into the annals of history at the beginning of the 1990s, the American Right needed new scapegoats. Although the New Right had campaigned on traditional morality issues for years, these concerns had always been subordinate to the threat of communism. With the latter's demise, traditional morality became the new bonding force that would unite the New Right and the Christian Right.[8] For many on the Right the greens also became the new scapegoat, the new evil, to be stigmatised and attacked.

Republican William Dannemeyer wrote in 1990:

> The greatest enemy for the United States has always been the enemy from within – those who would sacrifice American independence on the altar of global interdependence. No foreign enemy could threaten our national security more than we have done to ourselves by keeping us mostly dependent on foreign sources of energy. The enemy I speak of in this case is a group of powerful special interests I refer to as the Environmental party . . . The environmental party is the enemy from within.[9]

The same year, Accuracy in Academia, a right-wing think-tank warned that, 'With the death of Marxist regimes in Eastern Europe and Central America, groups seeking control of the economic and political agenda have been without a philosophy. Many see the growing environmentalist movement as the saviour of left-wing politics.'[10] As millions celebrated Earth Day in April 1990, *Human Events: The National Conservative Weekly* cautioned that, 'Earth Day is a creation of the left, a pagan holiday devoted to fashioning a Socialist/Marxist world, even while Marx is being discredited across the globe.'[11]

The Right were reiterating not only the communist threat, but the religious argument too. 'Today we face an ideology as pitiless and messianic as Marxism,' wrote Llewellyn H. Rockwell Jr in an 'Anti-environmentalist manifesto', in the publication *Buchanan from the Right*. He continued:

> And like socialism a hundred years ago, it holds the moral high ground. Not as the brotherhood of man (since we live in post-Christian times), but as the brotherhood of bugs. Like socialism, environmentalism combines utopianism, statism and an atheistic religion.[12]

Calling it a 'battle for America', Craig Rucker from the Committee for a Conservative Tomorrow labelled environmentalists 'watermelon marxists', in 1992.[13]

THE FRONT LINE

In the last three years, right-wing and Wise Use activists have often repeated the watermelon jibe about environmentalists being green on the outside but red inside. But the backlash is more than words and name-calling. The current Republican agenda, one to roll back legislation, is one that the Right has been working on overtly in both local and federal government for some time. There have been distinct sections of the Right that have been directly and indirectly instrumental in the green backlash sweeping America: the Christian Coalition, the New Right, and right-wing think-tanks.

THE CHRISTIAN COALITION

The Christian Coalition is now the fastest growing grassroots movement in the USA, and its stated goal is to 'build the most powerful political force in American politics'.[14] Since President Clinton's election they have been growing at a rate of 10,000 new members a week, and soon the size of the mailing list and budget will exceed that of the Republican Party.[15] They already represent the largest single veto in the GOP and it was riding on the back of Christian Right support that the Republicans achieved such political gains in the November 1994 elections. Three months later, the Christian Coalition sent Chairs from fifty states to Washington to lobby in favour of Gingrich's anti-green 'Contract With America'.[16] Although they do not work directly on environmental issues, they could be tending that way. In early 1994, the Coalition's leader, Ralph Reed, addressed the annual public relations conference organised by the Public Affairs Council. Whilst discussing 'grassroots' organising, Reed 'suggested that the Christian Coalition could ally itself with the business community around "environmental issues" especially "if a corporation is involved in getting a lot of harassment"'.[17]

THE NEW RIGHT

Paul Weyrich is considered the guru and key strategist of the New Right. He told the *National Journal* in 1978 that, 'We are different from previous generations of conservatives . . . We are no longer working to preserve the *status quo*. We are radicals, working to overturn the present power structure in this country.'[18] Weyrich has been working towards that goal ever since. As part of his strategy he admits, 'I believe in rollback.'[19] Newt Gingrich's 'Contract With America' is just the end result of an ideological struggle embarked upon by Weyrich in the 1970s.

Weyrich, through the Free Congress Foundation (FCF) also gets directly involved in ecological issues. During the Earth Summit in 1992, the FCF co-hosted an 'Earth Summit Alternatives Conference', with the Society for Defense of Tradition, Family and Property (TFP). One of the delegates reiterated a favourite theme: the environmental movement is dominated by communists who are using it to promote Marxism.[20]

Both Alan Gottlieb and Ron Arnold have become increasingly associated with Weyrich. 'There has been a realignment of a whole spectra of a constellation of right-wing organisations, including the Center for the Defense of Free Enterprise, from the Unification Church network towards Paul Weyrich's New Right orbit,' alleges Tarso Ramos. Weyrich is president of the cable and satellite network, National Empowerment Television (NET), 'whose clear mission is to unite different reactionary movements in this country, different sorts of strains of reactionary action', says Tarso Ramos.[21] During a series of programmes championing private property rights, the editor remarked how people needed to 'spread the alarm and alert America to the growing oppression by rule and regulation which over-zealous bureaucracies and "EEEGs" – extremist environmentalist elitist groups – have imposed on basic and supposedly inalienable citizen rights'.[22] Wise Use activists such as William Perry Pendley from the MSLF, David Howard from the Alliance for America, and Ann Corcoran of the Land Rights Newsletter helped with the series.[23]

Weyrich wrote on the back cover of William Perry Pendley's book, published by Alan Gottlieb's Free Enterprise Press, *It Takes A Hero*:

> In an era without heroes this book gives us a number of outstanding examples of domestic freedom fighters who battle tyranny. In that sense, although the context is profoundly disturbing, it is also truly encouraging because such authentic Americans still exist and are willing to fight for their country.[24]

THINK-TANKS

Weyrich and Coors set up the Heritage Foundation in 1973, now considered the premier right-wing think-tank in the USA.[25] Modelled on Heritage, scores of smaller think-tanks have sprouted up around the country and there are now over a hundred of them in Washington DC alone. A former vice president of the Heritage Foundation calls the tanks the 'shock troops of the conservative revolution'.[26] 'The deluge of position papers the tanks rain down might be considered the movement's smart bombs, the understaffed state legislators who receive them targets of opportunity,' wrote Larry Hatfield and Dexter Waugh in the *San Francisco Examiner*.[27]

The agenda is one of radical free market conservatism. Hatfield and Waugh continued:

> With increasing frequency, legislation, proposed and enacted, can be traced directly to think-tank position papers on such conservative agenda items as welfare cuts, privatisation of public services, privatise options and parental choice in schools, deregulation of workplace safety, tax limitations and other reductions of government, even selling off the national parks.[28]

Economics America, a conservative organisation in America, publishes *The Right Guide: A Guide to Conservative and Right-of-Center Organisations*, which lists over 8,000 organisations and periodicals around the world, but mainly focuses on the United States.[29] Of the thirty-three subject areas, one is listed as working on the environment. In this category, twenty-four organisations are catalogued. Although there will be a great cross-over of subject area and very few institutions are purely single issue, some think-tanks are working on 'free market environmentalism', the conservative panacea to the world's problems.

'There are about ten or twelve very important think tanks that are actively working against the environmental cause, putting out position papers, that are trying to influence politicians and so on,' says Sara Diamond. 'The real activity comes from the Political Economy Research Center (PERC), the Competitive Enterprise Institute (CEI), and a whole flue of very well funded think-tanks, which do things like opinion articles trying to influence public opinion, trying to influence members of Congress.'[30] Many of these 'very well funded' think-tanks receive corporate and conservative foundation funding, a subject expanded on in the next chapter. PERC distributes papers which blame environmentalism for inventing crises and publishes op-ed pieces such as 'The Endangered Species Act: a perverse way to protect biodiversity'.[31]

Whilst launching a new project, the Center for Private Conservation, in December 1995, the CEI's President Fred Smith, remarked 'there is a need to develop new approaches to environmental policy to complement, not supplant, the current reliance upon the federal government for environmental protection'.[32] Furthermore, the CEI, along with organisations such as the Christian Coalition, Concerned Women for America, The Conservative Caucus and twenty other conservative organisations, are part of a right-wing legislative effort to 'defund the left' being coordinated by Republican members of Congress. Their target is to cut federal funding of non-profit advocacy organisations. Environmental groups under attack for receiving government funding include the American Farmland Trust; Center For Marine Conservation; Conservation Fund; Conservation International; Defenders of Wildlife; Environmental Defense Fund; National Audubon Society; National Wildlife Federation; Natural Resources Defense Council; The Nature Conservancy; Rainforest Alliance; Tides Foundation; Izaak Walton League; and World Wildlife Fund.[33]

EARTH DAY ALTERNATIVES

Both PERC and CEI are members of Earth Day Alternatives (EDA), a coalition of think-tanks who work specifically and generally on environmental issues. They offer their own solution to environmental problems with which to counter the environmental movement – free market environmentalism, which fits into the range of the corporate right-wing principles of privatisation, free enterprise economics and limited government. Much like its Wise Use counterparts, the CEI labels environmentalists 'anti-human'.[34]

The CEI is the coordinating body of the EDA and although the majority of think-tanks are based in the USA, there are also think-tanks in Australia, Canada, France and the United Kingdom.[35] The EDA, formed in 1990, is clear evidence of informal networking between key right-wing institutions around the world, sharing political ideology that ultimately would like to discredit the worldwide environmental movement, replacing it with advocates of free market environmentalism. The President of the CEI, Fred Smith, outlines what he means by free market environmentalism. It is 'an alternative approach to addressing environmental issues, recognising that those parts of the world that have the freest economies have also been most successful in protecting the environment,' says Smith. 'Those areas of the world that have had the most regulated, government controlled economies have also had the greatest environmental disasters,' he argues. 'There is a strange failure to address what the evidence shows, which is that market orientated societies do a very good job in the environmental arena.'[36]

While Smith is correct about the appalling environmental legacy left by the communist countries, he forgets that market economies can have disasters of their own, too. Nor does Smith take into account the legalised and non-legalised pollution from companies that takes place, every minute, every day, with chronic environmental and health consequences. In short, free market environmentalism could mean total privatisation, total utilisation, in an unregulated free market world. The EDA's free market manifesto also advocates the following:[37]

> Develop property rights in resources that are now controlled by bureaucrats. When workable systems are identified, return these resources to private stewards.
>
> Restructure pollution laws so that where possible, those who pollute pay for the damages . . .
>
> Encourage the responsible use of our rich scientific and technological resources to discover the true nature of possible threats to the environment. Condemn in the strongest possible terms those who would use distorted fraudulent science as weapons of panic and fear.

As their first goal, the EDA push for the complete privatisation of resources. All property, including National Parks and protected areas, would be privately owned, with the owners free to exploit the land. All creatures on land and sea could be privatised too. As the second goal, they advocate an unregulated market whereby regulation is replaced by pollution credits, so that pollution can be traded as just another commodity between companies. 'The problem with pollution permits,' writes Paul Hawken, in his best-selling book, *The Ecology of Commerce*, 'is that they do just that: permit pollution.'[38] Furthermore, pollution credit schemes, such as the one that is up and running in California, have been criticised by air quality advocates as being biased in favour of business.[39]

Moreover, in this unregulated economy, Fred Smith argues that, 'we should find ways of allowing people to address pollution by playing a positive ownership role'.[40] Basically the argument is that the public demands for protection against pollution are a better protector of the environment than government regulation. In response to this, Chip Berlet says:

> You can go to any major history and see the effect of unregulation. The very point of developing regulation around industrial society was that they were not only exploiting the workers to death they were befouling the planet, so regulation came because of that. What the right wing wants is for the public to have this role in the societal debate over balance of these issues and no power. The public power to confront these errors of industry is government regulation.[41]

As their third goal, Earth Day Alternatives hope to discredit environmental science, which they see as a threat to their political agenda. These think-tanks provide the intellectual acceptability to the green backlash movement, especially in the field of environmental science and counter-science. More will be said about EDA's attempts to discredit science later.

Another leading right-wing think-tank has also exploited Earth Day. The Heartland Institute in Illinois issued an *Earth Day Guide to Saving the Planet* for Earth Day 1996. The Guide explained how environmentalists had become leftist extremists who were anti-human, anti-science, anti-trade and anti-free enterprise, whilst dismissing global warming, ozone depletion, and environmental problems associated with chlorine, cars and dioxins. One of the authors of the Guide was Patrick Moore, an original founder of Greenpeace, but who now works for the forest industry in Canada. The Guide also listed many of EDA's think-tanks as sources of information as well as Arnold and Gottlieb's CDFE.[42]

THE HERITAGE FOUNDATION

Heritage, one of the EDA members, promotes a *laissez-faire*, free market approach to environmental issues. 'Privatisation is the highest priority for the environment,' Heritage's Policy Analyst for Environmental Affairs, John Shanahan told journalist David Helvarg. 'By denying ourselves material wealth today, by slowing the accumulation of wealth, we are denying our children. You deny the future by not using resources now.'[43] This language would certainly resonate well within the ears of the corporate donors to the Heritage Foundation and the Republican Party.

The Heritage Foundation itself was instrumental in providing the policy platform for the Reagan administration. In fact, an estimated two-thirds of Heritage's policy recommendations to the Reagan adminstration were adopted by the in-coming Republicans.[44] One recommendation that would be worked on would be environmentalism. Mark Dowie writes:

> After the election of Ronald Reagan, foundations funded by conservative business leaders like Joseph Coors and Richard Mellon Scaife mounted an ideological jihad against environmentalism. It was led by the Heritage Foundation, which has found ways to blame environmentalists for almost every social problem plaguing the country.[45]

In 1990 the Heritage Foundation's *Policy Review* outlined the conservatives' vision for the 1990s. One of the priorities was to 'Strangle the environmental movement. It's the greatest single threat to the American economy. It doesn't just include a few extremists. It is extremist. Even the mainstream environment groups are. An intellectual war must be waged – and won – against these upperclass Luddites.'[46]

THE PRO-GUN LOBBY

Alan Gottlieb is the President of two pro-gun foundations: the Second Amendment Foundation (SAF) and the Citizens Committee for the Right to Keep and Bear Arms (CCRKBA). CCRKBA's National Advisory Committee has included ex-Vice President Dan Quayle and ex-Secretary of Defense Dick Cheney.[47] According to Rudolph Rÿser from the Center for World Indigenous Studies, the CCRKBA was a participatory organisation in the Anti-Indian movement before it became incorporated into the larger anti-environmental/anti-government and pro-gun movements.[48] Furthermore, Gottlieb's book the *Things You Can Do to Defend Your Gun Rights* is routinely used by the militia.[49] The link between the militia and the anti-environmental movement is expanded on later in the chapter.

Gottlieb has been on the Board of Governors of the Council for National Policy since 1985. The CNP is a clandestine clearinghouse for key right-wing activists and funders and one which brings together people from the complete spectrum of the American Right.[50] Other key right-wing activists such as Weyrich and Pat Robertson the founder of the Christian Coalition also come together in the CNP. 'The militia and the Wise Use movement meet in the CNP, with Alan Gottlieb, Howard K. Phillips (National Taxpayers

Union), Robert K. Brown (*Soldier of Fortune*) and Helen Chenoweth,' alleges Sheila O'Donnell, who has spent years documenting violence against American citizens.[51]

Given that Gottlieb is such a pro-gun advocate, why is he promoting anti-environmentalism? The reasons are both political and financial. In a sense Gottlieb's involvement in anti-environmental activity is a microcosm of the culture wars. 'My impression, and this is a widely held impression, is that Gottlieb had no particular ideological interest in environmental issues *per se*,' says Tarso Ramos. Ramos believes that the attractiveness of environmental issues for Gottlieb was a combination of financial interests but also the political intent of not only rolling back environmental regulations, but also to create a pro-industry, anti-environmental counter-culture.[52] Furthermore, both Gottlieb and Arnold are actively trying to cross-pollinate Wise Use with anti-gun control.[53] Whilst Gottlieb argues that people who favour gun control are attacking liberty and freedom, Arnold and Gottlieb send out the same message about environmentalists too, giving reason for gun owners to oppose environmentalists. However, Arnold sees the 'anti-hunting, anti-gun movement advocating "animal rights"' as part of the environmental movement.[54] Arnold does not work solely on anti-environmentalism. He is also a contributory editor for the Second Amendment Foundation's *New Gun Week*.[55]

There is no doubt Alan Gottlieb is one of the best right-wing fund-raisers around, who also made a mint for Ronald Reagan.[56] Since Arnold joined the CDFE, Gottlieb has also turned his fund-raising hand to anti-environmentalism, with one simple aim: to make money. He does that by spreading fear. As with his gun literature, fear works. 'Last night I was raped . . . Where were the Police?' runs one Second Amendment Foundation advert. 'We all know of friends or family who have been raped, beaten, robbed or burglarised by thugs who don't think twice about hurting someone. You might be the next victim.'[57] Just as Gottlieb scares women into buying guns, so he seeks to persuade people into buying that the environmentalists are the enemy. It helps, Gottlieb told *The New York Times*, 'to have an evil empire: the spectre of an enemy to raise potential contributors' fears and open their wallets . . . For us, the environment has become the perfect bogeyman.'[58] 'Fear, hate, and revenge are the oldest tricks in the direct mail book,' Arnold told Jon Krakauer of *Outside Magazine* in 1991.[59]

Gottlieb told journalist Dave Helvarg:

I've never seen anything pay out as quickly as this whole Wise Use thing has done. What's really good about it is it touches the same kind of anger as the gun stuff, and not only generates a higher rate of return but also a higher average dollar donation. My gun stuff runs about $18. The Wise Use stuff breaks $40.[60]

As a fund-raising concept, Wise Use works. For as long as it does, both Gottlieb and Arnold will push anti-environmentalism to make a profit. 'Wise Use is a profit-making enterprise for those in control of the various organisations,' remarks Paul de Armond, who along with Jim Halpin interviewed Gottlieb at length.[61] De Armond continues:

He mails out twenty-five million direct-response letters every year. Recipients mail back $24 million. His costs, at 27 cents per letter, are $6.75 million, which means his mailers net $17.25 million. Put another way, $2.25 comes back for every dollar invested in direct-response letters.[62]

Of this, some $2 million goes to various anti-environmental clients.[63] Gottlieb also regularly holds fund-raising sessions at Wise Use conferences, and has held fund-raising seminars for key activists Chuck Cushman, William Perry Pendley and Clark Collins, so they can then fund-raise against the green threat.[64]

There are other exploitative forces around on the fringes of American politics, that of the conspiracy. The most frightening part is that it is on the increase.

LYNDON LAROUCHE: THE CONSPIRACY BEGINS

'Greenpeace: shock troops for a New Dark Age' ran the headline of *Executive Intelligence Review* (EIR) in April 1989. The article alluded that 'Greenpeace, indeed, is the ecological version of the Nazi SA, or what today might be called "eco-spetsnaz commandos."' This was just one of a number of startling accusations made in the article.[65] EIR is a magazine associated with extreme political chameleon, Lyndon LaRouche, someone renowned for his anti-semitic and racist views as well as his wild conspiracy theories such as the British royal family being behind the global drug trade and the environmental movement.

LaRouche, a former Marxist, did a severe political backflip in the early 1970s to the Right, where he has remained ever since. Dennis King, who has written an exposé on LaRouche, outlines what happened next.

> Organisers for his [LaRouche's] National Caucus of Labor Committees (NCLC) began contacting everyone they and their fellow radicals of the anti-Vietnam War movement had reviled – the CIA, and FBI, the Pentagon, local police red squads, wealthy conservatives, GOP [Republican] strategists, and even the Ku Klux Klan. Their announced objective was to build a grand coalition to rid American politics of the Enemy Within – the evil leftists, liberals, environmentalists and Zionists.[66]

Funding for his operations has been partly derived by pressuring supporters into taking out huge personal loans which are never paid back, credit card fraud and through a private political intelligence-gathering service.[67] Sometimes LaRouche goes too far, even in the eyes of the law. In December 1988, Larouche and six top aides were convicted on fraud and conspiracy charges and were sent to jail.[68] Finally released from prison on 26 January 1994, his global empire was kept running by his wife Helga Zepp LaRouche, whilst he was inside.

Although the LaRouche organisation is headquartered in the USA, mainly run as the National Caucus of Labor Committees (NCLC), it basically spans most of the world, run either through the International Caucus of Labor Committees (ICLC), the European Labor Party, the Schiller Institute or a LaRouche publication.[69] LaRouche publications are sold all over the world, and apart from pushing conspiracy theories, concentrate on two things. First they advocate a 'new world economic order', as the world is heading for economic collapse. 'Only we have the knowledge and methods for teaching a new elite the necessary historical, scientific and above all, economic knowledge the world needs for its survival,' says LaRouche.[70] Second, Larouche's followers vehemently promote nuclear power and high-tech industry, peddling fusion and fission power as the panaceas to the world's problems, whilst at the same time castigating anti-nuclear critics. 'Vote for me and I'll build 2,500 nuclear power plants,' LaRouche told voters in his 1980 presidential bid.[71]

More often than not Larouche articles mix fiction with conspiracy theories. The *Executive Intelligence Review* article alleged that Greenpeace could cause a major environmental disaster by sabotage to 'bring into operation a global "crisis management apparatus", that will be the *de facto* interim government of a "green fascist" new world order'.[72] The article also exaggerated the accusations from a 1989 film, *Survival in the High North*, by the Icelandic film-maker, and arch-Greenpeace critic, Magnus Gudmundsson.[73] Lyndon LaRouche, and publications associated with him have publicised Gudmundsson's work all over the world. Gudmundsson's relationship with the LaRouche people is examined more closely in chapter 13.

The EIR article also alleged that it is the British royal family and the Soviets that are Greenpeace's backers.[74] This is entirely consistent with LaRouche's beliefs. An editorial in *21st Century Science and Technology*, another LaRouche publication, claimed at the same time as the EIR article that 'In the Federal Republic of Germany, the Soviets have also used the Greens as a cover for covert military manoeuvres, involving their own *spetsnaz* special-forces troops in acts of sabotage and even, on occasion, assassinations.'[75]

Carrying on the conspiracy conundrum, the EIR article contended that Greenpeace is one of the 'officially patronised groups of the Lucis Trust, the umbrella organisation for the New Age movement, which was originally known as the Lucifer Trust'.[76] The Lucis Trust, according to LaRouche, 'is the leading, putatively respectable Britain-based Satan cult (it worships Lucifer)'.[77] Furthermore, the Trust opposes 'the materialism of science and every form of dogmatic theology, especially the Christian religion . . . and promotes a pagan form of Theosophical religion'. Other prominent front organisations for the Lucis Trust, apart from Greenpeace, are the following: the United Nations Association, the World Wildlife Fund UK, the Findhorn Foundation, Amnesty International, the Rudolf Steiner School, UNESCO, and UNICEF.[78] Feeling confused? What is the connection with Greenpeace, Soviets, paganism and wildly rampant conspiracy theories?

'LaRouchians accuse "dark forces" of being behind whatever happens in the world,' writes journalist Jerry Sommer, who has spent time studying LaRouche's activities in Germany. He continues:

> The conspirators are not always the same, and the conspiracy theories are not always logical, but conspirators are none the less almost always at work. The conspiracy theories are often so abstruse that it is simply incredible how people can cling to them. But they are often cleverly intermingled with facts and half truths, or legitimate political positions, to form an apparently inextricable tangle.[79]

According to LaRouche, everything can be traced to Babylonian times, and the forces of order – epitomised by Plato – and the forces of chaos – symbolised by Aristotle and the evil oligarchists. 'LaRouche claims that his followers represent a 3,000-year-old faction of "Neoplatonic humanists" locked in mortal struggle with an equally ancient "oligarchy",' says Dennis King.[80] Political analyst Chip Berlet adds:

> For the LaRouche people, if you accept the idea that there is a secret cabal that has been operating since the fall of the temple of Babylon, that was behind the Aristotelian thinking,

if you really believe that that is true, of course when you look at the environmental movement you see in it this conspiracy.[81]

Along with environmentalists and the Rockefellers, LaRouche often singles out Jews, especially Jewish bankers, as being behind the global conspiracy. Henry Kissinger is a particular LaRouche favourite, too, often being referred to as a 'Soviet Agent'. So too are the British bankers and the British royal family. In LaRouche's mind, Britain can be blamed for about anything, even Hitler was a British agent, according to LaRouche.[82] According to the LaRouchians, the British monarchy are also behind the international drug trade, as was outlined in his book *Dope Inc: Britain's Opium War Against the US*, published in 1978.[83]

Delving deeper into LaRouche's world we find that he considers many of the key conspirators part of a Malthusian plot to take over the world. Before the UN Conference on Population in Cairo, in 1994, the LaRouche organisation was rampant in its opposition. The 29 April edition of *Executive Intelligence Review* ran a headline 'Hitler in Blue Helmets: The Case for Halting Cairo 94', in which the conference is described as the 'direct heir to the 1932 New York eugenics conference which set Nazi policy'. The *New Federalist*, another LaRouche publication, quoted LaRouche as saying, 'There is no difference between those in the UN who are convening and supporting this population conference, and Adolf Hitler.' The article alleged that UN Secretary Boutros-Ghali was 'Britain's Brown-Skinned Hitler' who was installed by the British in 1992, because the British consider it easier to kill hundreds of millions of Africans and Asians under the direction of a 'brown-skinned' agent, rather than in their own name.[84]

The UN is, according to LaRouche, attempting to conquer the world with a police state or a UN-controlled 'New World Order'. *Executive Intelligence Review* called the Earth Summit in 1992, 'the Mother Earth cult festival in Rio de Janeiro, which is intended to spread mass psychosis and institutionalise a global police state in the name of saving the environment'. There was, according to EIR, 'fascism being promoted in Rio: the most evil threat, in sheer scale, which has ever faced humanity. For that, we need a herculean effort to educate people to overcome the brainwashing of the environmentalist media.'[85]

Asked just how significant a player LaRouche was in propagating the international anti-environmental message, Chip Berlet responds:

He has always been important [on both the national and international level], because he is like a deranged bee, cross-pollinating various flowers. His people are relentless in their pursuit of networking and even though people in the anti-environmental movement will swear up and down that he is crazy and that they do not work with him, in fact many of their staff do.[86]

One such busy bee is Roger Maduro, an Associate Editor of *21st Century Science and Technology* and rising anti-environmentalist. He is a visible bridge between the LaRouche organisation and the Wise Use movement, and the LaRouche magazine *21st Century Science and Technology* is increasingly becoming the mouth-piece for anti-greens too. 'Even a limited review of LaRouche-related publications makes clear that his organisation has found fertile ground among "wise use" and "property rights",' wrote Dan Barry and Ken Cook from the Environmental Working Group in 1994.[87]

Maduro is also a leading anti-environmental scientist, who co-authored the book *The Holes in the Ozone Scare. 21st Century* is peppered with articles claiming the fraudulence of environmental science and promoting Dixy Lee Ray, another leading science sceptic, up until her death in 1994. The process of debunking environmental science is discussed in chapter 5. But there are other links between the Wise Use movement and *21st Century*. Hugh Elsaesser is on the board of both the Environmental Conservation Organisation (ECO), a Wise Use network of over 400 groups and the Scientific Advisory board of *21st Century*. An article by Dr William Hazeltine, another ECO board member appeared in the Summer 1994 edition of *21st Century*.[88] Other prominent Wise Use activists have also had articles published in *21st Century*, such as William Perry Pendley, Kathleen Marquardt, and Michael Coffman.[89] An article by the hardline anti-environmental group, the Sahara Club, appeared in the Summer 1991 edition.[90] Teresa Platt from the Fishermen's Coalition, one of the 'Wise Use Heroes,' was interviewed by the EIR. Barry Clausen, a private investigator who infiltrated Earth First!, also has an article published and book advert in the Spring 1994 edition of *21st Century*. He has also teamed up with Roger Maduro to write a publication *Eco-Terrorism Watch* (see chapter 5).

What is frightening about LaRouche is his organisation's ability to collect and trade intelligence information all around the world. In the early 1980s, LaRouche and Helga met with the serving CIA deputy director to discuss Germany's environmental and peace movements.[91] Admiral Booby Ray Inman is said to have received 'enticing information' from LaRouche on the German Green Party. 'At the time, nobody in intelligence was covering them at all,' said Inman.[92] But there are other LaRouche attacks on the environmental movement in Germany. Also in the early 1980s the European Labor Party (ELP) attacked leading green activist Petra Kelly, who was leader of the German Greens until her death. Due to the harassment, Kelly sued for libel. According to her attorney, the 'LaRouchians had engaged in a "vicious campaign that made it difficult for her to appear in public".'[93]

The ELP had claimed that Greens were both fascists and were communist-controlled, a slight contradiction in terms. For example the party had, on the one hand distributed leaflets against the 'green environmental fascists',[94] whilst on the other, Helga Zepp LaRouche called for the German Green Party to be banned in the 1980s because it was run by the KGB.[95] Other magazines associated with LaRouche have also targeted Greenpeace. *Fusion* magazine in Germany has remarked that 'Greenpeace suggests using organic fertilisers to help save dying forests. Question: how many "Greens" do you need to fertilise a tree? Answer: Five. Actually one is enough, but you need four to persuade him to get into the bone crusher.'[96]

Because of his pro-nuclear stance, LaRouche has links to the nuclear industry, in particular the nuclear establishment, who although they must know he is an extremist are prepared to support him because of his pro-nuclear views. LaRouche's '1980 presidential campaign committee solicited donations from executives of nuclear power and aerospace corporations,' according to Dennis King.[97] Dozens of scientists and engineers signed a full-page *Fusion* advertisement backing LaRouche for President.

One thing is certain though about LaRouche, and that is that his conspiratorial rhetoric is currying favour with many people in the USA, not just in the nuclear industry, but on the extreme Right.

THE MILITIA: CONSPIRING OUT OF CONTROL

The most violent manifestation of the culture war and a reflection of the feeling of the despair, distrust and anger that are now endemic in many sections of the Right, is the rapid growth of the militia movement in the USA. The expediential growth of the militia, coupled with the depth of anti-government feeling, had caught many people and politicians totally off-guard.

A mass movement the size of the militias could not have grown as fast as it has without there being a very large pre-existing group of people willing to be organised around some extremely real grievances, argues Chip Berlet:

I think that is what people do not understand is that there is a very large sector of the population that has simply given up any faith in the government. Are they a paranoid lunatic movement? Absolutely, but even a paranoid lunatic movement has to be built on real grievances.[98]

By mid-1995, estimates of the number of members who were actively involved in the militia ranged from 10,000 to 40,000. Militias had formed in at least forty states, but militia organising was believed to be ongoing in at least fifty states.[99] Although their growth has been rapid, the militia would probably be confined to the concerns of a few leading researchers, academics and journalists who monitor the Right, if it were not for the bombing of the Alfred P. Murrah building in Oklahoma on 19 April 1995. The date of the bombing, it would become clear, was no accident.

In the worst episode of domestic terrorism ever witnessed on US soil, the bombing killed 168 people, including many children. In the immediate aftermath of the bombing, as journalists searched for solutions in the insanity of the chaos, many accused the Middle East of the atrocity, without one shred of evidence. They were made to eat their hasty words with the slow, painful realisation that this particular bomber had been home-grown. He was as American as apple pie. Fed on a diet of violent anti-government rhetoric, fuelled by conspiratorial madness, and ignited by hatred and a desire for revenge, the Oklahoma bombing was a disaster waiting to happen. Adding fuel to the fire had been the host of conservative radio talk-show hosts. For example, G. Gordon Liddy's instructions to his listeners was to shoot 'Head shots. Head shots' at members of the Bureau of Alcohol, Tobacco and Firearms (BATF). After Oklahoma, he amended his instructions for people to 'shoot in the groin area'.[100]

'Armed confrontation between the government and the militia members seem increasingly likely,' wrote Dan Junas, weeks before the bombing.[101] 'Some people connected with this movement advocate killing government officials,' wrote Kenneth Stern, from the American Jewish Committee, also just weeks before the bombing, 'They may attempt such an act.'[102] It is not that the warnings were not there, it is just that they were not heeded.

To some, Oklahoma was a justifiable act: 'As I watched the news I thought of all the working men who have been put out of business and the families made homeless by Federal regulations . . . ,' wrote Roger Hathaway of the Eagle Constitutional Militia, the day after the bombing. He continued:

I thought of the spirit of hope and the dreams of young people who find those dreams killed by environmentalists and political regulations. The vision that passed through my mind was an endless line of God's children who hold values of integrity, honesty, work ethic, and morality; there they are, walking so slowly in this unending line, thousands by thousands, heading where? There's no longer any place for them to go . . . They've been crushed by a liberal tyranny that takes delight in the destruction. Why has it taken so very long before some frustrated Americans got up the courage to do an Oklahoma City?[103]

William Pierce, author of the *Turner Diaries*, the fictional bible of the Far Right in which a federal building is blown up as an act of revenge, responded by remarking that:

When a government engages in terrorism against its own citizens, it should not be surprised when some of those citizens strike back and engage in terrorism . . . You are the real terrorists . . . the ones responsible for the bombing, for the deaths of these children.[104]

Bo Gritz, a renowned Christian Patriot with links to the militia, called the bombing 'a Rembrandt – a masterpiece of science and art together'.[105] A posting on the Nazi Bulletin Board on the Internet declared that although Timothy McVeigh, the alleged bomber, was 'widely condemned today', 'in the future he may be seen as a hero, as a fighter for his people'.[106]

There are many different reasons why people have joined the militia. Overriding factors in its growth are a deep distrust or fear of government, fear of losing their guns, and racial or political beliefs. It is a movement where a certain section of American society is restating their rights and beliefs, angered by recent social trends. 'The militias are a movement that for some time has been waiting to happen, born in the backlashes against civil rights, environmentalists, gay rights, the pro-choice movement and gun control,' argued Marc Cooper, in *The Nation*,[107] a sentiment reflected by Chip Berlet and Matthew Lyons. 'The militias are now only the most violent reflection of the backlash against the social-liberation movements of the 1960s and 1970s,' they argue.[108]

Berlet and Lyons write:

Overlapping right-wing social movements with militant factions appear to be coalescing within the militias. These include:

- Militant right-wing gun-rights advocates, anti-tax protestors, survivalists, far-right libertarians;
- Pre-existing elements of racist, anti-Semitic or Neo-Nazi movements such as the Posse Comitatus, Christian Identity, or Christian Patriots;

- Advocates of 'sovereign' citizenship, 'freeman' status, and other arguments rooted in a distorted analysis of the Fourteenth and Fifteenth Amendments. Among these groups are those who argue that African Americans are second-class citizens;
- The confrontational wing of the anti-abortion movement;
- Apocalyptic millennialists, including some Christians who believe that we are in the End Times;
- The dominion theology sector of the Christian evangelical Right, especially its most zealous and doctrinaire branch, Christian Reconstructionists;
- The most militant wing of the anti-environmentalist Wise Use movement;
- The most militant wing of the County movement, the Tenth Amendment movement, the states'-rights movement and the state-sovereignty.[109]

To understand the militia, it is necessary to understand the significance of 19 April, the day of the Oklahoma bombing: 19 April is considered to be 'Militia Day', where 'all able-bodies citizens are to assemble with their arms to celebrate their right to keep and bear arms'.[110] Not only is 19 April the day of the Battle of Lexington and Concord in 1775, between colonist Minutemen and the British which helped to spark the American revolution but there are two far more recent reasons that caused the militia to mobilise on that particular date. In April 1992, the White Supremacist Randy Weaver, wanted for a minor firearms charge, had holed up with his family in his cabin on Ruby Ridge in Idaho. In a botched attempt to arrest him, Weaver's wife and son were killed by the FBI. The shooting took place on 19 April, and the militia had its first martyrs. Then on 19 April 1993, the BATF stormed the Branch Davidian Cult in Waco, Texas, killing most of the inhabitants. Many saw this act alone as the government crossing one bridge too far on the road to tyranny. 'The government's refusal to take responsibility for those events as errors was a major factor in large segments of society of the US population thinking that the government had lost all credibility,' remarks Chip Berlet.[111]

Other factors had also brought the militia into direct conflict with the government. The Brady Bill mandated a waiting time for automatic weapons, an act that 'was like someone poured jet fuel on the movement', according to Ken Toole of the Montana Human Rights Network.[112] To add further fuel to the fire, the Crime Bill, passed in September 1994, outlawed nineteen kinds of semi-automatic weapons and accessories.[113] Many people believed that the government were just one step away from taking away the sacrosanct right to 'keep and bear arms'. 'For some members of the Patriot movement, these laws are the federal's first step in disarming the citizenry,' writes Dan Junas, 'to be followed by the much dreaded United Nations invasion and the imposition of the New World Order.'[114]

Many in the militia believe in some contorted concoction of the conspiracy theory, that takes many different forms. But most believe that the government is part of a greater conspiracy to take away their liberty and guns. The government are either controlled by a cabal of world elitists or Jewish bankers or the UN, who want to install a New World Order. Variations of the same theme may warn of the ZOG (Zionist Occupation Government), or the Freemasons, the Trilateral Commission or the British royal family. The conspiracies have reached epidemic proportions, fuelled by mysterious sightings of black helicopters and shadowy pictures of Russian attack craft seen on various state highways. These myths are perpetuated through magazines such as *Spotlight*.[115]

Many in the militia see the environmental movement as being in bed with the government and environmental regulations as just another example of a government severely out of control. Others see environmentalists as part of a greater conspiracy. Environmentalists have long been depicted as part of the 'New World Order' by some sections of the Right and conspiracies such as these are being believed by more and more people. In some counties, such as Ohio and Virginia, the militia are said to be stockpiling weapons for the imminent invasion by the UN.[116] The Militia of Montana, one of the leading militia organisations, talks openly of war, and its co-founder, John Trochmann openly accuses the US government of treason. It was formed directly after the Randy Weaver shooting.[117] 'What we are watching is that countries whose populations are under stress are unravelling and grasping at these theories out of desperation and that is the dynamic that people need to pay attention to,' argues Chip Berlet. 'It's not enough to look at LaRouche's palpably lunatic conspiracy theories and laugh at them if major sections of the population are not laughing.'[118]

MILITIA MEETS WISE USE

There are now open militia supporters in government, who are openly anti-environmental, too. One such person is Helen Chenoweth, who partly won against an incumbent Democrat by telling her voters that the only endangered species was the 'white Anglo-Saxon male'.[119] Chenoweth, a Council for National Policy member, Wise Use advocate, and anti-environmentalist, when speaking about Arnold and Gottlieb at the 1993 Wise Use conference, said, 'These two men are heroes. We must stand behind them and behold them.'[120]

She talks about the spiritual war between God-fearing Americans and environmentalists. 'We are in a battle today that is far more insidious and far more dangerous as conquering our people, their soul and this great nation.' Chenoweth's speeches are distributed by the Militia of Montana. In the tape she denounces environmentalists as communists and that the decision to protect the spotted owl habitat, was 'breaking down sovereignty and possibly leading to one-world government'.[121]

There are links in anti-government and scapegoating rhetoric between the anti-environmental movement and the militia itself. The same individuals also appear in both groups. 'The language is all the same and the theory is all the same, which I think is the most frightening part of it,' argues Sheila O'Donnell. 'We are not seeing individual people, we are seeing this theory being expanded and accepted by property rights people and the populace in general.'[122] Many people believe that the vitriolic rhetoric used by Wise Use could spur people to take up arms against the government and environmentalists. Dan Junas says:

> I think you sow a certain amount of discontent and you can expect that a certain amount of people can go in a violent direction: You are going to continue to see people whipping people up for their own political purposes, and that will continue to lead to the growth of the armed right wing.[123]

Other researchers have observed the synergy between the militia and Wise Use. 'There's no question that, at least in portions of the West in which we work, that Wise Use is one of the social forces that has created support for some of the stated objectives of some of these militia,' says Tarso Ramos, from the Western States Center. 'As evidence for that, we find militias that support some of the tenets of the Wise Use movement quite explicitly.'[124] Paul de Armond has identified individual links in the Washington States area between a local property rights organisation, the Snohomish County Property Rights Alliance (SNOCO PRA), and the Militia of Montana.[125] A joint meeting occurred in 1994 between the two groups. 'Militia of Montana literature keeps showing up at Wise Use property rights groups around here,' adds de Armond.[126]

However, some members of the group take their beliefs beyond just the militia. Two SNOCO PRA members have filed papers in the local court to make Snohomish a sovereign white state.[127] Dick Carver, the champion of the County movement from Catron County has also addressed the SNOCO PRA. Eric Ward and Devon Burghart from the Northwest Coalition Against

Malicious Harassment have also documented ties between property rights activists and Christian Patriots.[128]

Militia leaders themselves openly talk on environmental issues. John Trochmann from the Militia of Montana, who has links to the racist Aryan Nations, and who spews out conspiratorial rhetoric about a one world government and the United Nations wanting only 2 billion people, believes that the militia and Wise Use groups are headed in the same direction. 'All of it's about the same thing,' says Trochmann, 'This is our homeland. What's left after the United Nations takes all our land will be cities and concentration camps.' Trochmann has started warning his audiences of biospheres, where humans will be excluded for animals, whilst promoting the views and book of Michael Coffman, a leading Wise Use activist.[129]

Furthermore, Norm Olson from the Northern Michigan Regional Militia believes government departments such as the Environmental Protection Agency and the Bureau of Forestry are intruding on people's sovereignty.[130] Concerning the enforcement of the Endangered Species Act, Samuel Sherwood, the leader of the United States Militia Association, told an assembled audience in Challis, Idaho, that 'all it's going to take is for this crazy judge . . . to actually shut down the forests, and there will be blood in the streets'. Sherwood reportedly urged people in attendance to 'get a semi-automatic assault rifle and a revolver and a uniform'.[131]

In fact the militia are linked to the alleged perpetrators of the Oklahoma atrocity. Both James and Terry Nichols apparently belonged to the Militia of Montana for a while. Timothy McVeigh may also have attended their meetings.[132] It is also alleged that McVeigh acted as a bodyguard for one of the most notorious of militia spokespeople, Mark Koernke, more commonly known as Mark of Michigan.[133] Terry Nichols was a member of the Michigan Property Owners Association, formed by the prominent Wise Use activist Zeno Budd, who has also spoken at militia functions.[134]

THE COUNTY MOVEMENT

Furthermore, the 'County' or 'county supremacy' movement can be seen as an arm of the Wise Use movement and is another merging of anti-environmental, anti-federal government and anti-regulatory sentiments. The first county to attempt to overrule the federal government was Catron County that ruled that they had a historic 'civil right', and one based on 'custom and culture' to overrule federal regulation and work the land as they had always done. Any federal employee who violated the rights of local citizens could be arrested.[135] It was essentially a land-grab by the county over federal control, where no wilderness could be allowed.

The word play on 'custom and culture' is important. Sheila O'Donnell says:

> There is a great deal of discussion about custom and culture and those words are critical because of our founding documents, the Constitution and the Bill of Rights and the Declaration of Independence. They are all based on what was postulated as our custom and culture – civil rights litigation all through the years has been based on culture and custom. Now that language is being turned around and is being used as the reason and rationale behind stopping environmentalists.[136]

Those behind the County movement see it differently. 'This movement is a grass-roots response to an ever-growing federal bureaucracy which has strayed well beyond, and even violates, the provisions of the Constitution established by our Founding Fathers,' advocate the County Commissioners of Catron County.[137] 'In essence,' argue Paul de Armond and Jim Halpin, 'Catronions claimed the right to overgraze, overcut and pollute just as they had before the feds started passing environmental protection laws in the 1970s.'[138] The new ordinances would weaken the Endangered Species Act, the Clean Water Act, the Wild and Scenic Rivers Act, the Wilderness Act and the National Forest Management Act.[139] Furthermore, Catron County has also passed a measure requiring heads of households to own firearms to 'protect citizen's rights', and passed a resolution predicting 'much physical violence' if the government persevered with its 'arrogant' grazing reforms.[140]

The Counties movement was initiated and coordinated by the Wise Use organisation, the National Federal Lands Conference in Utah. Its Executive Director is Ruth Kaiser, considered by William Perry Pendley as one of his 'Wise Use Heroes'.[141] Furthermore, such Wise Use stalwarts such as Ron Arnold, Wayne Hage and Karen Budd were on the Advisory Board, but Arnold resigned due to time pressures rather than philosophical differences between the two organisations. The anti-regulatory Catron ordinances were written by Karen Budd, a Wyoming attorney and former assistant to James Watt.[142] The Counties movement is also supported by other Wise Use organisations such as People for the West! and the Center for the Defense of Free Enterprise.[143]

The NFLC claim to have some 200 counties enlisted in the Counties movement, although there was trouble ahead as Catron County were sued by the

Justice Department in 1995 for declaring state ownership.[144] When the case was settled the following year it was a real blow for the NFLC and the thirty-five counties who had passed anti-federal ordinances. In March 1996, a federal judge ruled that the County ordinances were in fact illegal.[145] Other counties have repealed federal environmental laws and shown complete contempt for international law as well. In early April 1995, the state of Arizona voted to legalise the manufacture and use of ozone-destroying CFCs. The measure contravened both US law and the Montreal Protocol. 'We resent people sitting in Washington DC, who have never visited Arizona, making laws and dictating what we should do,' said a supporter of the Bill, state Representative Don Aldridge.[146]

The NFLC also published an article in their October 1994 newsletter, 'Why there is a need for the militia in America'.[147] There was a need because 'we cannot trust the federal government to look out for our individual freedoms'. The Environmental Protection Agency was one of the government departments close to plunging the nation into an 'absolute dictatorial, martial law mode of repression'. The article concluded:

> At no time in our history . . . has our country needed a network of active militias across America to protect us from the monster we have allowed our federal government to become. Long live the militia! Long live freedom! Long live government that fears the people.[148]

During the debate over the Biodiversity Treaty, Ruth Kaiser warned that 'Americans will be disarmed through the Biodiversity Treaty! Eco-fascists have mounted a major effort to shut down gun ranges across the United States.' She also likened Acts by Bill Clinton to those of Adolf Hitler.[149] Rhetoric like this is going to stir up the considerable paranoid forces amongst the gun lobby even further against the environmental movement and the government. Kaiser also reiterated the militia favourite phobias of black helicopters being sighted along with large number of Russian military vehicles.[150]

Those people behind the County movement are prepared to enforce their beliefs. In Catron County, one local official behind the County movement chased two forest service employees off a road he was illegally bulldozing through protected federal forest land, backed up by an armed posse. When a federal biologist visited the region, he was simply told by a rancher 'If you ever come down to Catron County again, we'll blow your fucking head off.'[151] The same month the forest service office for the region was painted with a hammer and sickle.[152]

Dick Carver, a Nye County Commissioner has attended and spoken at both Wise Use events as well as Christian Identity functions.[153] 'White supremacist elements add a degree of militancy and experience in conflict with the Federal government that folks in the Wise Use movement and militias appreciate,' argues Jonathan Mozzochi, from the Coalition for Human Dignity.[154] In fact, members of the Aryan Nations had started a recruitment drive in 1991 in the resource-dependent communities of the Northwest, aimed at people who were threatened by economic downturn and seemingly hostile environmental legislation, according to the Coalition for Human Dignity. 'The coming warfare will literally be a war for survival not just of the logging industry or a way of life, but the very survival of the white race,' said one letter to the editor in *Log Trucker*.[155]

In a final ironic conspiratorial twist to the tale, only capable by a magazine associated with Lyndon LaRouche, *21st Century Science and Technology* reported that Nye County officials, like Carver were actually funded by the same people who were behind the environmental movement. It was also, according to Marjorie Hecht, the author of the article, Lord William Rees-Mogg and British Intelligence who were attempting to incite 'local militias to violence'.[156] Hecht also alleged that the British Crown and 'its wealthy friends and agents including top-level British intelligence figures like Rees-Mogg, is also manipulating *anti*-environmentalists'.[157] So the British monarchy are behind both the environmental and anti-environmental movement, and even the militia. The reason, according to Hecht, was to 'finish off what is known as the American system, so that an industrial American giant can never again threaten the British colonial system'.[158]

Another article in *21st Century* also alleged that 'the other British-party designers of the "Wise Use" movement, would destroy the United States faster than the Greens could do, by breaking up the Government and the union'.[159] You see everything is possible in the conspiracy. For further details, computer kids can scan talk.politics.guns, alt.conspiracy and misc.activism.militia on the Internet. But watch out for the black helicopters.

3

THE DEATH OF DEMOCRACY

The Twentieth Century has been characterised by three developments of great political importance; the growth of democracy, the growth of corporate power, and the growth of corporate propaganda as a means of protecting corporate power against democracy.[1]

The intensity of the corporate counter-attack against a burgeoning environmental and consumer rights opposition has been so powerful that in countries like America, it has, at best, derailed, at worst, destroyed, democracy itself. If democracy is meant to signify a representative government for all the people, in which everyone has an equal chance of being heard, of being able to influence their local politician, then democracy is dead, killed by the monoliths of the modern age – transnational corporations.

Overtly and covertly, by stealth and by design, big business has perverted the democratic process by buying politicians, by bribing them, by funding 'independent' think-tanks, by forming 'corporate front groups', by bullying citizens, by lobbying and by lying – all in the name of profit. At the same time, they have told us how much they care. The way companies have co-opted the environmental message and colonised the debate is examined in the next chapter.

CORPORATIONS AND POLITICS

REGULATORY POLITICS

Much of the focus for business activity has been around corporate opposition to regulation. As companies have expanded overseas, this anti-regulatory dogma has grown to incorporate moves to pre-empt international accountability. The case for less regulation is based on economics. Although most companies hanker after a free, unregulated, market in which to conduct their business, they need certain regulation to define the boundaries of the market and for public confidence. On the whole, though, once these boundaries have been set, corporations aim for a free market in which to operate. Free from environmental controls, free from worker protection legislation, and safeguards for society at large. Free to maximise profits.

Corporations should be left to regulate themselves, argue business leaders, although the problem with this argument is that corporations, through their pursuit of profit cannot be trusted to look after workers and the environment at the same time. 'Which arose first, the regulations or the violation of societal standards that called them forth?' asks green businessman Paul Hawken. 'It is the anti-democratic nature of business that has brought upon itself the minutiae of government regulation.'[2]

Journalist Mark Dowie writes of business's 'three bites of the apple' strategy towards environmental regulation:

> The first bite is to lobby against any legislation that restricts production; the second is weaken any legislation that cannot be defeated; and the third, and most commonly applied tactic, is to end run or subvert the implementation of environmental regulations.[3]

'Regulatory capture', whereby regulators are subverted rather than obeyed, becomes the strategy.[4] The pure pursuit of profit and anti-regulatory drive can lead some businesses to move to where there is the lowest common denominator of regulation and wages, which is often poor Third World countries desperate to attract outside investment at a dire cost to the workforce and the environment.

But this anti-regulatory attitude and adoration of the free market also unite many sections of the corporate world and the political Right. Many right-wing activists strive for a free market, not only on an economic basis, but also on an ideological one as well. 'When you talk about regulation, it is not difficult for ultra-conservatives to argue that all regulation is a denial of liberty,' says Chip Berlet of the Political Research Associates, adding that:

> What that does not take into account is that there is a balance between the needs of the individual and the needs of society. And it's that balance that is being debated right now. What environmentalists are demanding is that in the debate between individual liberty and societal needs, that environmental issues have to be factored in. What the Right is arguing is that all regulation is theft, all regulation is a denial of liberty. A very anarchic view

of society. It's a form of economic anarchy called capitalism. Unrestricted, unregulated capitalism, is a form of anarchy.[5]

An archaic belief in corporate anarchy that business leaders strive for on a daily basis. Transnationals head this drive for no national or international regulation, responsibility or accountability. Global, unrestricted, unregulated, unaccountable capitalism is the business end game, a goal to be fought for.

THE CORPORATE COUNTER-ATTACK

It was back in the 1970s when faced with fledgling environmental and consumer advocacy movements that the corporate world launched its counter-offensive. Industry had to act – it had been forced to modify its operations by new incoming regulations and been driven to clean up its image because of an increasingly sophisticated opposition. Environmental legislation was just the last in a long list of targets. 'Conflict between corporate and public goals has been a constant factor in regulatory politics,' commented Richard Kazis and Richard Grossman in their book *Fear at Work*. They wrote:

In the 1940s and 1950s the business community fought to undo worker gains of the 1930s. In the past decade there has been a similar push to reverse environmental protections and worker and consumer rights . . . This campaign began to take shape as soon as the first environmental, health, and workplace protection laws of the 1970s were passed.[6]

Business had to confront the environmental movement. 'The realisation that environmentalism was now in a position to cripple a major industry led to some serious reflection among the leaders of Corporate America,' writes journalist David Helvarg, in his authoritative book, *The War Against the Greens*. Whilst some companies worked on pollution prevention, or public relations, others, 'primarily in resource extraction industries such as oil, coal, timber, and beef, decided it was time to fight back against the environmentalists', continued Helvarg.[7] 'What is the goal the petroleum industry should strive for in the Decade of the Environment?' asked Bob Williams, a senior oil and gas advisor in his book *US Petroleum Strategies in the Decade of the Environment*. 'To put the environmental lobby out of business . . . There is no greater imperative . . . If the petroleum industry is to survive, it must render the environmental lobby superfluous, an anachronism.'[8]

Business leaders had to get involved in politics and learn how to become as ruthless in political spheres as they were in the business community. This is not to say that the corporate world purely started participating in the political process because of environmentalism, it was just one of a number of contributory factors. Company leaders had to understand how to become activists, learning from their opposition. As one businessman put it:

> We need a businessman's liberation movement and a businessman's liberation day and a businessman's liberation rally on the monument grounds of Washington, attended by thousands of businessmen shouting and carrying signs. We need a few businessmen to chain themselves to the White House fence.[9]

Although business leaders would be reluctant to be so obvious in their counter-offensive, corporations would learn much from their opposition movements. They also had to act in a more coherent and unified way. Coincidentally this process was already happening. In the 1980s, Patrick Useem examined in detail the political activity of the largest corporations in Britain and America. He found that the economies in both countries were increasingly becoming dominated by a relatively small number of large corporations who were linked in 'inclusive and diffusely structured networks'.

Useem found that a significant trend was one of 'interlocking ownership', whereby corporations were increasingly being owned by large financial institutions. Second, the boardroom was the second area of increasing cohesiveness, whereby a growing number of directors were sitting on the boards of fellow corporations, and had people from other corporations or financial institutions sitting on their own boards. The result of this network, argued Useem, was that there was now an 'informal organisational vehicle for the aggregation of pan-corporate political concerns'. Instead of acting as individual corporations, business leaders were increasingly making decisions which benefited the business community *per se*.[10]

Although Useem's book is now eleven years old, when asked how the findings of his book had stood the test of time, Useem replied that all evidence suggested that interconnected networks were still very much alive. If anything, Useem argued, they had become stronger and more international, reflecting the globalisation of business that has occurred over the last decade.[11] Adding weight to this assumption was the revelation in 1995 that over a third of Britain's biggest company boardrooms are connected through directorships.[12]

Becoming politically active poses a significant dilemma for corporations. Because what is best for maximising economic profitability of the company is often in direct contrast to what is socially and environmentally responsible or acceptable in how a corporation operates. 'The discrepancy between the acceptable public persona and the private hidden shadow of a corporation or government creates a vulnerable gap,' argues Joyce Nelson, the Canadian award-winning journalist, referring to it as 'the legitimacy gap'.[13]

This 'legitimacy gap' either forces companies to change public and political expectations so they reflect the corporate agenda, or makes the company itself change to meet society's expectations of the corporations. In fact, corporations are doing both. Sara Diamond, the political analyst, has outlined how corporations channelled money both into *electoral politics*, through funding politicians and political action committees as well as *ideological politics*, through the funding of think-tanks and other organisations.[14]

THE COLOUR OF MONEY

Business would influence both the electoral and ideological political arena, with the one commodity they had at their disposal : money. 'Only those who have accumulated lots of money are free to play in this version of democracy,' wrote William Greider in his exposé of how American democracy had been subverted by the corporate élite.[15] 'Money now creates the milieu in which debates are framed, voices heard, decisions made,' adds green businessman Paul Hawken, who adds that:

> Corporations have created a multi-billion-dollar industry of lobbyists, public relations firms, scholarly papers prepared by conservative think tanks, artificially generated 'people's' campaigns, 'expert' witnesses at public hearings who work for, or are paid by, corporate interests, and lawyers based in Washington DC, whose sole purpose is to influence lawmakers and regulators in their offices, in four star restaurants, at lavish receptions, on overseas junkets.[16]

'For many people, money explains almost everything about why democracy is in trouble,' writes William Greider, concluding that, 'Now, however, it is not an exaggeration to say that democracy itself has been "captured".'[17] Captured by an increasingly small number of corporations, who very rarely, if ever, have the public's interest at heart.

Over the last thirty years the global market has become dominated by an ever decreasing number of massive corporations, who make unbelievable sums of money and therefore wield immeasurable amounts of power. For example, in 1990, the *Fortune 500*, the world's largest 500 corporations, were responsible for 42 per cent of global GNP.[18] These same companies control two-thirds of world trade.[19] In fact, over 40 per cent of world trade is carried out by transnational companies. The top 500 manufacturing and top banking and insurance companies have a combined wealth of over US$10 trillion, twice that of the USA's GDP.[20]

These gargantuan giants are economically more powerful than many countries. For example, take the six largest private oil companies, the Anglo-Dutch company, Royal Dutch/Shell, British Petroleum and the American corporations, Exxon, Mobil, Texaco and Chevron. Individually, the sales for both Royal Dutch/Shell and Exxon, for example, were greater than the GNP for all but the twenty-seven most productive countries in 1991. BP and Mobil, the next largest oil companies, had sales greater than all but the top thirty-seven countries, Texaco and Chevron, all but the top fifty countries.[21]

Transnationals are, in effect, ungovernable organisations, out of control. A report by the US Office of Technology Assessment in 1993 found that transnationals were growing in power and authority so fast that the rate exceeded the ability of nation states to control them.[22] Individual shareholders have little or no effective say in the running of the business in which they have a financial interest, except by selling their share. Furthermore, a survey of 1,000 American corporations found that most undermined or denied the principle of 'one share, one vote' and other forms of shareholder rights.[23]

In Britain the situation is not much better. Evidence emerged of the impotence of individual shareholders at the 1995 Annual General Meeting of British Gas. After a concerted shareholder campaign to reduce the salary of top executives, including its Chief Executive who had awarded himself a 75 per cent pay rise to £475,000, 4,000 angry shareholders attended the meeting. The shareholders found that, despite a unanimous vote at the AGM not to re-elect the board, this was reversed by the large block votes of the city institutions.[24] 'Yesterday's squalid manipulations may have shown how the world works,' *The Guardian* editorial wrote. 'If so, they also showed how that world must now be radically changed.'[25]

For the moment, the situation for the individual shareholder is getting worse. In the aftermath of British Gas and RTZ AGMs in 1995, many companies were reviewing their share options with the view to preventing future disruption.

Shell, Barclays, Wimpey and British Aerospace had all been targeted by ethical, environmental and human rights campaigners that year. One of the proposals being considered by RTZ was to consolidate shares into blocks of 100, and then trade fractions of shares on the stock market. This would mean that only people who had bought 100 shares could attend an AGM – leaving out the small investor.[26]

It is large financial institutions that hold the majority of shares in a corporation and wield the power. Over half the equity held in British companies is held by big investors such as the Prudential, Standard Life and Scroedes.[27] However, Ward Morehouse, from the Council on International and Public Affairs and International Coalition for Justice in Bhopal, concludes that even large shareholder accountability is largely illusionary, as financiers shift their investments in a near continuous shuffle in pursuit of profit. It is estimated that about a trillion dollars *a day* is transferred on the international markets. Only about 10 per cent of this is investment, the rest is pure speculation. One vast unaccountable global gamble, that few people benefit from.[28]

'If accountability to consumers and shareholders is largely illusionary, multinational corporate accountability to communities, workers, the environment and even the countries in which they charter is virtually non-existent and declining rapidly,' declares Morehouse. He also outlines the inevitable consequences of such economic might coupled with such unaccountability.[29] 'Lord Acton predicted the inevitable consequence of this situation when he observed power corrupts and absolute power corrupts absolutely. The vast accumulation of power out of democratic control has led to abuses of that power, sometimes on a massive scale.'[30]

Nowhere is that abuse of power more evident than in politics.

CORPORATE CORRUPTION OF THE POLITICAL SYSTEM

After the successes in the early part of the 1970s of the environmental, labour and consumer rights movements, business started to fund politicians. This happened in both the United States and the United Kingdom. 'In both countries money generated by corporations is increasingly flowing into politics,' wrote Professor Michael Useem in 1984. 'Indeed, the flow has been transformed in the 1970s from a trickle to a torrent. The scale of the increase is so great that corporate money has now become a fundamental force in both British and American politics.'[31]

Useem believes it was a period of intensifying corporate activism. Few large British companies gave money to the Conservative Party in the early 1970s, but most did by the end of the decade. 'More than coincidentally, the decade ended with conservative, pro-business governments firmly in power,' says Useem. 'Radical reductions in spending on social programmes and ardent support for free enterprise were the dual banners of both the Reagan administration and the Thatcher government.'[32]

One of the mechanisms by which American business achieved this was through Political Action Committees, or PACs as they are known for short, which are entities by which money can be given to support a particular political candidate or cause. Through PAC funding, donors can influence the receiving politician. Although the ideas for PACs first originated in the 1930s with the labour unions, PACs have now become one of the vehicles by which corporations exert an inordinate amount of power over the American political system, because more than any other sector they have the cash to corrupt. In contrast, the average individual voter is left effectively powerless to influence their politician. 'PACs raise important questions about how money manipulates the political system,' argues the Council on Economic Priorities.[33]

Soon after the successes of the environmental movement in the early 1970s, changes began in the balance of PAC funding. For example, in 1974, the labour movement accounted for half the PAC money in the USA, by 1980, they contributed less than a quarter.[34] Corporations have increased their share ever since. In 1972 there were fewer than 100 corporate PACs in operation, by 1980 there were over 1,100.[35] Furthermore, PAC activity is concentrated amongst America's biggest companies and by the early 1980s specific industries were beginning to flex their corporate muscles.

The oil industry was one of the major funders that brought the anti-environmental Reagan administration into power. A survey by David Rogers from *The Boston Globe* in 1981 showed how:

> political committees financed by oil interests more than doubled their contributions between the 1978 and 1980 elections, and far more than other corporations, the industry used its wealth to support not incumbents but challengers, whose victories changed the face of Congress this year.

Federal Election Commission records reviewed by *The Globe* showed that oil contributions increased by at least $2.38 million in the last election, 'a 111 percent increase that was more than four times the rate of growth in general campaign spending'.[36] The increase in PAC donations had been due to a coherent campaign by the oil industry. 'We came to a decision some time ago that the only way we could change the political fortunes of the petroleum industry was to change Congress,' Harold Scroggins, a Washington lobbyist for the Independent Petroleum Producers Association, was quoted by Rogers as saying.[37]

Apart from appointing the anti-green James Watt as Interior Minister and Anne Gorsush as Head of the EPA, nearly sixty major existing environmental, health, safety, and other regulations were slated by Vice President Bush's Regulatory Relief Task Force. These ranged from controls on automobile pollution and safety to the testing of food and drugs, handling of hazardous wastes, noise reduction and health standards for asbestos, chromium, and cadmium.[38] Kazis and Grossman concluded that 'the Reagan administration has intentionally crippled the governmental machinery for protecting health, environment, and worker rights'.[39] This is essentially the intention of the current Republican revolution under Newt Gingrich's 'Contract With America'. By backing Reagan, if the corporate world had wanted less regulation on these critical issues, their investment paid off. The same is true today, and will hold true if a Republican president is elected this year. In the 1987–88 presidential election, the corporate world was up to its tricks again, donating more than $1.8 million or over 50 per cent of the $3.4 million of all PAC money donated to presidential campaigns.[40] By 1990, some 70 per cent of all contributions to major political parties was from corporations.[41]

Despite this, traditional mainstream environmental groups are still preoccupied with participating in this theoretical democracy. Outgunned by big business in terms of PAC contributions and lobbyists, business outmanoeuvres the environmentalists 10 to 1. 'No matter how large, clever, and sophisticated in the ways of Washington the environmental movement has become,' writes Mark Dowie, 'it has remained a mosquito on the hindquarters of the industrial elephant.'[42]

In 1995 a survey in *Business Week* also found interesting investments made by politicians in companies they had some political control over. Although ethics rules supposedly prohibit politicians from voting on or pushing legislation that benefits themselves, this does not include investments. Senator Frank Murkowski, a Republican from Alaska, reported owning between $15,001 and $50,000 of Louisiana-Pacific stock in 1994. In July Murkowski proposed legislation that could spur logging in the Tongass National Forest, where a Louisiana-Pacific subsidiary is one of the largest operators. A spokesperson for

Murkowski reported that he had apparently sold his stock before the decision was made.[43]

As mentioned in chapter 1, it transpired that the Republicans' 'Contract With America', and other anti-environmental legislation were being bank-rolled by a coalition of 115 corporate and industrial lobby organisations that donated $10.3 million to Republican Congressional campaigns, called Project Relief.[44] The investment had certainly paid off for industry as its experts and lawyers actually got to draft the new legislation, which rolls back environmental and health and worker safety gains.

EUROPE : THE SAME STORY

The corporate pollution of the political system is not isolated to the United States, either. The revolving door between business and government is going around and around in other places on the globe. Take, for example, the United Kingdom. 'Britain has been overrun by a seemingly endless succession of scandals at the interface between business and politics,' wrote Vincent Cable, the head of the International Economics Programme of the Royal Institute of International Affairs in April 1995.[45] But corruption seems to strike at the very heart of UK politics. 'The corruption of our politics can take many forms,' wrote the Liberal Democrat MP, David Alton, 'but the most insidious form of corruption is that which breaches no law but is part and parcel of the system.'[46]

This 'insidious form' of corruption Alton was talking about is the intricate link between big business and government. Alton explains how the 'merging of the interests of the big corporations and the parties reaches deep into Parliament itself'.[47] He adds:

> Past Ministers soon find solace in directorships and consultancies outside government. On the backbenches the same holds true. One hundred and thirty-five Conservative MPs hold 287 directorships and 146 consultancies between them, and the other parties are not immune. Twenty-nine Labour members share sixty directorships and forty-three consultancies; while Liberal Democrats hold a total of fifteen.[48]

In total, according to the 1995 Register of Members' Interests, some 389 of the 566 MPs had financial relationships with outside bodies, directly related to being an MP, and thirty held consultancies due to their position.[49] Of the big

British businesses listed in the MPs' Interests are companies such as Safeway, British Gas, ICL, Cable and Wireless, British Airways, Nissan UK, Fiat UK, British Aerospace, Albright and Wilson, British Petroleum, Nirex, Johnson Matthey, National Power, Racal, Rolls-Royce, Virgin Atlantic, Guinness, BAT Industries, and Sainsburys. Transnationals such as Boeing, BASF, American Express, Elf, Rothmans, Lockheed, Rhone-Poulenc, Amway Corporation, and ARCO are also represented in the Members' Interests.[50]

'The boardrooms of big British companies are brimming with politicians, both practising and retired. Most are Conservative,' wrote Patrick Hosking in *The Independent on Sunday* in September 1995. Hosking quotes a study by Labour Research that shows that sixty former Tory ministers who had left office since Mrs Thatcher came to power in 1979 had among them 407 paid posts. Altogether the ex-Ministers earned a total of £7.15 million or an average £119,200 each. 'Most large companies,' wrote Hosking, 'boast at least one figure from Westminster, or Whitehall in the boardroom. Usually it is an ex-minister.'[51] So it seems the revolving door just keeps on spinning in a synergetic relationship that both parties profit from, but at what cost to democracy? 'Whatever the attraction for the firms,' adds Hosking, 'the appeal for ex-ministers is usually for money.'[52]

By also representing corporations, MPs who are elected to represent constituents must suffer at least some conflict in interest. The dilemma that faces many an MP and many an ex-Minister, is that, as it has already been pointed out, what is normally good for a corporation is rarely directly beneficial to the public. Politicians are elected to speak on behalf of the people and if that process has been so badly corrupted by corporate interests, then somehow the system must surely change.

Just as in the USA, therefore, Britain's political system is influenced by business concerns and financed by corporate capital. The Conservative Party receive practically all of this corporate finance, although business has significantly reduced the money it has paid to the Party in recent years. For the financial year ending March 1994, the independent journal, *Labour Research*, traced payments of £2,500,000 from 168 companies, down from £3,762,000 in 1991.[53] In 1993/1994 only a quarter of Conservative income came from British limited companies, compared to 52 per cent in 1989, 55 per cent in 1990 and 36 per cent from 1991. Overseas donations do not have to be declared.[54]

Up to this point in time, companies had donated generously to the Thatcher and Major regimes. Many British corporate household names have aided and abetted the Conservative revolution: for example: Allied Lyons, Argyll Group

(Safeway supermarkets), Racal, United Biscuits, BAT, Black and Decker, Forte, Glaxo, Hanson, Inchape, John Menzies, Kingfisher, Kleinwort Benson, Legal and General, and Marks and Spencer, all donated monies, amongst others.[55] It is difficult to find concrete examples of where high party donors are rewarded with political favours. One such case could be where United Biscuits and two other companies that gave generously to the Conservative Party successfully lobbied the government to overturn a popular lorry ban, which limited heavy lorry movement in the capital. Furthermore, a Trade and Industry Minister, Tim Sainsbury also had a significant stake in the supermarket company that pressed for the end of the ban.[56]

Moreover, according to the MP David Alton, 'A recent study showed that the directors of companies donating more than £500,000 to Tory [Conservative party] funds had a 50 per cent chance of receiving an honour.'[57] If this honour is a knighthood, then the company directors find themselves gratuitously rewarded by their corporate gift, by being placed in the House of Lords, the upper house in Britain's two-tier parliamentary system. It is here that they could, if they so wish, block legislation that adversely affects the corporation in which they have a vested interest.[58]

The subject of corruption in British politics, or 'sleaze' as it is affectionately known, and rows over the 'revolving door' between business and ex-government ministers, and the exposure that some MPs had accepted money to ask questions in the House, forced John Major to set up an inquiry in 1994. Led by the respected Lord Nolan, the committee recommended seven principles of public life, when its findings were published in May 1995. These included 'Selflessness': whereby 'holders of public office should take decisions solely in terms of the public interest. They should not do so in order to gain financial or other material benefits for themselves, their family and friends'; 'Integrity': whereby 'holders of public office should not place themselves under any financial or other obligation to outside individuals or organisations that might influence them in the performance of their official duties'; and 'Honesty': whereby 'holders of public office have a duty to declare any private interests relating to their public duties and to take steps to resolve any conflicts arising in a way that protects the public interest'.[59]

It is difficult to see how the 70 per cent of MPs who have had the *honesty* to declare their private interests, can carry on being MPs with any *integrity* or *selflessness* if they continue to keep these outside interests, which do ultimately provoke a conflict of interest. In an unprecedented development in November 1995, the House of Commons voted to declare all the details of outside interests, against the wishes of the Prime Minister and the Conservative government.

But does actual corporate manipulation actually influence the way a politician will vote? A 1991 report by the American Center for Public Integrity discovered that they do. Charles Lewis, the Center's Chairman and Executive Director, said:

> We found that on critical matters which affect our daily lives, for the Congress of the United States, money talks and the public interest walks. In some instances, we have documented how some members of the House and Senate actually switched their votes to support special interests whose money they had received in the thousands of dollars.[60]

One such example the study quoted was the federal sugar price support programme which means that consumers pay an extra $43 billion in elevated sugar prices. In the House of Representatives every single member who received over $15,000 from sugar interests voted against reducing the price support mechanism. Furthermore, 85 per cent of senators who received $15,000 or more from sugar interests also voted against any reduction.[61]

Another example is where senators actually changed their votes in favour of the automobile industry after receiving PAC money from the industry, in a debate that would have made the car industry increase fuel efficiency standards on vehicles.[62] A measure that would have saved consumers money, reduced pollution, greenhouse gas emissions and the need to drill in environmentally sensitive areas such as the Arctic National Wildlife Refuge, which industry has been pressing for for years and is one of the key points on *The Wise Use Agenda*. Furthermore, through the use of lobbyists, 'corporate front groups' and think-tanks, industry misled the government and the American people in the fuel efficiency debate. The argument they used was that improved fuel efficiency would compromise safety, a statement flatly refuted by independent research.[63]

PERVERTING DEMOCRACY

William Greider, in his book *Who Will Tell the People?* illustrates how corporations, though their propaganda and through intermediaries, manipulated the truth so that senators were bombarded by seemingly 'independent' or 'white hat' organisations, who, believing industry fears about safety and urged on by lobbyists, backed the industry's position. The use of so-called independent grassroots supporters is a favourable PR tactic, explained in the next chapter. All this was coordinated by public relations firms such as Bonner & Associates, operating out of K Street, Washington DC.

Greider illustrates how firms such as Bonner & Associates will produce facts to support industry's lobbying claims. Or how Bonner will commission an opinion piece from a think-tank that his client just happen to underwrite. Polling firms are hired to produce favourable polling statistics. People can be found to support the industry's position, too. 'On the clean-air bill, we bring to the table a third party – white hat – groups who have no financial interest,' Bonner told Greider:

> It's not the auto industry trying to protect its financial stake. Now it's senior citizens worried about getting out of small cars with walkers . . . It's farm groups worrying about small trucks. It's people who need station wagons to drive kids to Little League games.[64]

This leads Greider to conclude, that:

> in the text book version of democracy, this activity is indistinguishable from any other form of democratic expression. In actuality, earnest citizens are being skilfully manipulated by powerful interests in a context designed to serve narrow corporate lobbying strategies, not free debate.[65]

The truth can be manipulated. When numerous states were considering banning disposable nappies in 1990 in the USA, Proctor and Gamble, the largest manufacturer of disposables in the country, commissioned the environmental consultants, Arthur D. Little, to undertake research into the issue. The states' action was being considered after a concerted campaign by environmental groups to highlight the ecological impact of disposable nappies. The consultants' report, completely contradictory to any other, however, showed that disposable nappies caused no greater environmental damage. The environmental campaign was effectively torpedoed by the results. Such 'tactical' research is more affordable to the more affluent – big business.[66]

In recent years, business in America has come up against the fastest growing grassroots environmental movement: the environmental justice movement, consisting mainly of people of colour who have highlighted the racial inequity of many corporate practices. In recent years, over sixty surveys have shown that there are racial inequalities in America's environmental laws and practices. Research like this has added great momentum to the environmental justice movement and its calls for reform to stop environmental racism. However, a report published in 1994 flatly refuted the findings of earlier research, concluding that it could not discover cases of racial inequity in the siting of toxic waste dumps. In fact, if anything, the report said these dumps were more likely to be located in white working-class areas.[67] Curious as to why this survey differed so differently from their own, activists soon discovered that, although it was written by the University of Massachusetts, it was actually sponsored by the Institute of Chemical Waste Management, an industry trade association. Furthermore, WMX, the world's largest waste company had donated $250,000 to the study.[68]

'Corporate-directed research, with its built-in conflicts of interest, poses demonstrable threats to public health and safety,' writes Ron Nixon in his article 'Science for sale'. 'It can serve as a tool in corporate efforts to derail movements attempting to impose even minimal restraints on their behaviour.'[69] Corporate research in the USA actually helps determine certain important regulatory guidelines, such as those for the Occupational Safety and Health Administration.[70]

There is also increasing corporate control over scientific research. For example, in the growing field of biotechnology in the UK, some 74 per cent of biotechnology companies have links with university research. For one-third of these, the links are central to the research effort.[71] The closeness between the oil industry and post-graduate research is shown by the fact that 47 per cent of graduates taking geological jobs between 1985 and 1990 in Britain joined oil companies or contracting firms.[72]

Serious ramifications for the future of the independence of British science occurred when Britain's Office of Science and Technology, the government's science body, was moved from the Cabinet office to the control of the Department of Industry in July 1995. It was a move that naturally angered Britain's scientific community. 'Try as you might,' wrote the leader writers at the *New Scientist*, 'it is impossible to find a logical reason for last week's decision . . . And what of research on the environment – the results of which are often painful for industry?'[73] The journal *Nature* called the decision 'potentially disastrous'.[74]

The situation in the USA also worries activists. 'The extent of corporate influence on scientific research today is staggering,' continues Ron Nixon. 'While universities have long had a cozy relationship with corporate research funders, the increasing weight of private interests is beginning to cause concern in the halls of academia,' he continued. Nixon pointed out over 1,000 university–industry cooperative centres have been set up on over 200 US campuses. In some sectors of research, industry grants outstrip government money.[75]

'Truth,' concluded Cynthia Crossen in her book *Tainted Truth: The Manipulation of Fact in America*, 'has come to those who commission it.'[76]

'CORPORATE FRONT GROUPS'

Corporations, though, will go even further to manipulate the truth. Industry is playing corporate grassroots politics, a tactic modelled on public interest organisations and taken from the environmental movement. PR firms have advised corporations to set up 'front groups', with smooth eco-sounding names, with reassuring brochures and designer labels designed to fool the public.

The increase in the use of 'corporate front groups' is 'a direct response to the burgeoning consumer, citizen and environmental movements', according to Mark Megalli and Andy Friedman, in *Masks of Deception: Corporate Front Groups in America*. They continue:

> Before these movements took hold in the late 1960s, big business corporations delivered their messages through their traditional lobbyists in Washington. The names of these old-fashioned corporate lobbies told the stories – Beer Institute, National Coal Association, Chamber of Commerce, American Petroleum Institute . . . But as public interest groups began to win widespread public support, it became clear that new mechanisms were needed to deliver the corporate message.[77]

Take, for example, the Clean Air and Fuel Efficiency debates in the USA. Not content with independent lobbying, Ford, General Motors and Chrysler, as well as the Motor Vehicle Manufacturers of America (MVMA), the National Automobile Dealers Association and the Association of International Automobile Manufacturers (AIAM) formed the Coalition for Vehicle Choice (CVC). This 'corporate front' group was actually created by the public relations company E. Bruce Harrison.[78] Ron DeFore, Vice President at E. Bruce Harrison and communications director of the Coalition, said Harrison set up the group but denied that it is a front for automakers. 'CVC is a broad-based coalition of nearly 1,200 automotive, insurance, safety, farm and consumer organizations,' he said, that represents millions of Americans from all walks of life.[79]

CVC's basic job was to convince the public that safety would be compromised by increases in fuel efficiency, although scientific evidence did not support this claim. It became an acrimonious war of words. A former administrator of the National Highway Traffic Safety Administration (NHTSA), Joan Claybrook, who went on to direct the Consumer group, Public Citizen, criticised Diane Steed, the head of Coalition for Vehicle Choice. 'Throughout the 1980s, Diane Steed personally led the fight against air bags, rear-seat shoulder belts, pick-up and minivan safety, rollover and side-impact protection,' said Claybrook, adding:

> Now, with a straight face, in the employ of the auto industry's Coalition for Vehicle Choice, she is trying to make us believe that improving fuel efficiency will put safety at risk. She is lying. Everyone of us who has spent a lifetime working to make cars safer knows that cars can be designed to be both safer and more fuel-efficient at the same time.[80]

CVC's adverts, which claimed that small cars were less safe than larger ones, were described by a coalition of environmental, consumer and public-health groups as one of the most misleading adverts of 1991.[81]

The Coalition for Vehicle Choice is nothing really new, as corporate citizen action has been going for a while, long before the first Wise Use conference in 1988. Since its inception in 1983, the Citizens for Sensible Control of Acid Rain has spent over $7.5 million attempting to defeat acid rain legislation, without a sensible citizen in sight, just the large electricity companies.[82] The utilities, including many nuclear companies, were also funding the United States Council for Energy Awareness to the tune of $20 million a year. It has spent the last sixteen years trying to promote nuclear power in the name of energy security. The Council's board include representatives from General Electric, Bechtel, and Westinghouse.[83] So, despite their names, the acid rain producers were promoting policies which would produce more acid rain and the nuclear companies more nuclear power. The same applies for other industries. Take the Alliance for Responsible CFC Policy, whose 400 members consist of companies and trade associations that produce CFCs and want to keep it that way,[84] or the Global Climate Coalition (GCC), whose oil and gas producer members want to keep changing the climate.

As global industries fight global environmental problems, industry front groups will increasingly be used on a worldwide basis. Transnational companies are now working together with some governments in international coalitions fighting transcontinental treaties which hope to solve these global problems. Not only interested in gutting domestic control and regulation, transnational companies are out to admonish any international oversight or accountability. This harsh reality will be clothed in language of 'scientific consensus', 'sound science' and a 'need for further research'. The way the language is changing in the environmental debate is examined more in chapters 4 and 5. Nowhere have international front groups sprouted up faster than in the debate over climate change. However, other international front groups have also appeared in the timber, chemical, and carton industries, which is expanded on in later chapters.

The Global Climate Coalition (GCC) is a coalition of the largest oil, gas, coal, utility, automobile and chemical companies and business trade associations. According to the GCC, it was established to 'coordinate business participation in the scientific and policy debate on the global climate change issue'. Congressman George Miller disagrees, commenting that the GCC's single purpose is the 'unimpeded production of oil, gas and coal'.[85] Membership includes such corporate giants as Amoco, ARCO, BP, Chevron, Dow Chemical, Du Pont, Exxon, Shell, Texaco and Union Carbide, and such heavyweight associations and institutes as the American Automobile Manufacturers Association, American Mining Congress, American Iron & Steel Institute, American Petroleum Institute, Chemical Manufacturers Association, and the US Chamber of Commerce.[86] All these companies and institutions are extremely energy-intensive and immense CO_2 emitters and therefore have a vested interest in seeing any international treaties on CO_2 reduction weakened. It is no coincidence that the GCC employs E. Bruce Harrison, one of the leading anti-environmental PR firms, to assist in this process.[87]

Despite the fact that there is now a scientific consensus that global warming is actually occurring and that it is caused by man-made emissions, the GCC is attempting to obstruct this serious environmental debate. Over the last three years, the GCC has grown increasingly hostile to the findings of the IPCC – the Intergovernmental Panel on Climate Change – made up of some 300 of the world's leading climate scientists and the leading scientific authority on climate change. In turn, the IPCC's scientists draw on the expertise of a further 1,500 scientists. Set up in 1988 by the UN's General Assembly, the IPCC's remit was to evaluate whether the Earth was warming up due to man-made emissions of greenhouse gases. In 1990, when they first reported back, they had a simple message. The greenhouse effect was real, and urgent action was required to control emissions, namely carbon dioxide, the primary greenhouse gas. By 1995 the IPCC had established for the first time that industrial emissions of greenhouse gases had contributed to a one-degree Fahrenheit warming of the planet over the past century. The IPCC concluded that, unless emissions are reduced substantially, an additional two to eight degree Fahrenheit warming is expected over the next century.[88]

Although there are inherent uncertainties about predicting something as complex as the world's climate and how it will change, scientists agree that urgent action is needed. In the pursuit of profit for its members, the GCC has set out to make sure that action does not happen. To this end, by lobbying the US government, the GCC was instrumental in undermining the inclusion of emission reduction targets at Rio in 1992.[89] They have been working against mandatory emission targets ever since. Three years later, in the run up to the biggest climate meeting since UNCED, the Framework Convention on Climate Change in Berlin in March 1995, the GCC infiltrated the preparatory climate talks. Ravi Sharma, associate director of the New Delhi-based Centre for Science and Environment (CSE), lamented that 'the polluters of this world have the leverage of control over the negotiations'. The GCC was widely credited with lobbying the US government into inaction and watering down the scientific case for action on reducing greenhouse gas emissions.[90]

The GCC has also started to shift the blame of climate change to the developing countries and has used the threat of job losses to argue against emission controls. After Berlin, the executive director of the GCC, John Shlaes, remarked, 'It's clear the agreement reached by UN negotiators in Berlin gives the developing countries like China, India, and Mexico a free ride.'[91] Greenhouse emissions would 'create a competitive advantage for our international trading partners at the expense of US jobs'.[92] Furthermore, in direct conflict with the IPCC's findings, the GCC concludes 'it is an open question whether man-made contributions of greenhouse gases have contributed, or will ever contribute, to an "enhanced greenhouse effect"'.[93]

The Berlin Summit saw the emergence of two new worldwide industry lobby groups with eco-sounding names: the Climate Council and the International Climate Change Partnership or the ICCP. The Climate Council was formed by Don Pearlman from the law firm, Patton, Boggs and Blow, whose clients have included the Haitian dictator Duvalier, the Guatemalan military regime and the disgraced BCCI bank. For the Berlin meeting, Pearlman's clients also included Du Pont, Exxon, Texaco and Shell. Pearlman also has intricate links with Middle Eastern countries who want to exploit their vast oil reserves without any hindrance from any binding climate change agreements. Described by Jeremy Leggett from Greenpeace International as the 'high priest of the carbon club', Pearlman used scientists from the oil-rich Gulf States to attempt to discredit the IPCC. He was accused of 'endless hair-splitting' by a respected Dutch climatologist and dictated precise tactical manoeuvres to the Arab delegation at Berlin, stalling any set timetables for emission reductions.[94]

The ICCP is surely some kind of humorous pun on the IPCC, but if the world warms up, partly thanks to the lobbying efforts of these organisations, the global population will not be laughing at the fossil fuel industry's joke. Ironically, the ICCP or the International Climate Change Partnership, seems just as its name suggests, a partnership to change the climate. Although its promotional literature views it differently, the ICCP is, in its own words, 'a coalition of companies and associations committed to responsible participation in the climate change policy process'.[95]

Although ICCP's Executive Director, Kevin Fay, commended the decisions made at the Berlin conference, he remained 'concerned that the timing of future negotiations may be too rapid, the parties agreed to a process grounded in scientific, technical and economic assessment'. This effectively means that there should be no emission reduction targets. Moreover, the ICCP singled out developing countries to have more of a 'realistic role',[96] which means, in untwisted jargon, that their role should admit that they are to blame. Due to the efforts of the fossil fuel lobby, the US government only favoured voluntary emission reduction targets at Berlin.

Ironically, another business organisation surfaced at Berlin – the Business Council for a Sustainable Energy Future (BCSEF). Its name sends alarm bells ringing and it is just another vested 'industry front group'. Made up of companies who are due to benefit from emission reduction measures (solar panel manufacturers and makers of energy conservation equipment, as well as from gas companies), the BCSEF is at loggerheads with the GCC.[97]

Big oil is heavily represented in another so-called grassroots front group called the National Wetlands Coalition (NWC), operating solely in the United States. What are Amoco, ARCO, BP, Chevron, Conoco, Enron, Exxon, Marathon Oil, Mobil, Shell, Texaco and Unocal all doing preserving wetlands, one might ask? The answer is, quite simply, that they are not. The NWC exists to weaken legislation that protects wetlands and promote drilling for oil and gas on wetlands. Office space and staff for the NWC are provided by the Washington DC-based law firm, Van Ness Feldman and Curtis.[98] Some $7.8 million was raised by the NWC from British Petroleum, Georgia Pacific, Kerr-McGee and Occidental. The chairman of NWC is Leighton Steward, the CEO of Louisiana Land and Exploration, the largest owner of coastal wetlands in the country.[99]

Van Ness Feldman and Curtis are also working to gut other legislation on the Endangered Species Act. 'Form a broad-based coalition with a simple name and a grassroots orientation. Incorporate [it] as a non-profit, develop easy-to-read information packets for Congress and the news media and woo members from virtually all walks of life,' Van Ness *et al.* told some south-western utilities who subsequently formed the National Endangered Species Act Reform Coalition.[100] And their clients are having success: working with Dan Quayle's Council on Competitiveness, the NWC succeeded in opening up half the USA's wetlands for development.[101] This is a case of legal firms taking up PR anti-green antics for their clients.

CORPORATE FUNDING

CORPORATE FUNDING OF WISE USE AND ENVIRONMENTAL GROUPS

Although the majority of the above organisations are purely corporate front groups, run out of the offices of PR companies and the like, some Wise Use organisations also receive corporate support and back-up. This said, the majority of Wise Use groups do not receive corporate funding. 'Of the roughly fifteen hundred or so of organisations in the United States which claim to be part of the Wise Use movement, a very, very, small fraction of them actually directly receive money from the corporate world,' says Dan Barry, the Director of CLEAR.[102]

When Wise Use activity started gaining media attention, many environmental groups dismissed the movement as a front for industry because so much corporate money was supposedly being received by these groups. This was wrong. It was also ironic because the majority of large mainstream environmental organisations in the USA also receive corporate and/or foundation funding. This is, of course, something which the Wise Use movement points out with great regularity. Wise Users believe that they are the underdogs up against the corporate-backed environmental goliath.

Donating money to environmental groups is a tactic favoured by PR companies to improve the environmental image of their clients. 'Cash rich companies, PR people say,' reported the PR trade journal *O'Dwyer's PR Services*, 'are funding hard-up environmental groups in the belief the imprimatur of activists will go a long way in improving their reputation among environmentally aware consumers.'[103] But some environmental organisations actively solicit industry money from some renowned polluters. For example, in 1993, Eastman Kodak gave the Wildlife Fund for Nature USA $2.5 million – its largest-ever single donation – and its chief executive, Kay Whitmore, is now on WWF's board. The Audubon Society gets donations from General Electric, Waste Management Inc. and Proctor and Gamble, and the National Wildlife Federation receives funds from Du Pont, Dow, Monsanto, 3M and Shell.[104] Co-option of environmental organisations is examined in greater detail in the next chapter.

Although some corporations initially turned up at the first Wise Use conference at Reno, with the likes of Du Pont, Exxon, Georgia Pacific Corp, Louisiana-Pacific Corporation, MacMillan Bloedel in attendance, visible widespread direct support for Wise Use from individual corporations seems to have decreased. This is due, no doubt, to adverse publicity about right-wing extremists and Moonies, and also the increasing occurrence of anti-environmental violence, all of which offer potential significant public relations problems for industry. Therefore, for the most part, if there has been funding it has been through more of a covert channel, say, through industry associations. For example in 1994 the American Forest and Paper Association, American Forest Council, American Mining Congress, California Mining Association, California Forestry Association, and Northwest Mining Association supported the Alliance for America.[105] Now all of the above corporations who were at Reno may be represented by the above trade associations but there is no visible corporate link.

However, there are still some industries that visibly support Wise Use groups. Traditionally this funding has come from the resource use industries and off-road vehicle manufacturers. The anti-environmental movement has actively sought corporate cash for their campaigning. For example, at the 1992 Alliance for America conference, Tom Synhorst, a professional fund-raiser, told the audience that the emphasis of the Alliance's activities should be on 'cultivating high donor corporate donations'.[106] The same year, and a month later, one of the main themes of the 1992 Wise Use conference at Reno was to gain credibility with major CEOs of all industries, in order to solicit funds from them.[107] Some specific Wise Use groups had corporate funding from the start though, for example, People for the West! (PFW!) and Blue Ribbon Coalition which have relatively intimate links with industry.

PFW! was formed to fight the amendment to an 1872 Mining Law which still allows anyone who finds valuable mineral deposits on federal land the right to purchase that land for a maximum of $5 per acre. The law is invaluable to mining companies, but robs taxpayers of their property and destroys the environment. PFW! is an outgrowth of the Western States Public Lands Coalition, before it changed its name to the National Coalition on Public Lands and Natural Resources. Run by the Grannells, who were brought in to run the first major Wise Use campaign of the timber industry (see chapter 1), but who then moved to help the mining industry with PFW!.[108]

PFW! has a full-time staff and annual budget which is 75 per cent funded by mining, ranching and timber companies. The Board of Directors is dominated by mining executives, taking twelve of the thirteen places, including those from Pegasus Gold and Homestake Mining.[109] Other corporate money comes from Chevron, who have donated over $92,000 to PFW! and other Wise Use groups such as the National Wilderness Institute.[110] It is easy to see why Chevron has been funding PFW!. Under the 1872 law, the company bought 2,000 acres of public land for $10,180, or $5 an acre. It will pay no royalties for the minerals it extracts which are estimated to be worth billions of dollars.[111]

At PFW!'s first annual conference in 1994 some 500 people attended. The themes discussed were how to appear more mainstream and how to become an important national political force. Also discussed was how to start networking with other 'Wise Use' groups around the world. PFW! is now forging close working links with The Forest Protection Society, a front group for the timber industry from Australia (see chapter 9).

In January 1996 it was announced that PFW! would merge with the Western States Coalition (WSC), an organisation comprised of elected and appointed state and local government officials, industry representatives and other individuals who have 'an interest in the management of public policy as it relates to issues impacting the social, cultural and community stability of the Western United States'.

The WSC has held six summits since its inauguration in 1994 with many Wise Use leaders in attendance, including Rep. Helen Chenoweth, Chuck Cushman, Dick Carver, Bruce Vincent from the Alliance for America, William Perry Pendley from the MSLF, Ruth Kaiser from NFLC, and Roger Maduro from *21st Century Science and Technology*.[112]

Another Wise Use group that receives large amounts of corporate funding is the Blue Ribbon Coalition. Financial contributions come from American Honda Motor Company, American Suzuki Motor Company, Motorcycle Industry Council, Yamaha Motor Corporation, Kawasaki Motor Corporation, and the American Petroleum Institute.[113] Representatives from Kawasaki,

Yamaha and Honda sit on the Advisory Board,[114] whilst another Wise Use activist, Grant Gerber, from the Wilderness Impact Research Foundation is also represented.[115] The Coalition is upfront about funding and publishes lists of supporters and contributors. It also tries to portray itself as a grassroots organisation, and in order to do this, the Coalition adds up all the members of all the associations who show support for it, thus inflating the actual figures of true Coalition members.[116]

The Coalition preaches the mainstream angle, one of moderation, too. Clark Collins, its Executive Director writes, 'the Blue Ribbon Coalition have a reputation among opponents of "Wise Use" as being against environmentalists. Nothing could be farther from the truth.'[117] Collins, like many within the Wise Use movement, considers himself the true environmentalist, meanwhile he calls environmentalists 'extremists', 'hate groups' and 'Nature Nazis' and Greenpeace 'nothing more than con-men and shysters'.[118] The Blue Ribbon Coalition also networks with the hardline off-road dirt bikers from the Sahara Club.[119] Like PFW!, they too are also networking internationally, establishing ties with the Motorcycle Action Group in England.[120]

CORPORATE FUNDING OF THINK-TANKS

The vast majority of right-wing think-tanks, legal foundations and front groups also receive funding from corporations and conservative foundations. The simple reason is to push forward an ideological policy agenda which favours the corporation economically. Funding an intermediate organisation to publish research that is favourable to the corporation gives that particular research more credence than if published by the company itself. In essence, think-tanks acts as buffer organisations for companies to get their message into the public domain.[121]

Much of the money for think-tanks comes from corporations or conservative foundations, whose fortunes are in turn usually based on corporate family fortunes. The Coors Foundation's fortune is from beer, Sarah Scaife from publishing, John M. Olin from chemicals, Lilly Endowment from Eli Lilly pharmaceuticals, and Smith Richardson from pharmaceuticals and Bradley from electronics.[122] For example, the Coors family and Foundation fund the following organisations which could be considered pillars of the general green backlash: the American Council on Science and Health, Center for Defense of

Free Enterprise, National Legal Center for the Public Interest, Pacific Legal Foundation, Mountain States Legal Foundation, the Heritage Foundation and the Free Congress Foundation, the American Legislative Exchange Council and the Council for National Policy.[123]

Corporations also fund two of the most active anti-green right-wing think-tanks, the Competitive Enterprise Institute (CEI) and the Political Economy Research Center (PERC). The CEI is funded by such corporations as Amoco, ARCO, CSX, Dow Chemical, Ford, General Motors, and IBM, as well as the American Petroleum Institute, and such conservative foundations as Scaife, JM, and Kreible.[124] Funding for PERC comes from Amoco, and conservative foundations such as Carthage, JM, John M. Olin, Bradley, Earhart and Sarah Scaife. Executives from the oil, chemical and financial industries sit on its board of trustees.[125] Other members of Earth Day Alternatives receive corporate funding too.

The Right's vanguard think-tank, the Heritage Foundation, has received funding from, amongst others, the following companies and foundations: Amway Corporation, Coors Foundation, Federation of Korean Industries, *Reader's Digest*, Taiwan Cement Foundation, Union Petrochemical Corporation, Alcoa Foundation, Amoco Foundation, Ashland Oil, Chevron, Dow Chemical, Exxon, Ford, General Motors, IBM, Eli Lilly, Lockheed Foundation, Mobil, Nestlé, Philip Morris, Proctor and Gamble, Pharmaceutical Manufacturers Association, RJ Reynolds Tobacco, Shell Companies Foundation, SmithKline Beecham, Texaco, and Union Pacific.[126]

THE LOWEST COMMON DENOMINATOR

Increasingly, companies are performing a kind of shifting regionalisation, where they are moving from region to region, country to country, attempting to avoid stakeholder responsibility or regulation, looking for the lowest common denominator of environmental protection, worker safety, rights and wages. These areas often happen to be in poor communities of colour in the USA, or in the impoverished regions in Asia, Latin America and or Africa. Ironically, the success of the environmental movement in securing regulation in America and Europe could have caused or encouraged this flight.

Operations are then performed to a different set of standards than if the plant was situated in a more affluent region or country, which amounts to corporations practising a set of double standards. One set for poor people, one set for the rich, one set for people of colour in the South, another set for the North. This trend, as has already been mentioned, has been called environmental racism.[127] In the move South, transnationals are taking the products and practices that have been banned in the North and employing people in working conditions that are outlawed where they have their head offices and are operating to environmental standards that are obsolete back home. They would be criminals back home.

Left behind are decimated communities and the legacy of resentment stirred up by corporations who blame environmentalists and environmental regulation as the reason for companies leaving. For example, when US timber firm, Louisiana-Pacific moved part of its operations to Mexico to exploit cheap labour, it told its workers that environmentalists were to blame for the job losses.[128]

Third World governments, lured by the illusion of trickle-down capital and tempted by the dreams of technology transfer, have actually encouraged this corporate and capital flight into the deregulated zone, by setting them up in their own countries. The effect of this, according to Paul Hawken, 'is that the world becomes a large, non-unionised hiring hall, with poorer countries lining up for plum investments, willing to donate land, resources, environmental quality, and cheap labour as their cost of achieving economic "development".'[129]

The reality is different though. 'The technology transferred is often inferior and often hazardous,' writes Vandana Shiva, from the Research Centre for Science and Ecology in India. 'In addition, these technologies threaten the jobs of workers.'[130] But the temptation is normally too strong and numerous countries have established free trade zones (FTZs) or export processing zones (EPZs) which offer lax or non-existent environmental legislation, lenient or non-existent tax regimes and ridiculously low wage costs from underpaid non-unionised workers. Mini business utopias, basically.

Over 120 EPZs have now been set up worldwide, in Bangladesh, Bombay, Brazil, Cameroon, China, Colombia, Cyprus, Dominican Republic, Egypt, Ghana, Guatemala, Haiti, Hong Kong, Hungary, Indonesia, Jamaica, Kenya, Kuala Lumpur, Malaysia, Malta, Mauritius, Mexico, Philippines, Puerto Rico, Romania, Sierra Leone, Singapore, South Korea, Sri Lanka, Sudan, Taiwan, Thailand, Tunisia, Venezuela, Yugoslavia and Zambia.[131] M. Iba, writing in the *Third World Guide* outlines the detrimental activities found in EPZs:

> These include: unfair labour practices such as paying low wages or refusing to recognise trade unions; the imposition of tough and inhumane working conditions; the dumping of

hazardous wastes; the destruction of natural resources and ecological cycles; taking control of the national business cycles of the recipient countries, unfair trade practices; and non-compliance with international and national standards for environmental protection, safety and labour practices.[132]

Take, for example, a knitwear factory in an EPZ in Sri Lanka, which employs 2,000 people. The *Workers' Health International Newsletter* explains the conditions found in the plant:

About 200 of them are exposed constantly to liquid chemicals and fumes in the dyeing section of the factory . . . Even though workers have complained of skin diseases and lung problems, the management has ignored them. Since there are no effective mechanisms for implementing the laws related to occupational health, the management gets away with their violations of workers' rights to compensation and other welfare measures.[133]

Conditions are appalling. A working week can be over 110 hours. After about four years, an employee's health will have deteriorated to such a degree that they are unfit to continue working. 'Some six million Third World women below the age of thirty have been summarily "used up" and discarded by the multinational clients of the zones,' writes journalist Joyce Nelson. Topping the transnational clientele, according to Nelson, is the electronics industry, 'one of the most highly polluting industries of our time'.[134]

Nelson alleges that in order to avoid higher labour costs and more stringent environmental regulations in North America, the electronics industry has:

hopped from country to country, exporting its most labour-intensive and polluting processes, while head office in California's Silicon Valley dictates production quotas. Japan's electronics firms have also followed this procedure. In fact the development of EPZs and FTZs historically parallels that industry's footloose migrations.[135]

Three other industries exploiting the difference in regulatory standards are the biotechnology, pesticides and oil industries. Take the biotechnology industry, for example, an industry dominated by an ever decreasing number of massive corporations, working to control the development, application and regulation of genetic engineering on a global basis.[136] Concerning the safe handling and use of genetically manipulated organisms (GMOs), most industrialised countries have had regulatory control since the 1980s, however, most developing countries have no regulation. As a consequence of this, transnationals often with the consent of host nations, are testing the release of GMOs in the South with little public health or environmental controls.[137] For example, between 1989 and 1992, the multinational Monsanto released genetically engineered soybeans in Argentina; soybeans, cotton and maize in both Belize and Costa Rica; as well as soybeans in the Dominican Republic and Puerto Rico. Ciba-Geigy released genetically engineered maize, canola and sugar beet in Argentina, Calgene released genetically engineered tomatoes in Mexico and Chile as well as cotton in Bolivia and Argentina. Campbell/Sinaloa released genetically engineered tomatoes in Mexico, whilst ICI/PetroSeed released tomatoes in Chile. All of these releases were unregulated.[138]

Transnational agrochemical companies, such as Hoechst, Shell, Bayer, Dow, Du Pont, Ciba-Geigy, Rhone-Poulenc, Sandoz, Sostra, Uniroyal and Zeneca are either expanding existing plants or building new ones in developing countries where regulations are less strict.[139] Many of the above named companies are still producing certain pesticides which are banned or restricted in some countries. For example, American Cyanamid still produces 5 products that fit into this category, BASF 18 products, Celamerck Gmbh and Co 19, Ciba-Geigy 24, Dow 6, Du Pont 17, Hoechst 7, Hokko Chemical 8, ICI 36, Monsanto 9, Rhone-Poulenc 23, Shell 17, Sandoz 34, and Schering 25 products.[140] Stockpiles of obsolete pesticides are now building up in the Third World, posing urgent environmental and public health risks.[141]

The results of operating to double standards in the pesticide industry are severe. The Bhopal disaster, the world's worst chemical accident, has been described as a classic example of double standards and would not have happened had the plant been located in the North, or operated to standards found there. It was on the fateful night of the 2 December 1984 that noxious fumes containing methyl-isocyanate, hydrogen cyanide, and cyanide leaked from the Union Carbide plant in Bhopal.

Up to 16,000 people are thought to have been killed as a result of breathing in the toxic gases. Even now, people are dying still, at the rate of a few a week. Spontaneous abortions and stillbirths tripled after the accident. Another quarter to half a million suffered some health damage, at least 30,000 of whom are described as incurably ill. An estimated 600,000 sued for compensation totalling US $10 billion. Union Carbide settled for $470 million in 1991 and quietly left the country. No-one from the company has ever been, or is ever likely to be, prosecuted for one of the corporate crimes of the century.[142] The disaster was a far cry from the world Union Carbide promised India in its advertisements in the 1960s: 'The people of Union Carbide welcome the opportunity to use their knowledge and skills in partnership with the citizens of so many great countries.'[143]

Tota Ram Chouhan, an ex-Union Carbide employee, who still suffers from breathing problems and calls himself a 'one man protest group' has said that the:

> disaster was a development of the gross negligence at the plant. Everything built up to that day. Before the disaster there were problems: instruments malfunctioned, vital instruments were taken out of service, pipelines broke due to corrosion, and people were always changing jobs.

So faulty were the safety systems at the plant that they were switched off at the time of the accident. For speaking out against the company, Tota Ram Chouhan, or TR, as he is more commonly known, has been harassed at work. Following two visits to the USA which the authorities tried to block and where he spoke about the disaster, the Indian government has not renewed his passport.[144]

Whilst Union Carbide may have left India to avoid any further responsibility, other companies move to avoid regulation. For example, Shawqi Issa, a lawyer from the Law and Water Establishment in East Jerusalem presented evidence to the Annual Public Interest Environmental Law Conference in Oregon in the spring of 1995, of how an Israeli company moved to an area of less strict regulation, after having been closed down in Israel for violating environmental regulations.[145] Companies also threaten to move: Shell threatened to pull out of The Netherlands when the country was considering imposing a unilateral carbon tax, and so the plans were scrapped.

Companies are not only avoiding regulation but also attempting to avoid any liability as well. For example, Du Pont in its negotiations with the state of Goa in India, drew up a contract 'which specifically exonerates the company from all liability in the case of damage to the environment and worker health', according to Kenny Bruno and Jed Greer from Greenpeace.[146]

GATT

On the tenth anniversary of the Bhopal disaster in December 1994, the Permanent Peoples' Tribunal, set up after the disaster, sat for the last time. The six eminent judges heard how transnational corporations are still not held accountable for their actions and attempts to introduce a code of conduct have failed.[147] These corporations are aided by free trade and the liberalisation of trade as facilitated by the Global Agreement on Tariffs and Trade or GATT agreement. Furthermore, a study by *Public Citizen* found that trade advisory councils with intimate links with US trade officials were heavily packed by 'companies notorious as polluters of the environment and opponents of stricter environmental protection'.[148]

Vandana Shiva argues:

> Free trade is basically the deregulation of commerce when commerce is deregulated, and national barriers to trade are removed, capital moves where people are poor, wages are low and regulatory frameworks are weakest. GATT is the main instrument for deregulation of economic activity. While GATT creates a level playing field for capital, it threatens to create an environmental apartheid for people.

GATT, alleges Shiva, 'contributes to the erosion of democratic decision-making power of local communities. It creates a global structure in which hazards increase in the Third World, the rights of corporations increase globally and their responsibilities decrease everywhere.'[149]

Vandana Shiva is only one of the many commentators who hold these views. As a result of the GATT agreement, 'outlook for realisation and protection for human rights of millions of people throughout the world is very glum', says Ward Morehouse, as multinationals 'consolidate and codify their power through a super national agreement that will further undermine the sovereign powers of nation states'.[150] Overall, the main beneficiaries of GATT reforms will be the TNCs and the main losers will be the environment and the world's poor,' concluded Colin Hines and Tim Lang, in their book on the effects of free trade, *The New Protectionism*.[151] 'We argue in the book,' says Tim Lang, 'that GATT represents the erosion of the citizen, the erosion of democracy: the French word citizen – citoyen – now becomes obsolete.'[152] Lang believes that because the World Trade Organisation, the body that determines the future shape of the world's economy, is totally unaccountable, we are witnessing the 'demolition of the struggle for democracy'.[153]

Assisted by GATT, global democracy is slowly dying, as unaccountable organisations assist unaccountable corporations seeking out countries which have the lowest common denominator of wages, and health and environmental protection. Local communities, workforces and the environment are being sacrificed in the new corporate bonanza. 'The new GATT lowers the floor of international environmental, health and safety standards,' concludes Harris Gleckman and Riva Krut from Benchmark Environmental Consulting.[154] Gleckman was the head of the UN's own Centre for Transnational Corporations (UNCTC), before it was closed down.

With the internationalisation and liberalisation of trade and the increasing immobility of labour, therefore, the future for environmental and worker protection looks bleak. Job blackmail will be the ace card that industry will use time and time and time again in the global marketplace. It will be used to water down environmental, social and worker safety regulation that would improve the quality of life of workers and the environment for all, rather than offer profit for the corporate élite. There could well be a downward spiral of environmental protection and workers' rights as counties are forced to lower standards to entice corporations and compete internationally.[155]

Transnationals have polluted the political process in many of the western countries, now they are setting out, hidden inside the Trojan horse of GATT, to exploit the resources and labour of the developing world. Harris Gleckman has remarked that, 'As a minimum, the job of a democratic society is to ensure that one party's freedom is not at the expense of another's. Unrestrained freedom for one group, destroys another's democracy.'[156]

4

GET ON THE GLOBAL GREEN

> The full effect of the corporate propaganda apparatus will never be known. It is most successful when the PR professional leaves no tracks near the scene of a winning campaign.[1]

Corporations have also set out to persuade the public that the very *raison d'être* of commerce has changed, and to co-opt the environmental debate. Companies were no longer insidious faceless corporations only interested in profit at any cost, they were now caring corporations concerned about communities, consumers, children. They were committed to pollution prevention, to people, to the planet. There was only one problem with this strategy – on the whole, they were lying.

Joyce Nelson, a Canadian journalist offers an explanation to this corporate co-option strategy:

> Since environmental concern is no longer a 'lunatic fringe' issue, it has become a litmus test indicating the degree to which individuals and organisations, as well as politicians and corporations, have a stake in maintaining the *status quo*. Obviously, widespread radical change in the way we live our daily lives is a threat to both a power structure and an economy that needs us to be diligent consumers. As a result the 'greening' of corporate and governmental personas is probably becoming the standard PR strategy in order to hold out the promise of 'reform' and prevent radical change.[2]

After the growing environmental awareness that swept America in the late 1980s and the multiplying consumer pressure for ecological change, industry had to respond. 'Around Earth Day 1990 every major corporation in the US became environmentally conscious overnight. That year the number of eco-ads more than tripled,' write Steve Strid and Nick Carter in *Tomorrow Magazine*. 'There were ads that made responsible claims and showed real environmental concern while others gave new lustre to the word "greenwash".' Greenwash is now a term that has entered the environmental debate.

GREENWASHING

So what do people mean by greenwashing? 'A leader in ozone destruction takes credit for being a leader in ozone protection,' writes Kenny Bruno, in the *Greenpeace Guide to Greenwash*. He continues:

> A giant oil company professes to take a 'precautionary approach' to global warming. A major agrochemical manufacturer trades in a pesticide so hazardous it has been banned in many counties while implying that the company is helping to feed the hungry . . . This is greenwash, where transnational companies are preserving and expanding their markets by posing as friends of the environment and leaders in the struggle to eradicate poverty.[3]

Motor vehicles, for example, the fastest growing source of pollution on the planet, suddenly became environmentally friendly. Lead-free petrol became 'green' fuel, as the advertisers omitted to mention the other cocktail of pollutants such as carbon monoxide, hydrocarbons or carbon dioxide still being emitted by the cars. Or the fact that unleaded petrol actually had more benzene in it – a carcinogen with no safe limit of exposure – than leaded fuel. Cars with catalytic converters were advertised by some manufacturers to actually clean the atmosphere. General Motors, the largest multinational on the planet, the largest producer of vehicles and a major defence contractor, suddenly announced the company's '20 years of environmental progress'.

The truth, however, is somewhat different. For years the car companies had been lobbying against the introduction of unleaded fuel, catalytic converters or any increase in fuel efficiency, which are all environmental benefits. The companies had lobbied against the introduction of airbags, a safety benefit, too. The reason the motor manufacturers had been forced to change was through legislation not through voluntary action. But when the change was forced upon them, they would be the first to crow about environmental and social benefits of the alterations.

Aerosol manufacturers, not to be outdone, suddenly became ozone-friendly; washing powders suddenly became phosphate-free; aluminium cans and paperbags suddenly became not only recycled but also recyclable. Greenbabble had arrived big time. So too, had deception and consumer confusion.

The oil companies were up to their tricks too. Chevron's 'People Do' campaign is one of the longest running greenwashing strategies in the USA. 'People Do', which was launched in 1985 at a cost of five to six million dollars a year – roughly 10 per cent of the company's annual advertising budget – has been criticised by environmentalists and public interest groups as being misleading. 'The ads are a selective presentation of the facts with a lack of context,' argues Herbert Gunther, Executive Director of San Francisco's Public Media Center. Conceived as a public relations exercise, it highlights what the company has done to protect certain animals, among others, grizzly bears in Montana and the kit fox in south-eastern California. The underlying message is that Chevron cares about its customers and the environment and that the company is a good corporate neighbour.[4]

When the company started drilling in the Northern Rocky Mountains of Montana, an area of national forest adjoining the Glacier National Park and one of the last remaining untouched habitats of the grizzly bear, it ran a series of 'People Do' adverts. Appearing under the headline, 'the bears who slept through it all', with a picture of a brown bear asleep, highlighting the fact that the company stopped its operations when the bears were active. What Chevron omitted was that these procedures were a prerequisite laid down by the US Fish and Wildlife Service and the Bureau of Indian Affairs.[5]

Mobil, another oil company, was separately sued by seven states in the USA for making false claims that their 'Hefty Bag' range of plastic bags were degradable. The charges of deceptive advertising were based on the fact that their bags, being photodegradable (i.e. needing sunlight to break down) were sold as biodegradable and that they would 'continue to break down . . . even after they are buried in a landfill'.[6] It is highly improbable that in landfill conditions, which are characterised by a lack of light and oxygen, and where even paper can take decades to break down, that photodegredation will take place. Even if the bags did break down, it is questionable whether they would be re-assimilated into the environment.[7]

'Environmental ad campaigns represent the limit and extent to which corporations are presently willing to accept ecological truths,' argues green businessman Paul Hawken. He adds:

> Corporations do not perceive that present methods of production will deprive future generations . . . what corporations do perceive is that genuine environmentalism poses an enormous threat to their well-being. If you define well-being as their ability to continue to grow as they have in the past, they are correct.[8]

Other greenwashing tactics used by companies are designed to negate the threat of environmentalism. Words and terminologies once used by environmental activists suddenly became corporate jargon. Companies even decided to host their own Earth Day, and called it Earth Tech. At the 1990 Earth Tech fair over 100 companies rushed to show the public 'how the world can continue to develop for the progress of mankind without continuing to degrade the environment in which we live'. The event was dismissed by many environmental groups as a corporate 'fashion show'.[9]

But by the beginning of the 1990s, the green fashion floodgates were open. At the 1991 Seventh Annual Harlan Page Hubbard Lemon Awards for the worst advertisements in the USA, two of the ten were for deceptive green advertising. General Electric won an award for exaggerating the environmental benefits of its Energy Choice light bulbs and the Coalition for Vehicle Choice for deceptively claiming that large cars were safer than small cars.[10] A year later the US Council for Energy Awareness was one of the award's recipients for misrepresenting the environmental benefits of nuclear power, something the industry had been doing for over a decade.[11]

Furthermore, the US nuclear industry used PR efforts to try and eliminate public opposition to low-level radioactive sites in 1991. 'The nuclear industry is beyond desperation in its search for a credible solution to the problem of radioactive waste. They have given up on science,' wrote *Rachel's Hazardous Waste News*. 'They have abandoned the democratic process and rational decision-making. They are now resorting to secret campaigns of bribery, persuasion and deception to convince America that black is white, evil is good, and danger is safety.'[12]

In England, greening had begun in the late 1980s. Friends of the Earth had become so concerned at how green consumers were being ripped off that by 1989 it had devised its own 'Green Con Awards' – an award specifically designed to be given to companies that mislead the public on ecological issues. Winners were British Nuclear Fuels Ltd (BNFL), who tried to promote nuclear power as environmentally friendly. BNFL attempted to justify its claim by maintaining that nuclear power does not produce any carbon dioxide. However, they forgot to mention radioactive waste and radiation.

The following year, 1990, when green fever was hitting the USA, Eastern Electricity won the coveted 'Green Con' award for actually proclaiming that their customers should 'use more electricity rather than less' to combat global warming. Runners-up included Scott Paper for falsely claiming that their logging operations helped counter the greenhouse effect. British Rail won for proclaiming that their paper bags were 'recyclable' without providing any recycling facilities. Shell and ICI also won awards.[13]

ICI was awarded an award in 1991 for double standards for proclaiming that it was 'firmly committed to the protection of the environment and fully supports the Montreal Protocol' whilst still manufacturing ozone-depleting chemicals in India. They did not win the top con prize, though, that went to Fisons, a leading peat-cutting company for proclaiming that 'buying British peat-based compost in no way endangers our remaining wetlands of conservation value' despite the fact that 90 per cent of the company's peat-cutting operations were on protected areas – SSSIs – Sites of Special Scientific Interest.[14]

ICI has been found guilty of using deceptive green advertising in other countries too. In Malaysia, the corporation ran a pesticide advert in the *Malay Mail* in April 1993, under the headline: 'Paraquat and Nature working in Perfect Harmony'. The advert, depicting a scene of palm trees, birds and flowering plants in a rural idyll, went on to state 'In fact, Paraquat is environmentally friendly. For over 30 years Nature and Paraquat have been working in perfect harmony.'[15] Josie Zaini from the Education and Research Association for Consumers called the advert 'horrendous and outrageous'. The advert broke the UN Food and Agricultural Organisation's (FAO) International Code of Conduct on the Distribution and Use of Pesticides, which prohibited misleading advertising and requires that manufacturers draw health warnings to the harmful effects of the product.

Paraquat is far from a harmonious pesticide. It is one of Pesticides Action Network's (PAN) 'Dirty Dozen' chemicals, which PAN are seeking to outlaw because they are so toxic. It can cause severe health problems to humans as well as wildlife and has already been banned in Austria, Bulgaria, Burkino Faso, Finland, New Zealand and Sweden. In Malaysia 450 people were killed by paraquat poisoning between 1978 and 1985. ICI's advert was just one of many 'environmentally friendly' claims made by transnationals in the country, according to Josie Zaini.[16]

This is just one example of multinationals violating the FAO Code of advertising on pesticides on numerous accounts in countries such as Colombia, Costa Rica, Ecuador, Indonesia, Mexico, Paraguay and the Philippines. Companies such as Bayer, Ciba-Geigy, Dow, Hoescht, Monsanto, Rhone-Poulenc and Shell are the violators. Often adverts show scantily clad women promoting these deadly products.[17] The chemical industry is just one example of a greenwashing industry that is going global.

The Japanese Power Reactor and Nuclear Development Corporation has created a green cartoon, called 'Mr Pluto', whose job it is to teach children that nuclear power is safe. 'If everybody treats me with a peaceful and warm heart, I'll never be scary or dangerous,' says the Smurf-like creature.[18] The European Nuclear Society prefers a different approach. One presentation at its conference in 1994 was entitled 'How to encourage and train female research scientists to become ambassadors for nuclear energy'.[19]

The Rainforest Action Network has alleged that Mitsubishi spent millions of dollars on greenwashing its global rain forest activities.[20] Rhone-Poulenc, whilst promoting a green image under the title 'welcome to a cleaner world' was ordered by a Brazilian court to close one of its production facilities in Cubatao due to high pollution levels.[21] Rhone-Poulenc and Mitsubishi were two of the multinational companies singled out by Greenpeace in the run-up to the United Nations Conference on Environment and Development (UNCED), for greenwashing practices along with Aracruz, Du Pont, Shell, Solvay, Sandoz, General Motors and Westinghouse.[22]

GREEN BUSINESS NETWORKS

Executives from Shell, Du Pont and Aracruz were to join the Business Council for Sustainable Development, one of the green business networks (GBNs) that have sprung up since the beginning of the decade. There are now a myriad of GBNs and industry initiatives such as the chemical industry's Responsible Care programme, the International Charter for Sustainable Development, the Advisory Committee on Business and the Environment, Global Environmental Management Initiative, and the World Industry Council for the Environment or WICE. In fact a survey of GBNs by environmental consultants, Sustainability and *Tomorrow Magazine* in December of 1994 identified some forty GBNs, some global and some local. Ward Morehouse dismisses such business initiatives as the Responsible Care initiative of the US Chemical Manufacturers Association and the Business Charter for Sustainable Development adopted by the ICC as 'public relations exercises that have little, if any, discernable impact on corporate behaviour'.[23]

Voluntary codes of practice filled with vague ambiguous language, such as sustainable development, are the industry's tactics to avoid mandatory regulation. By advocating policing itself, with schemes of voluntary oversight, industry aims to pre-empt regulations. Industry lobbying under the umbrella of an eco-sounding business network is greenwashing that aims to prevent international agreements that oversee corporate behaviour. The most important GBN is the Business Council for Sustainable Development (BCSD), which was formed by the largest global public relations company Burson-Marsteller (B-M) and worked to such devastating effect at the Earth Summit, as will be explained later. B-M and other global PR companies have had a profound effect on the environmental debate.

THE PR PRACTITIONERS

With corporate backing, the PR industry has been extremely influential on the process of democracy in the USA. In the PR jargon of integrated communications, and with the combined tactics of traditional public relations, political lobbying and corporate grassroots organising, as well as corporate advertising campaigns and corporate-sponsored coalitions, American democracy is based on PR. PR, that is, standing for corporate public relations and not proportional representation of the people.

John Stauber, the editor of the independent journal *PR Watch*, conservatively estimates American's PR industry to be worth $1 billion a year.[24] This is probably an underestimate, but precise figures are understandably hard to come by. This said, corporate anti-environmental PR was estimated to be $500 million in 1990.[25] In the last five years the figure is likely to have increased, albeit slowly. Globally, no figure exists which shows how much money corporations spend on green or environmental PR, as it is so aptly called.

Ralph Nader, the worldwide-renowned consumer advocate has called environmental public relations practitioners masters of the 'fabricated, phoney, incomplete anecdote'.[26] PR companies have worked on environmental issues for over thirty years, however, *O'Dwyer's PR Services*, the leading PR insiders have labelled environmentalism the 'life and death PR battle of the 1990s'.[27] 'I see the PR companies as having played a critical, pivotal role in orchestrating the environmental backlash movement,' says Stauber. 'I think that to a very large part, it is the public relations companies and the public relations experts employed by industry who have developed and defined and conducted anti-environmentalism.'[28]

Stauber continues:

> The public relations strategies and tactics that have been employed are sort of two-face. They're employing a good-cop, bad-cop, or a carrot and stick approach, where on the one hand the companies and their PR agents try to co-opt and occupy the time of, and distract and smooch environmentalists. On the other hand, they are defining environmental activists as terrorists. They are drawing the line that absolutely no ecological reforms are working, hand in hand with the lobbyists. They are also employed by the same industries to kill any environmental reform legislation, and they are working with or promoting or encouraging Wise Use types of right-wing anti-environmental extremism. If you ask me who is behind anti-environmentalism – it's business, but if you ask me who is waging the campaigns, who is choosing the tactics, who is coordinating the fight, who are the field generals, it is PR practitioners.[29]

So who are these secretive spin-doctors waging the anti-green war? Jack Bonner Associates, mentioned earlier, whose clients include the likes of Alliance of American Insurers, American Bankers Association, Chase Manhattan Bank, Chrysler, Dow Chemical, Ford, General Motors, Monsanto, Motor Vehicle Manufacturers Association, Philip Morris, Prudential and Westinghouse, is a small company in the global public relations game.[30] So too is E. Bruce Harrison, which has also been alluded to.

The largest PR company in the world is Burson-Marsteller, itself a subsidiary of American advertising leviathan Young and Rubicam. Second is Hill and Knowlton (H&K), a subsidiary of another advertising giant, WPP, whose Chief Executive earns a reported £8 million a year.[31] B-M has a client list that comprises some 350 of the *Fortune 500* companies as well as many governments and trade associations.[32] It has 62 offices in 29 countries around the globe. Of any PR company, B-M receives the greatest income for work in the environmental area, with a worldwide income of $18 million in 1993, followed by Ketchum PR with $15.3 million, H&K with $10 million, Fleishman-Hillard $9 million and Shandwick $6.7 million. E. Bruce Harrison, comes in at number six with an income of $6.5 million.[33]

Whilst the former PR companies are engaged in PR across the board, E. Bruce Harrison can be seen as an environmental specialist.[34] The PR company E. Bruce Harrison, is actually named after the man of the same name, who is known as the 'dean' of PR. The company employs over 50 staff for 80 of the world's largest corporations and associations, which include American Medical Association, AT&T, the Business Roundtable, the Chemical Manufacturers Association, Chlorox, Coors, Ford, General Motors, Monsanto, the Motor Vehicles Manufacturing Association, Rhone-Poulenc, RJR Nabisco, RJ Reynolds, Uniroyal Chemicals, Vista Chemicals and Waste Management.

Harrison recently opened an office in Europe to assist 'its transnational clients work through the complexity' of environmental regulations in the EC.[35] Basically they are out to avoid European environmental legislation.

E. Bruce Harrison, the man that is, earned his spurs coordinating the chemical industry's counter-attack against Rachel Carson's *Silent Spring*, which, more so than any other book, is heralded as giving birth to the ecological awareness that resulted in the current environmental movement. No sooner had the conscience been awakened, however, than it was being countered by corporate PR. Harrison, under the auspices of Allan Settle of the Manufacturing Chemists Association and in cohorts with PR people from Du Pont, Dow, Monsanto and Shell helped put forward a 'rational, responsive programme', to counter Carson.[36]

John Stauber writes:

> They hit back with the PR equivalent of a prolonged carpet bombing campaign. No expense was spared in defending the fledgling agrochemical industry and its $300 million/year in sales of DDT and other toxins. The National Agricultural Chemical Association doubled its PR budget and distributed thousands of book reviews trashing *Silent Spring*.

Carson came under personal attack for being a spinster, although she was dying of cancer.[37]

PR TACTICS

Stauber elaborates some of the tactics used. 'They used emotional appeals, scientific misinformation, "front groups", extensive mailings to the media and opinion leaders, and the recruitment of doctors and scientists as "objective" third party defenders of agrochemicals,' he says.[38] These are tactics that have been used time and again, in different circumstances, by different PR companies for different clients to nullify different environmental problems. Tactics being used now in different countries and even different continents around the globe. Tactics that will be used until time immortal. Tactics perfected by the likes of Jack Bonner, E. Bruce Harrison, B-M, H&K and Ketchum, but also by other PR practitioners. Tactics that include co-option strategies and the denigration of environmental activists or scientists by labelling them as terrorists. According to Stauber:

There is really a fairly small but unfortunately very effective repertoire of tactics and techniques and language that is being employed to defeat environmentalism, and also to create the appearance of, and in many cases the reality of, a growing anti-environmental groundswell.[39]

Moreover, the PR gurus are networking with each other, communicating how best to get in on the act, or 'on to the green' as they call it. E. Bruce Harrison has perfected other techniques to counter the environmental movement elaborated in his book *Going Green: How to Communicate your Company's Environmental Commitment*. In the book, Harrison argues that the 'American brand of environmentalism – an activist movement – has slowed down as a vehicle operating outside the institutions of America. In a sense, this style of environmentalism died; it succumbed to success over a period roughly covering the last 15 years.' Harrison also claims that a movement has been replaced by a message – sustainable development. Business leaders are now the true ecological pioneers, he argues. 'Corporate environmentalism is now more lively than external activist environmentalism,' says Harrison, 'and this trend will continue to grow.'[40] Although, for business, 'getting on the green is not easy', 'environmental communication, or envirocomm', is now the way forward to establish 'sustainable relationships' with customers, or create 'public relationships' with stakeholders or your customer publics.[41] Welcome to the land of greenbabble where the traditional environmental movement has been replaced by 'envirocomm'.

Harrison argues that environmental groups are business ventures designed to create membership and money. Period. This means that those concerned with respectability will sit around with industry, such as the joint project between Environmental Defense Fund and McDonalds. The fast food giant has been repeatedly praised throughout the industry for its eco-partnership with the EDF, however it is also suing two environmental activists in Britain who questioned the company's health, and environmental record (see chapter 12). McDonalds has threatened to sue numerous groups over its ecological record, a tactic designed to silence people through legal intimidation.

Harrison continues:

As a number of activists move towards the center, competition for those members will become more keen and the extremists will insist on marketing strategies which rule out business coopting environmentalism . . . Greenpeace is proving this trend true with recent attacks on Canadian pulp producers with its anti-dioxin campaign, and on Du Pont's manufacture of CFCs, despite its continued phase out.[42]

In order to try and defeat groups like Greenpeace, co-option rather than confrontation is seen as the way forward. This is especially the case if industry wants to neutralise the effectiveness of the opposition. 'The cold war is over between environmental activists and companies,' reported *O'Dwyer's PR Services* in February 1994, 'Each side is willing to work with the other on projects designed to improve the environment, say PR pros eager to play matchmaker roles.'[43] 'One good thing about' corporate co-option efforts, according to a board member of the oil company ARCO, is that 'while we are working with them, they don't have time to sue us'.[44] Furthermore, links with environmental groups help industry establish or improve its 'green credentials', according to Dale Didion, Environmental Practice Director at H&K.[45] An endorsement from an environmental group carries more weight in the public eye than anything the company can say itself.[46]

Didion and H&K are 'active in setting up dialogue groups between key representatives of environmental groups, Chambers of Commerce and the federal Government'. Ways to set up relationships are for industry to help environmental groups raise money and to sit on their board of directors. 'That can open a good symbiotic relationship,' says Didion.[47] B-M, which sees a closer alignment between environmentalists and industry, also argues for a better dialogue between their clients and environmentalists. E. Bruce Harrison also gives advice on 'relating to the "active opposition"'. As do Steven Bennett, Richard Frierman and Stephen George in their book *Corporate Realities and Environmental Truths*, where they devote a chapter to 'Strategic Partnerships', in which a reader can learn 'how to form a partnership with an environmental group'.[48]

The PR companies are also enjoying the current political climate in Washington. *O'Dwyer's PR Services* recommended that so-called environmental PR companies, should 'ride the Republican fuelled anti-environmental backlash wave as far as possible'. 'The business community enjoys the upper hand,' says B-M's head of Worldwide Environmental Practice Michael Kehs, 'There is a new contract on the street. Although the word "environment" is never mentioned, many observers believe that it's less a Contract with America than a "contract on environmental busybodies" . . . There is no better time to extend an olive branch.'[49]

At the end of the day, when the greenspeak is seen for what it really is, PR companies are simply advising on a time-honoured theme, one of divide and rule. Insight on this strategy can be gleaned from a meeting that occurred in 1991. Ronald Duchin, from the mouthful PR company, Mongoven, Biscoe and

Duchin (MBD) addressed the 1991 Convention of the National Cattlemen's Association, a supporter of the Wise Use Movement. Duchin outlined a divide and conquer strategy on 'how corporations can defeat public interest activists'. *PR Watch* reports how:

According to Duchin, activists fall into four distinct categories: 'radicals', 'opportunists', 'idealists' and 'realists'. To defeat activists, says Duchin, corporations must utilise a three-step, divide-and-conquer strategy. The goal is to isolate the radicals, 'cultivate' the idealists and 'educate' them into becoming realists, then coopt the realists into agreeing with industry.[50]

Two years later Duchin addressed the Annual Conference on Activist Groups and Public Policy Making in the USA. Also speaking were Fred Smith from the right-wing anti-environmental Competitive Enterprise Institute and Ralph Reed from the Christian Coalition. Duchin talked about methods by which industry could build alliances with carefully chosen activists for mutual benefit 'in the regulatory and legislative arenas and in the shaping of public opinion'.[51]

It should come as no surprise that MBD is advocating such tactics. It is a secretive and small PR company whose speciality is not only co-opting activists but tracking them for corporate clients, who pay a retainer of some $3,500 to $9,000 per month for these services.[52] 'We monitor issues, specifically, environmental issues, biotechnology, and other areas . . . solid wastes and hazardous wastes,' says Ronald Duchin, 'We track the issues, we track the players in the industry, both individuals and groups.'[53] MBD's clients include Shell, Philip Morris and Monsanto and other *Fortune 100* companies, although MBD are loath to disclose this information. Monsanto and Philip Morris are, according to John Stauber from *PR Watch*, currently using MBD to report on and 'undermine consumer activists and family dairy farmers opposed to Monsanto's controversial new animal drug, bovine growth hormone (BGH)'.[54]

In 1990 MBD infiltrated SEAC, the Student Environment Action Committee. MBD attended 'Catalyst', the largest meeting of students on environmental issues ever assembled at the University of Illinois. Documents showed how MBD where 'unable to determine the positions and locations of a number of individuals described as organisers of the Catalyst meeting'.[55]

MBD was founded in 1991, but its roots lie in another PR company, Pagan International, which also engaged in covert activity and information collection. As far back as 1985, Pagan International had prepared a briefing document on Greenpeace 'to inform business, government, and other opinion leaders of the objectives and activities' of the organisation.[56] In the late 1980s Shell, under criticism for its continuing involvement with the apartheid regime in South Africa, hired Pagan International to run an anti-boycott campaign. Codenamed the 'Neptune Strategy' it involved recruiting well-known public figures to promote Shell's position, to investigate the 'personal characteristics' of key boycott supporters and to infiltrate boycott meetings. A front group called the Coalition on Southern Africa was also formed.

In the past, Pagan had also run campaigns for Nestlé which was targeted for a boycott due to its milk powder promotions in the Third World and Union Carbide after the Bhopal disaster.[57] When Nestlé resumed selling infant formula to the Third World in 1988, it was once again targeted for a further consumer boycott. This time it hired another PR firm, Ogilvy & Mather (O&M) Public Relations, for assistance. O&M's report advised on 'neutralising or defusing the issue by working with key interest groups' as well as 'grass-roots monitoring'.[58]

Ogilvy & Mather, like Hill and Knowlton, a WPP subsidiary, also recommended that it would be beneficial for the company to 'establish a "positive do good public service campaign on behalf of Carnation [Nestlé's subsidiary] to show the company's social responsibility"'.[59] This was not the first time Ogilvy & Mather had recommended such a strategy to a client – Shell's 'Better Britain Campaign', had also been set up by the PR company.[60] The scheme ran into criticism though. In 1991, the Campaigns' Director of Friends of the Earth UK, Andrew Lees, criticised it as an example of corporations trying to buy a green image, rather than actually earning one. 'During the 20 years the campaign has been running, the British countryside has gone to rack and ruin and Shell has itself been flogging organo-chlorine pesticides, which have had a highly damaging effect,' said Lees.[61] Shell had to postpone the 25th Anniversary of the Better Britain Awards in 1995, because it was caught up in the controversy on whether to dump the Brent Spar rig on land or at sea.

Another key tactic being used against the environmental movement is for PR companies to put across the cost of environmental regulations and concerns. It is a tactic industry has been using for years and one which has gained pace with the right-wing/Wise Use advocacy of cost/benefit analysis on regulation. 'It's the job of PR firms to make sure federal, state and local governments along with host communities understand the economic trade-offs involved in complying with environmental requirements,' says *O'Dwyer's PR Services*. It is also the job of the PR companies to shift the blame of pollution from their corporate clients on to the public at large. Irresponsible individuals, and society at large are to blame for pollution rather than corporations.[62]

BST

But most of all, environmental PR companies exist to undermine and effectively annul environmental campaigning organisations. For example, apart from MBD, there were many other PR companies employed to legitimise the use of bovine growth hormone, known as BST. A growth hormone, BST, can increase yields of milk in cows by up to 25 per cent. Critics of the drug, including many dairy farmers, citizen action groups and environmentalists claim that it could be unsafe for cows and humans, causes infections in cattle, and will saturate a market already overflowing in excess milk production. Its use raises many 'fundamental ethical, social and economic considerations', according to Professor Samuel Epstein, from the School of Public Health, University of Illinois,[63] but with international sales expected to reach at least $100 million in the first year, the commercial stakes are high.

Because of the controversy, B-M were employed by Eli Lilly and a subsidiary of Monsanto, to counter any criticism, especially those from the 'Pure Milk Campaign'.[64] H&K were also employed by Monsanto to lead the pro-growth hormone fight. Both H&K and B-M were instrumental in defeating state legislators' attempts in Wisconsin, Minnesota, California and Vermont to label the growth-hormone milk, a measure that would have been in the interests of consumer choice.[65]

B-M formed a 'grassroots' coalition to fight for the drug's introduction. Internal documents from B-M on 'grassroots mobilisation' to defeat proposals to restrict BST in Minnesota, Wisconsin and Washington reveal why the appearance of grassroots support is all important. It 'demonstrates broad-based support, shows position in public interest, complements lobbying efforts, counters competitors' efforts and draws attention to the issue'. B-M asked these so-called 'grassroots' activists to write letters to legislators and other decision-makers; to testify; to write op-ed pieces, letters to the editor; to attend meetings, in the community or with public and elected officials; to circulate petitions; to send telegrams to legislators; to phone or appear on talk-shows on radio and television; to undertake interviews for newspapers and broadcasters; to make speeches; to phone or fax decision-makers.

In short, everything normal community organisers would do but not under the auspices of a multinational PR company for a transnational client. The grassroots network included: 'Business executives and lobbyists; biotech company executives; farmers; veterinarians; university professors; high school teachers; Chambers of Commerce and elected and appointed officials'. Letters sent to the editor of the *Hudson Star Observer*, arguing for BST's approval, appealed that 'biotechnology will be the electric lights of the 21st Century. Let's not turn them off.'[66]

BST has been approved for use in the USA, but not yet in Europe, despite the lobbying efforts of B-M over there. In Britain, where the government maintain the drug to be safe, scientists involved in BST field trials have been unable to challenge the government's position for fear of being prosecuted under the Official Secrets Act and the Medicines Act. Professor Richard Lacey, a government advisor, has also claimed that the government's safety statement on BST was drawn up after meetings with the drug companies themselves and not the government's own scientists.[67]

Fresh from their North American victory on BST, the biotech industry will want to capitalise on it. 'Soon to follow,' writes John Stauber, 'are genetically engineered fruits and vegetables from Monsanto, Upjohn, Calgene and other firms; infant formula from Bristol-Myers Squibb produced from genetically engineered cows and containing human mother's milk protein; and meat from cloned cattle owned by W.R. Grace.'[68] Sure to be there to make genetically engineered foods more palatable will be B-M. In 1994, Sheila Raviv, who led B-M's pro-BST campaign, and whose speciality is developing industry coalitions to defeat social change activists, was named as B-M's new CEO in Washington.[69]

BURSON-MARSTELLER (B-M)

Burson-Marsteller is the one company that stands alone in its global anti-environmental activity. 'When you need Help on Environmental Issues. You need environmental professionals' reads their promotional literature:

> People who deal with these issues day to day. They have to hit the ground running. Burson-Marsteller offers a world-wide environmental team. Issue experts. Lobbyists. Community relations cancellers. Technical advisors. And media specialists. For insight, ideas and implementation, Burson-Marsteller provides the right environment.

B-M provides the right environment for its clients. Clients that include the powerful, the polluters and the powerful polluters. ARCO, British Gas, Boots, BP Chemicals, British Nuclear Fuels, Cadbury Schweppes, Chevron, Ciba-Geigy, Citicorp, Coca-Cola, Dow Chemical, Deutsche Telecom, Eli Lilly,

Fisons, Ford, Gallaghers, General Electric, Glaxo, Grand Metropolitan, Hoechst Roussel, Hydro-Quebec, IBM, ICI, Johnson and Johnson, Johnson Matthey, McDonalds, McDonnell-Douglas, Nutra Sweet, Ontario Hydro, Perrier, Philip Morris, Pioneer, Proctor and Gamble, Quaker Oats, Repsol, Rhone-Poulenc-Rorer, Sainsburys, Sandoz, Scott Paper, Shell Oil, Smithkline Beecham, Tetra Pak, Unilever, Visa, Warner Lambert and Zeneca are amongst its clients.[70]

Many of these B-M clients were also sponsors of the Wise Use movement, and actively engaged in anti-environmental activity. B-M is also leading the animal rights backlash, too. It represents the Fur Information Council of America in its multi-million campaign to combat 'animal extremists'.[71] The link between the Wise Use movement and B-M is extremely intricate. 'During the middle and late 1980s, when industry first started pouring money into the "Wise Use movement" in the US, thirty-six of the known corporate sponsors of that movement, which neatly pits workers against environmentalists, were B-M clients,' explains Joyce Nelson.[72]

Nelson continues:

> For example, by 1987, US timber giant Louisiana-Pacific (L-P) had been a B-M client for a decade, during which L-P busted its workers' union, began rallying its employees as a lobbying arm against environmentalists in the Pacific Northwest, blaming them for job losses. At the same time that L-P was bussing loyal workers to anti-environmental rallies, the company was building huge pulp mills in Mexico, where it could pay workers less than $2 per hour. By 1990, L-P was barging old-growth timber down to its state-of-the-art mills in Mexico.

The tactic of using employees as loyal 'grassroots' citizens who will promote the industry's cause is one being perfected by industry and PR companies. In Australia, Europe, Canada as well as in America, workers have become industry's front-line troops of advocacy, often given free trips and paid time off to go and demonstrate on the company's behalf. 'Don't forget the chemical industry has many friends and allies that can be mobilised,' says James Lindheim, B-M's director of worldwide public affairs, 'employees, shareholders, and retirees. Give them the song sheets and let industry carry the tune.'[73]

B-M sees itself as the global conductor of the anti-green orchestra. Company promotional literature publicises 'Burson-Marsteller is the new definition of "global public relations". We advise and counsel on what an organisation should *do*, position what the organisation should *say*, and *implement* total communications programmes against specific objectives.' One of those simple objectives is to nullify the environmental movement. For its clients B-M has formed 'front groups', mobilised 'grassroots opposition', undertaken covert activity, lobbied politicians and co-opted the environmental debate. B-M has also pulled off the biggest eco-trick of the decade at UNCED and diverted criticism from clients responsible for some of the most outrageous environmental and human rights abuses of the twentieth century. 'Burson-Marsteller is,' according to the company, 'the firm that has guided its clients through some of the worst crises that have beset business in the past several decades.'[74] What are some of the most revealing crises?

In 1976, shortly after a military junta led by General Jorge Rafael Videla seized power in a *coup d'état* in Argentina, B-M were hired to help improve Argentina's 'international image, especially for fostering foreign investment'.[75] During Videla's reign of terror, 35,000 people 'disappeared' and thousands of political prisoners were routinely tortured. Argentina hosted the 1980 conference of the Latin American Affiliate of the World Anti Communist League, at which Videla spoke alongside other death squad leaders from the region, including the notorious Roberto D'Aubuisson.[76] Joyce Nelson explains some of the torture used during this period in Argentina:

> 'el submarino' – holding the person's head under water or excrement until near drowning; 'la picana' – the electronic prod applied to the most sensitive parts of the body; rape – sometimes by police dogs; tearing out toe nails; and putting live rats to feed on fresh wounds.[77]

Meanwhile, leaked documents from B-M showed that the PR company hoped to 'generate a sensation of confidence in Argentina . . . through projecting an aura of stability for the nation, its government and its economy'.[78] But not even PR companies can hide the truth forever. The general whom B-M helped legitimise to the world, Jorge Rafael Videla, is now serving a life sentence for murder.[79]

No-one has been sent to prison for the world's worst industrial accident that occurred on 3 December 1984 in Bhopal India, mentioned in the last chapter. As thousands died and suffered, B-M was there, along with Pagan International, to do a damage limitation job for their clients, Union Carbide.[80] Eager to deflect blame away from the company, Union Carbide alleged that the disaster was a result of sabotage. A stance they still maintain to this day.[81] Whether this line was recommended by B-M is not known, but blaming someone else for your own problem is typical of industry's response to disasters or criticism.

Five years previously, in March 1979, B-M had been there to assist Babcock and Wilcox, when their reactor failed at Three Mile Island, causing America's worst nuclear disaster. In the early 1990s, over 2,000 lawsuits were still pending because of the disaster.[82]

Current clients being guided by B-M through the mire of international condemnation are the government of Indonesia which has been criticised for human rights abuses – causing genocide in East Timor – as well as ecological destruction.[83] Over one-third of the East Timorese population, some 200,000 people, have been slaughtered by the Indonesian government, and the UN does not recognise Indonesia's 'annexation' of East Timor. The only country to do so is Australia and that is due to the vast oil reserves found in the Timor Gap, which Australia is profiting from. B-M also defends rain forest destroyers, as it is the PR company for the Malaysian Timber Industry Development Council. Malaysia has faced severe criticism from western environmental groups in recent years for allegedly practising indiscriminate logging (see chapter 10).[84] A subsidiary of B-M's, Black, Manafort, Stone and Kelley was also working with the Nigerian military regime until recently. The atrocities committed by this regime against the Ogoni people are documented in chapter 11.[85]

B-M has also set up, or is involved in other 'industry front groups' to do with sustainability and forestry. One is the Business Council for Sustainable Development (BCSD), another the BC Forest Alliance (see chapter 7), and a third is the Forest Protection Society in Australia. The BCSD must be seen as one of the most serious attempts by industry to co-opt the environmental debate, but also to stop one of the few attempts ever made to seriously make corporations accountable. Over a year before UNCED, B-M issued a press release about the BCSD, one of the few times there has been any public association between the company and its creation. It read:

> In a major new initiative on the future development and use of the world's natural resources, over 40 top world business leaders have joined forces in the form of an international organisation to propose new policies and actions on the sustainable development of the Earth's environment.[86]

B-M made sure that the business side of the story would be heard and any measures to hold multinationals accountable or liable would be quietly slipped into the many paper recycling bins that now adorn the modern 'progressive' business office. BCSD's members include executives from Norsk Hydro, ABB Asea Brown Boveri, Rio Doce International, ENI, Chevron, Volkswagen, Kyocera Corp, 3M, S.C. Johnson & Son, Ciba-Geigy, Nippon Steel, Mitsubishi,

ALCOA, Dow Chemical, Du Pont and Shell.[87] Stephen Schmidheiny, a Swiss billionaire industrialist and a close friend of UNCED Secretary General, headed the BCSD. In the run-up to the conference, BCSD would have unprecedented access to influence the policy agenda being drafted for the conference. With B-M's assistance, 'a very elite group of business people was seemingly able to plan the agenda for the Earth Summit with little interference from NGOs or government leaders', wrote Joyce Nelson.[88] The BCSD managed not just to co-opt the debate but colonise it, and even create it on its own terms.

Within UNCED, for example, corporate interests effectively blocked discussion of the environmental impact of transnational corporations: recommendations drawn up by the UN's own Centre for Transnational Corporations (UNCTC) which would have imposed tough global environmental standards on TNC activities, were shelved,' wrote *The Ecologist* in 1992. 'Instead, a voluntary code of conduct, drawn up by the Business Council on Sustainable Development, a corporate lobbying group, was adopted as the secretariat's input into UNCED's Agenda 21. The UNCTC's proposals were not even circulated to delegates.'[89]

A voluntary code of conduct was put forward instead. 'There were two issues we missed out on: transnationals and the military,' a Brazilian delegate lamented on the closing day.[90] Of the 30,000 or so other participants at Rio, ranging from environmentalists to indigenous and community leaders from all over the world, as well as representatives from churches, academic institutions, development agencies, women's groups and a myriad of other organisations, essentially none supported BCSD's proposal. Once again the concerns of the world's majority had been destroyed by the vested interests of the few. 'It is precisely with these stark refusals to acknowledge the democratic process that business must come to terms,' writes Paul Hawken of the UNCED fiasco.[91]

Meanwhile the greenwash continued, as the vested interests declared their belief in a process that, to them, symbolises business-as-usual. 'As business leaders, we are committed to sustainable development, to meeting the needs of the present without compromising the welfare of future generations,' proclaims the declaration of the BCSD, in the introduction to *Changing Course*, BCSD's great global business manifesto, that has now been printed in eleven languages.[92] In its promotional literature, BCSD does not consider itself a lobbying group working on behalf of its corporate interests.[93]

Furthermore, the international community's last chance of any regulatory oversight of multinationals effectively disappeared when the one UN body that oversaw international corporate activity, the UNCTC, was quietly closed down because of pressure from industrialised nations. The one organisation with any power to oversee the activity of transnationals was shut down by companies and countries who so wholeheartedly embrace sustainable development. Peter Hansen, the Director of the UN Centre for Transnational Corporations, before it was closed, told a conference in 1991, that: 'There are no globally agreed upon rules of what's right and what's wrong for transnational corporations, no sense of global responsibility to match the global reach of corporations.'[94] In 1993, the remains of UNCTC were moved to Geneva, where it became part of the UN Conference on Trade and Development, but it has no power and any UN Code of Conduct on Transnationals is officially recognised as dead. Transnationals want to keep it that way, and with the help of a Swiss billionaire, a Canadian millionaire and the largest public relations company in the world, they could well have succeeded.

In late 1994, the BCSD merged with another green business network, World Industry Council for the Environment or WICE, to form the World Business Council for Sustainable Development (WBCSD), which is still based in Geneva.[95] So the BCSD became the WBCSD and its new chairman came from BP, the third largest global oil company.

B-M has set up other international 'front groups' for industry. In Europe the Alliance for Beverage Cartons and the Environment, was set up 'in defense of the beverage carton against environmental and regulatory pressures'. Launched at the Action for a Common Future Conference in Norway in May 1990, the Alliance represented twelve manufacturers of paperboard and beverage cartons: Elopak, Bowater, Tetra Pak, Billerud, Enso-Guzeit, Frovi, Korsnas, Champion, International Paper, Potlach, Westvaco, and Weyerhaeuser.[96] Internal B-M documents show that the campaign, coordinated by B-M's offices throughout Europe, involved using a Brussels lobbying firm, Robinson-Linton, as well as getting a 'white hat' or third party interest to lobby on their behalf. Also proposed for the campaign were a luncheon and 'placement of an article in a major publication by a recognised environmental authority'. Public relations companies often pay leading academics to write articles for the press, although the public is not told that they represent a vested interest in the debate. A recent example in the USA was where Edelman PR Worldwide offered to pay a professor at the Harvard School of Public Health $2,500 for his signature on a ghost-written editorial, which he refused but announced to the public.[97]

'Whilst their input will be of great value we must ensure that wherever possible, their time is charged to clients,' the internal B-M documents also reveal, about the 'third party' scientist. Media tours of beverage recycling plants, and the publication of a newsletter were also on the agenda. Objectives were to produce a 'green halo' around the beverage carton. Documents describe in a brainstorming session on 'what we want them to think', that the message should be 'do not worry. You made a sound investment. The beverage carton is an environmentally sound package.'[98]

By 1991 the Alliance had had some success, with coverage in over fifty major international publications. In Denmark, meeting with ministers had resulted in the 'successful prevention of the inclusion of the polycarbonate bottle in forthcoming study of carton collection and recycling'. Media awareness was also highly successful in Denmark, France, Germany, The Netherlands and the UK. Internal documents from the Alliance talked of 'environmentalism taken to the extreme' and that 'the environmental zealots have focused on waste as symbolic of today's throw-away society, and the beverage carton has become the symbol of the evil that must be eliminated'.[99] Other PR companies have also attempted to marginalise environmentalists as extremists (see next chapter).

The carton manufacturer Tetra Pak is also associated with a carton-producing PR effort in Australia (see chapter 9) and an environmentally sounding 'front organisation', called Waste Watchers. At first Waste Watchers seems to be an anti-glass bottle and pro-incineration, pro-cardboard packaging, environmental organisation. However, below the surface, there are some key personal links between Waste Watchers and Tetra Pak.[100] After Waste Watchers criticised a public awareness campaign into waste reduction by the Department of the Environment in Germany, the Environment Senator Dr Fritz Vahrenholt responded:

> I'm not against criticism by the packaging production industry nor against the non-recycling lobby. However, if they pretend and disguise themselves as an independent environment organisation, they confuse the public. I am somewhat grateful to them for their silly criticism, because of that, Waste Watchers has revealed whose side they are on and what their hidden agenda is. They made it blatantly obvious that they are not a genuine environment organisation.[101]

There are other examples of B-M co-opting or employing environmental specialists who work on environmental issues. In the UK, B-M employs Simon Bryceson as the senior counsellor of Public Affairs. A former board member of Friends of the Earth, and consultant for Greenpeace, Bryceson is chairman of the board of the environmental media charity, Media Natura,[102] which was set up by Chris Rose, currently Greenpeace UK's Campaign Director and who still sits on the board with Bryceson. Also employed by B-M for a short period was Des Wilson, described once as 'the scourge of big corporations' by some and named as environmentalist of the decade in the 1980s by Independent Television News. Wilson led the housing pressure group Shelter, ran the campaign for unleaded petrol in the UK, directed the Campaign for Freedom of Information, and was chairman of Friends of the Earth in that decade. In the 1990s, Wilson became the director of public affairs and crisis management at B-M for a salary of over £100,000.[103] Wilson's departure from B-M to the British Airports Authority to highlight the 'environmental benefits' of a fifth terminal at London's Heathrow airport, gave government officials the opportunity to attack environmentalists as opportunists out for personal financial gain.[104]

In Canada B-M also formed the 'front group' the Coalition for Clean and Renewable Energy to fight opposition to Hydro-Quebec's plans to build a massive dam in Quebec. Critics argued that the schemes proposed by the company would flood 4,000 square miles, release large quantities of mercury and relocate the native Cree people. John Dillon, writing in *Covert Action* wrote of the consequences of the PR firm's actions:

> B-M and companies like it have become matters of manipulation. If a pro-utility group calls itself by a nice, green-sounding name; if speakers at public forums are not identified as being on the Hydro-Quebec payroll; if supposed activists are really moles for the opposition; image triumphs and truth becomes a casualty.[105]

HILL AND KNOWLTON

Truth has been stretched to the limits by the other global anti-green PR company, Hill and Knowlton (H&K). Its clients include: American Iron and Steel Institute, American Paper Institute, American Petroleum Institute, Ashland Oil, Boeing, Chem Waste, Ciba-Geigy, Du Pont, Hoechst Celanese Corp, IBM, Johnson and Johnson, Louisiana-Pacific, Marathon Oil, Monsanto, Nuclear Electric, Olin Corporation, Pepsi Cola, Philip Morris, Proctor and Gamble, Sandoz, Smithkline Beecham, UNOCAL, Wellcome, Weyerhaeuser and Woolworths.[106]

H&K's tactics have 'raised grave concerns about the grip of lobbyists on democratic processes around the world', according to Judy Steed of *The Toronto*

Star.[107] One of its ex-employees even called H&K 'a company without a moral rudder'.[108] It also helped in the aftermath of the *Exxon Valdez* spill in Alaska and the Three Mile Island accident. Susan Trento in her book *The Power House* explains how H&K countered the environmental movement:

> H&K recruited students to attend teach-ins and demonstrations on college campuses at the height of the Vietnam war, and to file agent-like reports on what they learned. The purpose was for H&K to tell its clients that it had the ability to spot new trends in the activist movement, especially regarding environmental issues. After all, H&K's clients, large chemical, steel, and manufacturing companies, represented some of the worst polluters in the country. Many of its corporate clients were on all the 'dirty lists' of the emerging and increasingly powerful environmental groups.[109]

For example, Trento alleges that 'H&K worked with Proctor and Gamble to keep phosphates in laundry detergent, despite environmental concerns that the chemical was not biodegradable and did not significantly contribute to cleaner clothes.'[110] 'The big corporations, our clients, are scared shitless of the environmental movement,' says Frank Mankiewicz of H&K:

> They sense that there's a majority out there and the emotions are all on the other side – if they can be heard. They think the politicians are going to yield to the emotions. I think the corporations are wrong about that. I think the companies are too strong, they're the establishment.[111]

Mankiewicz and H&K were hired by the Body Shop after a critical piece by Jon Entine in *Business Ethics*.[112]

Internal documents show how H&K have actually generated work on environmental issues, too, by telling businesses Greenpeace was about to target them. A letter to Fidelity Tire Company in 1989 outlined how the company 'is facing the possibility of an unusual and difficult crisis' from Greenpeace, a 'militant activist group'. H&K were giving the company an opportunity 'to preempt them by putting its message across in a credible and effective manner'.[113]

Like B-M, H&K has tried to convince the world that countries with appalling human rights records – Egypt, Haiti, Indonesia, Kuwait, Morocco and Turkey – were good guys to do business with. It smoothed over the international condemnation against China after the massacre in Tiananmen Square.[114] Robert Gray, the Head of H&K, is a personal friend of the Rev. Moon and Oliver North. Gray set up 'front groups' to lobby on behalf of Moon and also worked for the Young Americans for Freedom.[115]

Plate 4.1 The Gulf War was to exert a heavy human and environmental toll.
Source: E.P.L./Michael McKinnon

All this pales into insignificance compared to the incredible campaign H&K led to convince the world to go to war. It formed the organisation 'Citizens for a Free Kuwait' that spent $8 million of Kuwaiti money to persuade the world to liberate the country after Saddam Hussein's invasion in 1990.[116] According to John MacArthur, the publisher of *Harper's* magazine, H&K's campaign amounted to 'subversion of democracy on a grand scale'.[117] H&K arranged for the daughter of the Kuwaiti Ambassador to the USA to appear as an ordinary Kuwaiti girl who had witnessed Iraqi soldiers taking babies out of incubators and killing them. It was a testimony that drove the USA to war. It was a testimony that was totally fictitious.[118]

SUSTAINABLE DEVELOPMENT

Truth is becoming a casualty as PR companies adopt global strategies to fight global ecological problems. As transnational companies move into developing countries around the world, PR guru, E. Bruce Harrison sees it as the PR companies' role to communicate their clients' 'green' policies on a global basis.[119] Nowhere in the next decade will the PR machinery churn so fervently as pushing their clients' commitment to sustainable development.

Sustainable development has become the development catchphrase of the 1990s.[120] Since its inception by the Brundtland Commission in 1987, it has become the promised panacea to cure all ecological problems and solve global pollution. United under the banner 'sustainable development is development that meets the needs of the present without compromising the ability of future generations to meet their own needs', everyone is committed to this social, cultural and environmental utopia. 'Sustainable development is a "metafix" that will unite everybody from the profit-minded industrialist and risk-minimising subsistence farmer to the equity-seeking social worker, the pollution-concerned or wildlife-loving First Worlder, the growth-maximising policy-maker, the goal-oriented bureaucrat and, therefore, the vote-counting politician,' writes one author in the *World Development Magazine*.[121]

PR people like it. 'For companies going green, sustainable development is the key concept for guiding business decisions,' argues E. Bruce Harrison.[122] The Business Council for Sustainable Development like it too. However, for them 'free trade is essential for sustainable development'.[123] Industry also like it, 'Growth in energy demand is almost a prerequisite of achieving sustainable development,' say Shell.[124] Unlimited growth and free trade surely mean business as usual with technological fixes thrown in as tasters for the public to swallow the bitter pill of the resulting ecological destruction.

Take as an example, agri-business, where the annual global market for pesticides is worth $US 20 billion. It is estimated that by the year 2000 the seed market will be worth $US 28 billion of which $12 billion will be made up of genetically manipulated varieties. A future where by the global seed market is manipulated and owned by a small number of mammoth corporations is hardly a sustainable future for the world's farmers and consumers, but biotechnology is seen as one of the growth areas in a sustainable transnational future in which the world will not go hungry.

Its universality, and the different ways in which it can be interpreted, are essentially sustainable development's problem.[125] 'No wonder "sustainable development" is rapidly becoming the buzzword of choice in corporate boardrooms across the world,' says Joyce Nelson. 'It really *does* mean sustained development and the continued altering of the ecosystem to suit human greed.'[126] Even E. Bruce Harrison admits that 'sustainable development is, even after years of debate, a theoretical concept. It's the green grail. It means growth plus greening, or greening plus growth. The definition depends on where you live.'[127] The definition also depends on who you work for to such a large extent, that to all intents and purposes, the definition is defunct. The dichotomy was neatly summarised by Mark Dowie. 'To a corporate economist sustainable development means development that will allow his company to remain in business forever. To an environmentalist it is development that will allow the earth to stay in business forever.'[128] The two are not necessarily compatible.

Green businessman, Paul Hawken, adds:

> There is still a yawning gulf between the kind of 'green' environmentalism that business wants to promote – one that justifies growth and expansionary use of resources – and the kind that actually deals with the core issues of carrying capacity, drawdown, biotic impoverishment, and extinction of species. Business, despite its newly found good intentions with respect to the environment, has hardly changed at all.[129]

The truth is that hardly any current business activity is sustainable, a fact which some businesses have recognised. Volvo admits that 'our products produce pollution, noise and waste'. The Body Shop, considered by many to be a progressive company challenged 'the notion that any business can ever be "environmentally friendly". This is just not possible. All businesses involve some damage.'[130]

If, through current business practices, we are already compromising the future, it is difficult to believe in sustainable development, as its definition has already been rendered defunct. Scientists estimate that the planet is already losing approximately 50,000 species annually, or about 140 each day.[131] A corrupted oxymoron, sustainable development could be the ultimate green-babbling greenwash, that translates into 'business as usual'. For sustainable development read 'multiple use' or 'wise use' or 'sound science' or 'market solutions' or 'free market environmentalism' or 'deregulation' or 'free trade' or 'growth'. The words also perform a paradigm shift by depicting business to be the greenest company on the global green.

5

THE PARADIGM SHIFT

> Things are moving a little faster than I thought they would . . . it's almost cool to be contrarian and okay to be a contrarian liberal.
>
> Michael Fumento, author of *Science Under Siege*[1]

TACTICS

So Rachel Carson was a communist, didn't you know? She may have sparked the modern ecological movement, but that too is like its mentor – communist, subversive and corrupt. Or so the story goes . . .

Carson was accused by the makers of DDT, Velsicol, of 'being in league with sinister influences', whose attack on the chemical industry had the purpose of creating 'the false impression that all business is grasping and immoral' as well as 'to reduce the use of chemicals in this country and in the countries of western Europe so that supply of food will be reduced to east-curtain parity'.[2] Monsanto also responded to *Silent Spring* by publishing *The Desolate Year*, a parody about a DDT-free America being overrun by a plague of locusts of biblical proportions. They sent it to over 5,000 media outlets. Carson's publishers, Houghton Mifflin, were also sued by Velsicol.[3]

The responses to Carson's book by both industry and the political Right are illustrative of the issues covered in the previous two chapters, and which are expanded on in this chapter. First, by attempting to label Carson as some kind of subversive/communist/sinister force, who would send living standards tumbling and society back to the dark ages, industry comes across as the voice of moderate reason fighting immoderate opposition. Second, by seeking to debunk her scientific arguments as unfounded and extreme, industry is the source of the truth, arguing against illogical and inaccurate antagonists.

If environmental activists, such as Carson, have the ability to threaten the corporate profit margin, they should expect retaliation, argues Brian Glick, who has documented government harassment of social change activists. Glick maintains that historically 'dissenting groups come under attack as they begin to seriously threaten the *status quo*'.[4] Attacks against environmentalists are on the increase worldwide, as activists are portrayed as manipulators of science, the press, the politicians, and ultimately the truth. Environmentalists can now expect a direct and severe backlash to any successful, or high profile, campaign they undertake, if they disrupt the *status quo* of governments, industry or the corporate media. The intense backlashes Greenpeace experienced in Europe over its Brent Spar campaign and in New Zealand during its anti-French testing antics in 1995, are a foretaste of the future for environmental campaigners.

The reply to Carson's book from the anti-environmental right wing is indicative too, of both the tactics used and the length of time these arguments have been circulating. Commenting about Carson and the anti-toxics movement, some thirty years after the book had been written, Patrick Cox, an associate policy analyst at the Competitive Enterprise Institute, dismissed them as 'hysterical ideologues'. Chip Berlet, from the Political Research Associates, analysed Cox's arguments and the rhetoric of DDT's opponents and proponents.

Berlet writes:

> Those who oppose pesticides and believe DDT is unsafe, reject science; are affiliated with 'environmental hypochondria'; circulate 'apocalyptic, tabloid charges'; have 'no evidence' for their 'hysterical predictions'; use 'gross manipulation' to fool the media; are 'unscrupulous, Luddite fundraisers'; suffer 'knee-jerk, chemophobic rejection of pesticides' and create 'vast and needless costs' for consumers.[5]

Those who support DDT, on the other hand are: 'Pro-science and pro-logic; have support from the "real scientific community – the community of controlled studies, double blind experiments and peer review"; and help US consumers and farmers save money.'[6]

This process of verbal subjugation against any movement is best described, as the '*paradigm shift*,' argues Chip Berlet. '*Paradigm shift* means a major negative change in the way the public perceives the political movement that is ultimately victimised,' continues Berlet.'[7] In this case, paradigm shift not only means scapegoating the environmental movement but greenwashing by industry. The tactics are simple: to de-legitimise and demonise the opposition, whilst legitimising and canonising yourself.

By reframing the rhetoric within which the public debate is taking place, by recasting the perception of players in the public's eye, by redefining the meaning of words, it is possible to close the 'legitimacy gap' which was explained in chapter 3. The process of greenwashing by industry is to legitimise what is often an ecologically or socially illegitimate process. Greenwashing creates the paradigm shift, making industry seem reasonable and caring and working for the public good. As we have seen, increasingly, industry is using the terminology of 'sustainable development' to promote a 'business as usual' scenario. 'Common sense', 'sound science', 'sensible economics', 'sustainable use' and 'environmental balance' have all crept into the corporate vocabulary. This is reiterated in the number of corporate front groups or so-called 'citizens coalitions', working for 'sensible' this, 'sound' that, or 'responsible' something else.

The other part of the paradigm shift is to make your opposition seem a manipulative, extremist and violent subversive movement, bent on destruction of employment and the way of life as we know it. There is a concerted effort underway to portray the peaceful environmental movement as violent. Doing this, of course, will lose public support and sympathy for the movement, whilst increasing support for yourself. 'Manipulation', 'misinformation' and 'propaganda' seem the in vogue words for the world's press in the 1990s, as they seek to justify the *status quo*.

Combining the two strategies of the paradigm shift is something the green backlash movement has become extremely adept at doing, and just as greenwashing legitimises industry, so moderate and ambiguous language portrays the anti-environmentalist mainstream. Just as industry uses 'greenwash', so 'whitewash' is being used by the Wise Use movement. The naming of an anti-environmental movement as 'Wise Use,' which promotes 'multiple use' of land, and advocates 'putting people back in the environmental equation' was the first step up the paradigm shift ladder, whereby Wise Use became 'mainstream' and the environmental movement becomes 'extreme'.

The main thrust of the Wise Use movement is that it balances the needs of people and nature, something radical environmental extremists do not. 'The Wise Use movement stands for man and nature living together in productive harmony,' says Ron Arnold.[8] Huddled under such caring rhetoric, the anti-environmentalists have set out to vilify their opposition. The tactics of Wise Use are far from caring or conciliatory, but confrontational, and do nothing to solve any problems caused; they only exacerbate them. Their tactics of single-issue scapegoating cut through the complex economic and social factors affecting a community and result in polarisation and alienation. For example, whilst there must be incidents where environmental regulation will have resulted in human hardship, regulation will not have been the only contributory factor to this situation. The real reasons for social problems in affected communities lie with the companies that have done nothing to diversify the workforce or the jobs available in certain areas. A policy of short-term profit has been pursued at the expense of long-term viability.

The end result of this type of paradigm process is to marginalise the environmentalist movement in the USA. This marginalisation process has yet to happen in other countries, but then it could only be a matter of time. As global industries internationalise their activities and governments entice these industries into their countries in a fiercely competitive global market place, people who stand up and question the ecological, social or cultural problems associated with industries will experience 'paradigm shift'.

In fact, for the first time what we are seeing is a 'blueprint' or 'template' of anti-environmental behaviour. What we find is a similarity of techniques being employed in different countries. In some regions such as North America, Australasia, and Europe, there is direct networking between anti-environmental groups in those countries and their Wise Use counterparts in the USA. In others there are no known linkages but in essence, the Wise Use movement is a small player in the anti-green game.

Global PR companies and global industries are also the protagonists of many of these techniques. PR companies have developed strategies for their multinational clients that can be adapted to attack environmental activists anywhere on the globe. Environmentalists now have to be aware that any successful campaign could have severe dangers for them, as these companies will attempt to demonise and marginalise them, advocating corporate environmentalism, instead. Techniques tried and tested in one country against one set of activists are now being used against activists somewhere else in the globe. People have to realise that whilst there may not actually be a coordinated anti-environmental conspiracy going on, there is increased networking of proven effective techniques against environmentalists that have one sole purpose. To effect a paradigm shift in public perception of the environmental movement so that it is marginalised at first and then effectively destroyed.

Outlined below are the main tactics of the paradigm process being used by anti-environmentalists in the USA.

GIVE THEMSELVES ECO-FRIENDLY NAMES

From the corporate front groups to Wise Use groups, many have, as already been stated, eco-friendly names. These have only one aim, which is to confuse the public into thinking they are actually working for positive environmental change. For example: Alliance for Environmental Resources, American Environmental Foundation, Committee on Wetlands Awareness, Coalition of Responsible Environmentalists, Conservation Coalition, Environmental Conservation Organization, Forests Forever, Friends of the River, National Council for Environmental Balance, National Wetlands Coalition, Our Land Society, Responsible Environmentalism Foundation, Save Our Lands, Soil & Water Conservation Society, TREES, and Wild Rivers Conservancy Federation.[9]

Public relations companies are now using this tactic around the world. There are eco-sounding corporate front groups in Australia, Europe, and Canada. As was discussed in chapter 3, the fossil fuel industry currently has three international eco-sounding coalitions promoting inaction on climate change: The Global Climate Coalition, The Climate Council and the International Climate Change Partnership.

CALL THEMSELVES THE 'TRUE' ENVIRONMENTALISTS; WHEREAS THE ENVIRONMENTALISTS ARE 'PRESERVATIONISTS'

Because environmentalists and environmentalism now have a broad base of support in the USA, it is necessary for anti-environmentalists to attempt to tap into this support, whilst at the same time alienating greens from their traditional backers. One way of attempting to do this is by saying you are the true environmentalists, whereas the environmentalists are now 'preservationists'. Some examples are as follows:

William Perry Pendley: 'We are the true environmentalists. We are the stewards of the earth.'[10] Alan Gottlieb: 'Wise use will *be* the environmentalism of the 21st Century.'[11] Clark Collins, Blue Ribbon Coalition: 'I see all of this as a new environmental movement. The characterisation of us as rapers and pillagers is not right. We want to protect the environment'.[12] Ron Arnold: 'We're not out to hurt the environment. We love the environment more than they do.'[13] 'We are all environmentalists,' insists Fred Smith, from the Competitive Enterprise Institute.[14]

CALL ENVIRONMENTALISTS RELIGIOUS FANATICS

For example: Ron Arnold: 'They're pagans, willing to sacrifice people to save trees.'[15] Chuck Cushman: 'Environmentalism is a new paganism. It worships trees and sacrifices people.'[16]. Helen Chenoweth, 1993 Reno Conference: 'We are in a spiritual war of a proportion we have not seen before . . . A war between those who believe that God put us on this Earth and those who believe that God is Nature.'[17]

ENVIRONMENTALISTS ARE LABELLED AS COMMUNISTS

'Environmentalists are like watermelons, red on the inside, green on the outside,' said William Perry Pendley at the Wise Use Conference 1993.[18] 'Watermelons': Kathleen Marquardt 1994 Wise Use Conference.[19] James Catron, Catron County, speaking at the ECO conference: 'We are not in a struggle with environmentalists but with tyranny. There is no moral difference between Earth First! and Stalin. If the radical environmentalists took power you would not live, you would not be left alive.'[20] Joan Smith, California Women in Timber, compares the environmental movement to the Chinese Communist Party.[21] Abundant Wildlife Society: 'The enemy within – environmentalists. We must note that the ideology of Environmentalism is very similar to Marxism.'[22] Bill Bennet, Oregon Citizens Alliance: 'This green tide is really the new red tide rolling over the face of America, the New Socialism.'[23]

ENVIRONMENTALISTS ARE CALLED NAZIS

Chuck Cushman: 'Eco Nazis', 1994 Reno Conference.[24] National Parks are, according to Cushman, 'scenic gulags'.[25] Michael Coffman, Wise Use activist, talks of 'green gestapo'. Pat Bradburn, Alliance for America: 'green gestapo'.[26] Bruce Vincent, from Communities for a Great Northwest, compares the environmental organisation, Earth First! to the Nazis, as does William Perry Pendley, who says 'they are no better than Hitler's Brownshirts'.[27] Other Wise Use literature talks of the 'wildlife gestapo'.

ENVIRONMENTALISTS ARE DEEMED ELITISTS

Arnold: Environmentalists are rich, educated people so high up the needs hierarchy that they actively seek to mock and disdain the lower needs, such as food and shelter;[28] People for the West!: environmentalists stink of elitism and arrogance.[29]

CREATE THE IMPRESSION THAT THE ENVIRONMENTAL MOVEMENT HAS A HIDDEN AGENDA

Michael Coffman, Alliance for America 1994: talked about the UN running the environmental movement. The UN's goal is the total transformation of our culture and society system.[30] 'Fossil Bill' Kramer, the 'Angry Environmentalist' talks of 'eco-totalitarianism', and 'dictatorial environmental bureaucracy' 'driven by hostility to humanity'. 'Leaders of these enormously wealthy, supposedly "non-profit" groups press ambitious government bureaucrats to control every facet of life which makes existence worthwhile – recreation, sports, employment, private property, even our bedrooms.'[31] Dixy Lee Ray: If you don't fight against mindless environmentalist regulations and irrationality, you'll find yourself living in a fascist police state.[32]

DEMONISE ENVIRONMENTALISTS AS EXTREMISTS OR ANTI-AMERICAN

This is not a new phenomenon, either. Any person campaigning for social change has long been regarded as anti-American. 'Environmental activism, according to one economist, is "marked by strong undertones of anti-Americanism",' wrote Richard Kazis and Richard Grossman in 1982.[33] 'People who fought for the 8-hour day, the minimum wage, Social Security, and the right to unionise were called anarchists, socialists, communists, and un-American troublemakers – anything to challenge their credibility without responding to their analyses and demands,' continue Kazis and Grossman. 'Today there are business leaders, government officials, elected representatives, academics – even some labor leaders – who use similar terms to describe environmentalists.'[34]

David Howard: Alliance for America: environmental extremists.[35] Magnus Gudmundsson, Icelandic film-maker, talking at the 1994 Alliance Conference: 'extremist groups'. Bill Wewer, Wise Use conference 1994: 'environmental extremists'.[36] Bill Grannell, People for the West!, talks of 'environmental extremism.'[37] Paula Easly, Municipality of Anchorage, 1994 Wise Use conference: suggests labelling environmentalists as GAGS, as they 'gag freedom'.[38] Clark Collins at the 1993 Wise Use conference 'Our opponents are hate groups.'[39]

ENVIRONMENTALISTS ARE OUT TO KILL YOUR JOB, THE ECONOMY AND TO DESTROY CIVILISATION OR CAPITALISM

Ron Arnold: Environmentalists are evil incarnate. Environmentalists invent environmental threats in order to recruit members and make money. Environmentalists 'are out destroy industrial civilisation'.[40] David Howard, Alliance for America: 'We are fighting for our lives, fighting for our survival, fighting for our country.'[41] Wise Use is a 'battle and a war for survival'.[42] Yellow Ribbon Coalition: 'If we do not unite, we will perish; the victims of a radical preservationist movement.'[43] Bill Bennet, Oregon Citizens Alliance:

> The environmental extremists who put animals above people, who proclaim that the planet is becoming overpopulated (to prime people to accept abortion as an environmental must), who rant about anything nuclear, whose sky is falling about some new threat each week, these people are covertly seeking the destruction of capitalism.[44]

DOWNPLAY THREATS TO THE ENVIRONMENT, BY DEBUNKING SCIENTISTS

Grant Gerber: environmentalists are 'anti-scientific'.[45] Alliance for America: Fred Singer 'has done a lot to demythologise crackpot scientific theory'.[46] 'There isn't such a thing,' says Ron Arnold, about the ozone hole. If CFCs cause ozone depletion, why are there not ozone holes above CFC manufacturing plants? contends Arnold. Acid rain is exaggerated.[47] Global warming, Alar, and species depletion are pure scare tactics, says Arnold, 'to create the illusion of crisis'.[48] Barbara Keating-Edh, Consumer Alert, 1994 Wise Use conference, 'Panic peddlers'.[49] Michael Coffman, Alliance for America Conference 1994: acid rain, global warming are issues of the 'extreme imagination' of environmentalists. Bud Houston, Wise Use activist: 'Mankind has not caused ozone depletion or global warming or acid lakes. They are natural conditions.'[50]

ENVIRONMENTALISTS ARE VIOLENT TERRORISTS

Lyndon LaRouche was one of the first people to brand anyone involved in any progressive protest as a 'terrorist'.[51] Following his lead many sections of society are now labelling environmentalists as terrorists. William Perry Pendley calls greens terrorists. MSLF has held conferences on eco-terrorism, and has been instrumental in having Earth First! labelled as a terrorist organisation.[52] In 1990, MSLF filed suit against the US Forest Service seeking to stop them giving a permit to the 'environmental terrorist organisation'.[53] Grant Gerber too has held workshops for companies such as Georgia-Pacific and Exxon on eco-terrorism.[54] 'Ecoterrorism: The Dangerous Fringe of the Environmental Movement', was the title of the briefing paper that the Heritage Foundation published to celebrate Earth Day 1990.

These last two issues, that of debunking environmental science and attempting to demonise environmentalists as violent are key parts of the paradigm process and warrant further analysis.

COUNTER-SCIENCE

The paradigm shift is coming from all sections of society who have reason to oppose environmentalists: industry, PR companies, the Wise Use movement, the government, the Right and increasingly the media. Furthermore, Lyndon LaRouche and publications associated with him have long defamed environmentalists and attacked environmental science and much of the paradigm shift process can be traced directly back to him. Wise Use activists are picking up on LaRouche counter-science and crackpot conspiracy theories. So too are the militia. More importantly, through Wise Use activists, right-wing groups and radio hosts, this vehement rhetoric is reaching the mainstream media.

LaRouche himself sounds very similar to the Wise Use movement, calling ozone depletion a fraud, and groups like Greenpeace and the Sierra Club 'nut groups, which are determined . . . to destroy industrial society for what they deem a post-industrial, depopulated planet'.[55] The LaRouchians are spreading their message and it is not just confined to the magazines *21st Century Science and Technology* and *Executive Intelligence Review*, now it is reaching into the heart of Washington. In 1994, the Biodiversity Treaty was not signed because of a campaign by LaRouche associates and Wise Use activists. At both major anti-environmental conferences that year, people reiterated bizarre conspiracy theories that the ultimate agenda of the environmental movement is the destruction of mankind. LaRouche's associate Roger Maduro warned that 'population is the enemy of the environmentalist. . . . The actual agenda is to rid us of human beings.'[56] Because of this so-called anti-human bias, Maduro recommended to the participants at Reno that they should join forces with the anti-abortion movement.[57]

LaRouche publications offer a forum for much anti-green theory whilst vehemently supporting the use of pesticides and nuclear power. Numerous anti-environmental scientists have had their work published in *21st Century*, mixing a strange concoction of conspiracy and counter-science. The only problem for the reader is distinguishing what is reality and what is fantasy. 'Save the planet's humans: lift the ban on DDT!' ran the headline of the *Executive Intelligence Review* in 1992.[58] Inside Marjorie Mazel Hecht, Managing Editor of *21st Century*, wrote that 'DDT was the "mother" of all the environmental hoaxes to follow, from saving the lousewort, to the ozone hole.'[59] According to Hecht, there were 'millions of lives lost as a result of the environmentalists' victory in banning DDT . . . DDT does not have harmful effects alleged by the scaremongers.'[60] Articles published in *21st Century Science and Technology* also liken Rachel Carson to Joseph Goebbels.[61]

Publications associated with LaRouche have long argued that there are no limits for growth because it can be accommodated by nuclear power. For example, an editorial in *21st Century*, entitled 'The World Needs Nuclear Energy' proclaimed 'Ironically, the very promise of nuclear energy – its capability for fuelling economic prosperity and population growth – is exactly what spurs its chief opposition. Nuclear fission is anathema to today's neo-Malthusians because it means that there are no limits to growth.' The article proclaims that, 'At least 115 million people died unnecessarily between 1965 and 1980, as a result of the obstruction of nuclear power development worldwide.'[62]

Roger Maduro is also one of the main ozone sceptics, who believes that 'there is no long-term ozone depletion . . . the ozone hole is a natural, seasonal phenomenon', and 'CFCs sink to the ground and go up in smoke; they don't get to the stratosphere'.[63] Maduro also contends that the greenhouse effect is a fraud,[64] a fact perpetuated in other LaRouchian magazines, which consider global warming to be a world federalist plot whereby 'Kremlin leaders and their Trilateral Commission friends are using "ecological emergency" as the pretext to destroy the sovereignty of nations and establish one-world rule.'[65]

Apart from LaRouchite scientists like Maduro, the main proponents of counter-science are a dozen or so extremely prominent 'independent' anti-environmental scientists whose works are regularly used by industry, governments, the Right and Wise Use activists to discredit environmental scientists. So just as industry uses 'third party' grassroots groups to enlist support for its cause, so 'third party scientists' exist who argue that a corporation's products are essentially harmless. Many accept heavy finance or support from industry and then promote scientific arguments that are favourable to those corporations.

Much of their work is a cross between science, scepticism, cynicism and scapegoating. Just as the Wise Use movement attempts to polarise any ecological debate into one of 'environment versus jobs,' 'us versus them', so the sceptics polarise the scientific debate by concentrating on the uncertainties inherent in predicting the outcome of many ecological problems into a debate of 'we are right, they are wrong' and anyway they are 'extremists'. Just as Wise Use activists exaggerate the uncertainties in job security, the sceptics exaggerate the scientific uncertainties in the debate. Healthy scientific research needs articulate well-argued debate, it does not need political polarisation intent purely on creating conflict and confusion.

Indeed, Rhys Roth from the Northwest Atmosphere Protection Coalition, calls such scientists 'atmosphere confusionists', who are 'sowing enough confusion in the minds of Americans about the science behind the greenhouse effect to diffuse our collective concern and outrage, rendering us politically mute'.[66]

These climate sceptics, contrarians or counter-scientists, as they are also known, are not only sowing confusion but spreading public paranoia towards environmental issues as well as empathy towards the anti-environmental agenda. 'If they can get the public to believe that ozone wasn't worth acting on,' says Michael Oppenheimer from the Environmental Defense Fund, 'then there is no reason for public to believe anything about any environmental issue.'[67]

There are now a myriad of books questioning environmental science. The majority seem to be backed by either right-wing publishers or think-tanks. There are some riveting titles: *The Apocalyptics* by Edith Efron, *Eco-Scam: The False Prophets of the Ecological Apocalypse* by Ronald Bailey, a project of the libertarian Cato Institute, as is *Apocalypse Not: Science, Economics and Environmentalism*. Elizabeth Whelan from the right-wing American Council on Science and Health has written *Toxic Terror: The Truth Behind the Cancer Scares.* Also, *Science Under Siege*, written by Michael Fumento, who is associated with the Competitive Enterprise Institute. *Protecting the Environment: Old Rhetoric, New Imperatives* by Jo Kwong Echard, by the right-wing Capital Research Center. Other counter-science books have recently appeared in the UK, which are analysed in chapter 12.

DIXY LEE RAY

Co-author of two prominent backlash books, *Environmental Overkill: Whatever Happened to Common Sense?* and *Trashing the Planet*, one of the most celebrated counter-scientists in the USA at least, was Dixy Lee Ray, until her death in 1994. Ray has been described as the 'queen bee of green-bashing' by the Earth Action Network.[68] She was 'a purveyor of what can only be called *pseudoscience*, in the service of global pillage', according to Stephen Leiper, writing in *Propaganda Review*, and was 'a chief source for disinformation about the state of nature, the environmental movement, and pollution-as-usual'.[69] Ray was a Wise Use favourite and was also on the Board of Directors of MSLF.[70] In fact

Ray was given a posthumous special tribute in the MSLF's President William Perry Pendley's book, *It Takes A Hero*.[71]

Ray was vehemently pro-nuclear and was the Chairman of the US Atomic Energy Commission for numerous years. One of the scientists to whom *Environmental Overkill* is dedicated is Edward Teller, the 'father of the hydrogen bomb'.[72] The other was Petr Beckmann, another vehement proponent of nuclear power and author of *The Health Hazards of Not Going Nuclear*.[73] Ray believed that:

> nuclear power is an unparalleled success. Nuclear generation of electricity is safe. In the more than a quarter century of commercial operation in the United States and the Western World, there have been no fatalities, no significant releases of radioactivity to the environment.[74]

Ray omits the evidence that the Irish Sea is now the most radioactive in the world because of discharges from Sellafield. Nuclear workers have a history of contracting cancer above what would be expected in the normal population. Nuclear proliferation is a serious concern of most world governments worried about terrorism. Ray's writings include the wording 'western world' so forgetting about the people who died and are still suffering from radioactivity due to the Chernobyl disaster (see below).[75] Solar power, according to Ray, is more dangerous than nuclear, because there is a risk of people falling off roofs whilst fitting solar panels.

Meanwhile, Ray's rhetoric is similar to LaRouche's warning of 'environmental extremists, alarmist environmentalists, doomsayers, scare substances' who want nothing less than 'world government'.[76] The paradigm process is prevalent throughout Ray's books, for example, 'calm reason and alarmist environmentalism do not co-exist'.[77] 'At a first glance *Trashing the Planet* seems to be an exposé of environmental problems, but it is really a call for nuclear power and technology,' wrote *The New York Times* of Ray's first book. It continues:

> At their best, the authors call for an objective analysis of environmental issues, a perception of environmental problems in terms of acceptable and unacceptable risks . . . but these positive qualities are lost in emotionalism, rhetoric and unfair selection of facts. In short, *Trashing the Planet* falls victim to the very tendencies it condemns.[78]

Dixy Lee Ray was also a frequent contributor to *21st Century* magazine, which she called 'one of the best science magazines being published in the US today'.[79] Indeed, the magazine called her 'one of the great women of this century'.[80] Ray uses Roger Maduro's book *The Holes in the Ozone Scare* as the main source for her information on ozone depletion. Sherwood Rowland, the 1995 Nobel Peace Prize Winner for chemistry, regarded by many as the father of ozone research, called Maduro's book 'a good job of collecting all the bad papers [in the field] in one place'.[81] In fact 'it's the standard bad science technique,' according to Garry Taubes in *Science* magazine. 'You take only the evidence that backs your case, over-interpret it and call all the contradictory evidence part of a plot.'[82]

Ray also cites other known sceptics who in turn cite her work. What happens, in effect, is a self-perpetuating circle whereby each researcher quotes another close colleague, who in turn is probably quoting them as their source of information. This is not to say that Ray is wrong in doing this, but these sceptics just do not cite the overwhelming majority of the world's independent scientists who disagree with them. A thorough independent peer review process, the normal safety net for sound science, is therefore avoided, as the peer review process is carried out by a close circle of sympathetic scientists. Bert Bolin, the Chairman of the Intergovernmental Panel on Climate Change (IPCC), has remarked that much of the sceptics' work has been presented in popular or semi-popular articles, rather than in peer-review journals.[83]

RATIONAL READINGS

Many of these sceptics appear in another book that is also a victim of the very tendencies it condemns. In the true spirit of paradigm shift, this lengthy tome, which is one of the most comprehensive counter-science books ever compiled, is entitled *Rational Readings on Environmental Concerns*, which it claims offers 'the best source of accurate and comprehensive data on environmental science'. Apart from Dixy Lee Ray, many of the most prominent counter-scientists are represented: Jay H. Lehr; Edward Krug; J. Gordon Edwards; Elizabeth Whelan; Petr Beckmann; Jo Ann Kwong; Hugh Ellsaesser and Fred Singer.

In this one 'rational' book on science, environmentalists are labelled as 'extremists', 'apocalyptics', 'alarmists', 'zealots', 'emotional extremists', 'ignorant', 'chemophobes', 'insincere environmentalists-for-a-weaker-America', 'fundamentally elitist', 'professional scaremongers', 'potential mass murderers' who are 'assaulting reason', full of 'environmental paranoia,' 'overzealous environmentalist rhetoric', 'environmental tyranny', 'toxicity' and 'nihilism' and who are fighting an 'ideological battle against economic growth' which threatens 'democracy' and has brought America to her 'knees'.[84]

Having vilified environmentalists, the book debunks many ecological concerns. Acid rain is advantageous: 'the acidity of lakes frequently contributes to their beauty, particularly the crystal clarity of acidic waters with little living in them. And swimmers do not have to worry about nuisances such as slimy green algae or leeches.'[85] Pollution is no problem: 'There is no persuasive evidence from epidemiology or toxicology that pollution is a significant cause of birth defects or cancer.'[86] 'Scientific studies have produced no evidence that PCBs in the amounts found in fish are harmful to humans'.[87] 'Dioxin . . . has never been shown to cause any deaths or serious harm to humans.'[88] 'After half a century of trying, no toxin or air pollutant has been found to have been present in concentrations demonstrated to be hazardous to health.'[89]

Global warming is also good for you:

> If global warming occurs to the degree and extent that the doomsday global warming models predict, it will be of great benefit to the world . . . science indicates that increases in temperature, moisture, and CO_2 inherent to the global warming scenario will transform the Earth into a Garden of Eden and not a den of death as we are led to believe.[90]

Nuclear power is safe: 'The problems cited against nuclear energy, such as disposal of spent nuclear fuel, are mostly political and ideological rather than technical.'[91] In fact radiation is good for you:

> We would all be better off – i.e., be healthier, live longer, have fewer genetic defects – if we had more exposure to radioactivity than we do now, up to 10 times more . . . exposure to the fallout from Chernobyl will actually prove to be a net health benefit.[92]

THE CLIMATE SCEPTICS

Fred Singer of the Science and Environmental Policy Project (SEPP) in Washington; Hugh Ellsaesser, from the Lawrence Livermore National Laboratory and the Wise Use group ECO; and Sherwood Idso, a research physicist for the US Department of Agriculture are three of the authors in *Rational Readings*. Along with Robert Balling, Director of the Office of Climatology at Arizona State University; Richard Lindzen, Sloane Professor of Meteorology at the MIT and Patrick Michaels, the Associate Professor of

Environmental Sciences at University of Virginia, they are the most prominent half dozen global warming sceptics.

Apart from global warming, Singer has broadened his attacks to include other ecological issues such as ozone, acid rain, automobile emissions and even whaling.[93] 'The CFC-ozone theory, such as it is,' wrote Singer in 1992, 'is simply not good enough to predict chlorine values or ozone depletion.'[94] Singer has also debunked acid rain for the Committee for a Constructive Tomorrow, which is a member of Earth Day Alternatives.[95] The Director of CFACT's environmental projects is Edward Krug, who is a noted acid rain sceptic, who also appeared in *Rational Readings on Environmental Concerns*. Krug believes that 'acid rain is only one of a dozen manufactured crises ranging from toxic waste to chemical poisoning of food and global warming'[96]. Global warming could be 'a smokescreen for another agenda?' as 'mounting evidence has convinced many scientists that greenhouse theory is just a lot of hot air'.[97] CFACT's Director is the Treasurer for the American Nuclear Energy Council and a lobbyist for General Atomics.[98]

Along with Singer's Science and Environmental Policy Project, Earth Day Alternative members mentioned in chapter 2 are prominent in debunking environmental science. So too is the George C. Marshall Institute, that specialises in the 'politicisation of science', which is intending to open a London office to counter this threat internationally and to debunk the myth of global warming. It was the Institute's report on the greenhouse effect in 1989, which was used by the Bush administration to justify a more lenient approach to CO_2 emissions. One of the authors of the report, Dr Robert Jastrow, argues that global warming advocates are 'motivated by an anti-growth, anti-business ideology'.[99] Six years later, another report by the Institute dismissed the chances of a major global warming as 'inconsequential', a fact used by the science-sceptics' new protagonist, journalist Gregg Easterbrook.[100]

For the scientists though, global warming is where they have made their professional stand and where they have staked their professional reputations, and where they are making a comfortable living off counter-science. Their arguments basically boil down to the fact that either global warming is not happening and is a fraud manufactured by greens; or if it is occurring, the problem is vastly over-exaggerated as observed warming does not reflect what is predicted in the climate models. But then, they add, even if global warming does actually occur, then it will be beneficial to the plants, as they thrive in CO_2-rich conditions.

For example, Richard Lindzen sees global warming as similar to other 'environmental "crises" including ozone depletion, acid rain, diminishing species diversity and contamination by PCBs, dioxins, asbestos and lead'. Basically, a manufactured problem solely designed to bring more money into the environmental movement.[101] Patrick Michaels is ambivalent: he believes that the greenhouse effect is not happening, but if it does it will be good for the plants.[102] This latter view is held by both Hugh Ellsaesser, who says that 'carbon dioxide will exert primarily beneficial effects on the biosphere' and Sherwood Idso who believes that a 'CO_2 enriched atmosphere will increase the biological carrying capacity of the earth by as much as an order of magnitude'.[103]

The beliefs of these scientists have brought them close to industry (especially large CO_2 emitters), the Right and the Wise Use movement. Ray, Singer, Lindzen, Michaels and Ellsaesser have all debunked global warming at press conferences organised by the Wise Use group Consumer Alert and Singer, Ellsaesser and Balling have spoken at Wise Use conferences.[104] Industry has provided the backbone of support and promotion for these sceptics and many receive industry finance, or work on industry's behalf. For example, Balling, Michaels and Singer have all represented the fossil fuel lobby's Global Climate Coalition (GCC).[105] Balling, Michaels and Idso have also acted as spokespeople for the 'Information Council on the Environment' (ICE). ICE, originally called 'Citizens for the Environment', was founded in 1991 with $75,000 from the National Coal Association. Leaked memos detail how the association planned test marketing of the idea to 'reposition global warming as theory (not fact)'.[106] ICE funding reads like a 'Who's Who' of the US coal industry.[107]

Balling was to later pull out of ICE, commenting that 'people do not like the idea of a respected scientist acting as a "mouthpiece" for a private group'. Subsequently, Balling has represented both the Global Climate Coalition and the Competitive Enterprise Institute, which is also a leader in global warming scepticism. Hardly surprising when contributions come from the American Petroleum Institute, Amoco, ARCO, Dow Chemical, Ford Motors and General Motors, amongst others.[108] Richard Lindzen was invited by OPEC to speak at their 'Conference on the Environment' in 1992 in Vienna and toured New Zealand in the run-up to the 1995 Berlin Climate Conference. Fred Singer has also made the trip down under (see chapter on Australia and New Zealand).

Michaels and Idso also receive funding from the fossil fuel lobby. *The Greening of Planet Earth*, a video on the 'benefits' of carbon dioxide, produced by Idso, costing $250,000, was funded by the coal industry and widely circulated by OPEC and coal companies. A thousand copies of his book, *Carbon*

Dioxide and Global Change: Earth in Transition were bought and distributed by Western Fuels.[109] A quarter of Michaels' research funding is reportedly received from companies such as Edison Electric Institute, the largest utility trade association in the USA. Michaels' magazine, *World Climate Review* is funded by the Western Fuel Association and a video produced by him was funded by coal companies and is distributed by the Denver Coal Club. When asked about his funding coming from companies which have such a vested interest in climate change, Michaels replies, 'Everybody has an agenda, let's face it.'[110]

'If it doesn't taint their science, it doesn't taint my science,' is Fred Singer's response to criticism of his industry funding, adding that 'every organisation I know of gets money from Exxon, Shell, Arco and Dow Chemical'. Fred Singer has worked for companies such as Exxon, Shell, Arco, Unocal and Sun. His organisation, the Science and Environmental Policy Project (SEPP) was founded as an affiliate of the Washington Institute for Values in Public Policy. Singer has admitted receiving free office space from them for a year. He is also on the executive advisory board of the Moonie-funded magazine, *The World and I*.[111] Candace Crandall, the Executive Vice President of SEPP is the former PR person for the Saudi Arabian Embassy,[112] the country that protects the largest fossil fuel reserves on the globe. Environmentalists 'have lost sight of the goal, which was to balance the needs of people with the needs of nature', says Crandall, reiterating a time-honoured Wise Use phrase.[113]

ACSH

Another person who has been severely criticised for taking industry funding and then attempting to appear as an independent scientist is Elizabeth Whelan, who often talks of 'toxic terrorists', 'self-appointed environmentalists' and 'voodoo statistics,'.[114] Whelan founded the American Council on Science and Health (ACSH) to 'bring reason, balance and common sense to public debates about food, nutrition, chemicals, pharmaceuticals, lifestyles, the environment and health', according to William Perry Pendley, as he named her one of his 'Heroes' fighting 'environmental oppression'.[115]

ACSH receives some 50 per cent of its funding from corporations and foundations.[116] 'The interests of her benefactors inevitably raise questions,' wrote Howard Kurtz in the *Colombia Journalism Review* in 1990.

Could there be any connection between Whelan's defense of saccharin and funding from Coca-Cola, the PepsiCo Foundation, the NutraSweet Company, and the National Soft Drink Association? Her praise for fast food and a grant from Burger King? Her assurance about a high-fat diet and support from Oscar Mayer Foods, Hershey Foods Fund, Frito-Lay, and Land O'Lakes? Her defense of hormones in cows and backing from the National Dairy Council and American Meat Institute?[117]

'I'm proud of my relationship with corporations,' says Whelan who has received *Rolling Stone* magazine's 'Hall of Shame: who is the foulest of them all?!! Award'.[118]

Whelan was one of the main critics of the Alar scare that rocked America in the late 1980s but some 10 per cent of her funding came from the makers of Alar at one stage.[119] The NRDC, who published the original controversial report on Alar, were dismissed by Whelan as a group whose 'target is the free-enterprise, corporate-America system'.[120]

Elizabeth Whelan also employed standard Wise Use tactics against a book entitled *Diet for a Poisoned Planet* by David Steinman, which highlighted pesticide contamination of certain foods, and gave consumers the chance to make a more informed and safe choice of foodstuffs. Before the book was published, Whelan wrote to the then White House Chief of Staff saying that Steinman, and people like him, were 'threatening the US standard of living and, indeed, may pose a future threat to national security'. The letter was copied to all the major government agencies. Whelan also pressurised for an introduction written by a senior science advisor to the US EPA, Dr William Marcus, to be dropped from the book. When Marcus refused he was later sacked by the EPA.[121]

ACSH are also clients of Ketchum Communications, who were also employed by the California Raisin Advisory Board to nullify any criticism in the book (Steinman had found 110 industrial chemical and pesticide residues in sixteen samples of raisins). According to a whistleblower from inside the company, Ketchum found out which talk shows Steinman was due to appear on and then tried to depict Steinman as an 'off-the-wall extremist without credibility'.[122] This is not the first time Ketchum have used such tactics to try and discredit environmentalists (see later).

Although Whelan maintains, 'I'm not politically driven',[123] it is undoubted ultra-conservative philosophy that drives many of the organisations and think-tanks who send material out to the press in the knowledge that little, if any, of the sources of the information will be checked. This means that the paradigm process, although starting from known extremists like LaRouche, can be legitimised by the use of so-called respectable think-tanks or scientists. 'I think they are very vital in taking what starts out as right-wing rhetoric and putting a veneer of intellectualism over it,' says Chip Berlet, and it 'would be dismissed if it was coming from someone who wasn't wearing a business suit'.[124]

RUSH LIMBAUGH

It is easy to follow the legitimisation process of poor science into the public domain. Rush Limbaugh is one of the premier right-wing radio hosts in the USA, reaching an estimated 20 million Americans every day. One of the celebrated shock-jocks that broadcast an emphatic right-wing viewpoint across America. 'So beware: there are people out there – Communists, Socialists, Environmentalist Wackos, Feminazis, Liberal Democrats, Militant Vegetarians, Animal Rights Extremists, Liberal Elitists – who will try to prevent you from reading this book,' writes Limbaugh in his first book, *The Way Things Ought To Be*.[125]

People should read Limbaugh's book to see how conspiratorial and incorrect science can be legitimised and infused into the consciousness of an unsuspecting America. Environmentalism and environmental science are two of Limbaugh's favourite topics for dour denunciation. Relying heavily on Dixy Lee's writings, he denies that there is ozone depletion or threat of global warming. 'The fact is we couldn't destroy the earth if we wanted to,' Limbaugh writes.[126] 'The best way you can arm yourself against the junk scientists is to read a book by Dixy Lee Ray,' say Limbaugh. By quoting Dixy Lee Ray, who was in turn quoting Roger Maduro, Limbaugh is just reiterating the extremist views of Lyndon LaRouche.

Limbaugh calls the destruction of the ozone layer by man-made sources 'poppycock' and 'balderdash', saying that, if chlorine from volcanoes have not destroyed the ozone layer in the 4 billion years they have been erupting, then there is no chance that man-made CFCs will do it.[127] The only problem with Limbaugh's argument is that naturally occurring hydrogen chloride from volcanoes is transformed into hydrochloric acid and is therefore washed out and does not reach the stratosphere, whereas man-made CFCs are inert and do.[128] Concerning the possible future extinction of the spotted owl, Limbaugh says, 'So what if they are no longer around to kill the mice. We'll just build more traps. Either that or we'll breed more dogs.'[129]

Limbaugh also demonises the environmental movement *per se*, talking about 'Environmentalist Wackos'. He writes:

the modern environmental movement is the major remaining hiding place for socialists now that communism has collapsed. Environmentalist Wackos are fringe kookbergers and are not to be confused with serious and responsible ecology-minded people . . . Environmentalists are dunderhead alarmists who are long-haired and maggot-infested religious fanatics.[130]

In his second book, *See I Told You So*, Limbaugh says that:

despite the hysterics of a few pseudo-scientists, there is no reason to believe in global warming . . . The real enemies of the radical environmental leadership are capitalism and the American way of life. There are more acres of forest land in America today than when we discovered the continent in 1492.[131]

'Rush Limbaugh's best-selling books,' write Leonie Haimson, Michael Oppenheimer and David Wilcove from the Environmental Defense Fund, 'are so full of statements on the environment that are misleading, distorted, and factually incorrect'.[132] So concerned was America's national media watch-dog, Fairness and Accuracy in Reporting (FAIR) over Limbaugh's inaccuracies on a range of issues, that their magazine *EXTRA!* ran a special edition called 'Rush Limbaugh's Reign of Error' which listed over six pages of Limbaugh's rhetoric versus actual reality. A whole book, called *The Bum's Rush*, by Don Trent Jacobs has been written purely debunking his environmental rhetoric.[133] The problem is, though, that millions of his listeners will neither read or hear a different viewpoint and to them environmentalists are socialists out to destroy capitalism with fraudulent science.

Paradigm shift, Rush-style, continues unabated on the airwaves across America. What Limbaugh has achieved for millions of Americans is a legitimacy for LaRouchite rhetoric that not even Lyndon LaRouche himself could have dreamed of. Furthermore, Dave Helvarg wrote, in *The War Against the Greens*:

It could easily be argued that Rush Limbaugh has done more to spread the anti-enviro message than the Alliance for America, Center for Defense of Free Enterprise, PFW, NIA, OLC, AER, ECO, and the whole alphabet-soup collection of Wise Use/Property Rights groups combined.[134]

It could also be argued that Limbaugh did more than most to assist Newt Gingrich and the Republicans ride so triumphantly back into power in late 1994. It remains to be seen if the feat will be repeated in 1996.

SCEPTICS GO MAINSTREAM

Through the likes of Maduro, Dixy Lee Ray and Limbaugh on ozone depletion and Fred Singer, Richard Lindzen, Patrick Michaels and Robert Balling on climate change, the science backlash was an important international phenomenon by the early 1990s. They were heavily backed by the fossil fuel companies as well as OPEC, and promoted by the networks of right-wing think-tanks, who were advancing their own political agenda. Counter-science was suddenly in vogue, with the media, looking for confrontation, seeking to shoot the 'doomongers' down. The only problem was that many journalists were not looking where it was coming from. Articles questioning ozone depletion and global warming began popping up all over the globe, as were the counter-scientists.

An indicator of how influential the sceptics had become was when ozone research was called a 'politically motivated scam' by the magazine *Omni*. Furthermore, the *Washington Post* reported that with the Montreal Protocol, 'the problem appears to be heading toward solution before [researchers] can find any solid evidence that harm was or is being done'. Despite the fact that the peak of ozone destruction is still to come sometime next century.[135] In the summer of 1993, *Science* magazine ran an article entitled 'The ozone backlash', linking Limbaugh, Ray and Singer.[136] '1993 was the year of the backlash about environmental hype and hoax,' says Singer. 'The quality papers finally, that is, *The New York Times*, *The Washington Post*, finally had articles that questioned some of the assumptions about environmental disaster.'[137]

The anti-green standard bearer in the print media was certainly Keith Schneider, a reporter for *The New York Times*. 'Journalists are coming around,' says Ron Arnold. 'Keith is only the first on the "let's-beat-up-the-environmentalists bandwagon".'[138] Journalist Dave Helvarg writes of Schneider's 'mainstreaming' of the Wise Use movement in the pages of *The New York Times* 'legitimising' the anti-environmentalists.[139] Schneider has been followed by another national journalist writing for *Newsweek*, Greg Easterbrook, who has started questioning the 'Green Cassandras'.[140] Although environmentalists 'are on the right side of history,' claims Easterbrook, 'they are on the wrong side of the present'.[141] 'Anyone who relies on exaggeration to exist politically runs the risk that their bluff will be called,' he says.[142] Easterbrook's book, *A Moment on the Earth*, a 'feel-good' story about the Earth's ecological problems, became a best-seller in 1995, but was criticised by the Environmental Research Foundation for being mistakenly optimistic.[143]

Environmentalists are now getting ripped apart in the media, but the media are not checking the validity, or the political or corporate agenda, of their sources. 'Environmentalists are on the run' reported *Fortune Magazine*,[144] declaring, 'A big part of the problem is that America's environmental policy making has increasingly been driven more by media hype and partisan politics than by sensible science.' What group did *Fortune* use to back this up? Elizabeth Whelan's industry-funded ACSH and her so-called 'independent' scientists.[145] In the spring of 1995 the *Ecologist* magazine ran an article complaining of the 'ozone backlash' and the *New Scientist* reported that 'the greenhouse backlash' was in full swing.[146] Showing how mainstream the anti-ozone depletion argument had progressed, the US Congress actually debated whether there was an ozone hole in September 1995.

Bert Bolin, the Chair of the Intergovernmental Panel on Climate Change had warned seven months previously that,

> the public press is anxious to seize on scientific controversies, and rightly so, but an accurate account is seldom given of how various differing views are founded in the scientific/expert community. An increasing polarisation of the public debate that has been developing in some countries is not a reflection of a similar change among *experts at work* on this issue. [emphasis in original][147]

The polarisation process and with it the resulting paradigm shift were working. The climate sceptics have achieved more prominence than either their numbers or scientific credibility warrant. Part of the blame must be laid at the media door, as journalists seek out this small minority, looking for the headlines that represent conflict not consensus, which is what sells newspapers.

When Fred Singer, Patrick Michaels and Sherwood Idso were featured on Ted Koppel's *Nightline* show on 24 February 1994, after all the furore of corporate and Moonie-funding, Koppel ended by saying, 'The measure of good science is neither the politics of the scientist nor the people with whom the scientist associates. It is the immersion of hypothesis into the acid of truth. That's the hard way to do it.'[148] So what is the acid truth, as it is generally regarded by leading expert scientists as regards ozone depletion, the legacy of Chernobyl and global warming, to take but three examples?

A report by United Nations Environment Programme (UNEP) concluded in 1994 that,

> the scientific evidence, accumulated over more than two decades of study by the international research community, has shown that human-made chemicals are responsible for the observed depletions of the ozone layer over Antarctica and likely play a major role in global ozone losses.[149]

Also at the end of 1994, NASA concluded: 'Three years of data from NASA's Upper Atmosphere Research Satellite (UARS) have provided conclusive evidence that human-made chlorine in the stratosphere is the cause of the Antarctic ozone hole. UARS instruments have found CFCs in the stratosphere.'[150] The destruction of the ozone layer hit a record high over the Antarctic in September 1994, according to the World Meteorological Organisation.[151]

In 1995, international research showed an ozone hole above the Arctic as well as above the Antarctic. Research published by the Department of the Environment in the UK showed that these results illustrated 'the importance of the internationally agreed phase out of ozone depleting substances'.[152] In August of that year British Antarctic Survey scientists announced that the 'springtime' ozone hole was deepening and that there was new evidence that the ozone layer was also thinning into the southern hemisphere's summer.[153] Scientists also found a thinning of the ozone layer over Europe.[154] In March 1996 the World Meteorological Organisation announced that the ozone layer had depleted by a record 45 per cent at times over the previous winter over a northern zone stretching from Greenland to Scandinavia and Western Siberia.[155]

In 1995, the UN also reported the social and environmental effects of the Chernobyl disaster. The minimum number of people forced to leave home was estimated at 400,000, however, the number of people affected in some way by the disaster was put at 9,000,000. According to the World Health Organisation (WHO), there are some 800,000 people who were involved in the clean-up operation. According to WHO, medical monitoring is already indicating growing morbidity (illness, disease and invalidity) and mortality among this group. The main diseases include cardiovascular and heart diseases, lung cancers, gastrointestinal inflammation and tumours and leukaemia. Some 7,000 Russian liquidators have already died since the accident from various causes, including suicide. In Byelorussia alone an area the size of The Netherlands has been rendered unusable.[156]

The Intergovernmental Panel on Climate Change (IPCC), which was convened by the UN General Assembly in 1988 to investigate climate change, reported back in 1990 and 1995. In 1990 they were not quite sure if human activities were producing global warming, but by December 1995 they concluded that 'the balance of evidence suggests a discernible human influence on global climate'. The new 'best estimate' for the increase in global mean temperature relative to 1990 was estimated to be about 2 degrees centigrade by the year 2100, about one third lower than the 'best estimate' made in 1990. The IPCC's 1995 report had faced strong opposition from the corporate front groups the Global Climate Coalition and the Climate Council, and countries such as Kuwait and Saudi Arabia.[157]

PORTRAYING ENVIRONMENTALISTS AS TERRORISTS

Perhaps the most serious angle of the 'paradigm process' is the attempt to tar the peaceful environmental movement as violent terrorists. There is undoubtedly a concerted effort to undermine environmentalists by doing so. 'Such labelling undermines public support and thus sanctions the use of aggressive surveillance and harassment by government agencies or private security firms,' argues Chip Berlet. 'There is also a self-fulfilling prophecy with labelling, as police are likely to respond with unjustified force when they have been trained to think of peaceful protesters as violent terrorists.[158]

The message is not just confined to the Wise Use movement. As far back as 1988 private security firms had long been labelling environmentalists as terrorists.[159] Public relations companies are also propagating the terrorist message. For example, in 1991 a memo was leaked to Greenpeace which had been written by PR giants Ketchum on behalf of the chemical company Clorox, a client for whom the PR company had drawn up a crisis management plan. The plan was promoted by fears that Greenpeace, who are running a chlorine phase-out campaign against the pulp and paper industry, might also target household use of bleach and call for its elimination.

The plan recommended labelling environmental critics as 'terrorists', and calling Greenpeace a violent organisation which produces spurious research. The plan called on an independent industry association to run a campaign 'Stop Environmental Terrorism', calling on Greenpeace and the press 'to be more responsible and less irrational in their approach'.[160] The whole approach was designed to promote a paradigm shift in the public perception that Greenpeace's tactics were not peaceful but openly violent.

The year before, a false memo on Earth First! letterheaded paper calling for acts of violence was distributed by PR companies working for the forest industry, even after they knew it to be fake. 'We intend to spike trees, monkey-wrench, and even resort to violence,' said the fake release, which was sent to the *San Francisco Examiner*.[161] That summer Earth First! was preparing a series of non-violent direct action (NVDA) protests against the cutting of old-growth forests in North California, called 'Redwood Summer', modelled on the non-violent civil rights demonstrations of the 1960s which had been called 'Mississippi Summer'. One of the main companies targeted by activists was also busy distributing false press releases purporting to be from Earth First! and calling on activists to 'fuck up the working of the mega-machine'.

Louisiana-Pacific also employed intimidation tactics by urging employees to attend environmentalists' meetings 'with rolled up sleeves, wearing their work boots and hard hats'. They then should go and sit right down next to environmentalists as they 'would not speak so freely if they knew pulp mill-workers were sitting beside them'. So outraged by the advocacy of such tactics, the union filed a grievance against the company.[162] These tactics seem common throughout the logging industry. Former logger Gene Lawthorn recalls the compulsory anti-environmental workshops given at his mill. 'For an hour they made you sit while they described environmentalists as preservationists, as tree-hugging pagans trying to overthrow Christianity,' remembers Lawthorn, 'They said the preservationists were going to take away your jobs. It scared the hell out of you.'[163]

But it is not just industry who hold anti-environmental workshops. So do the Sahara Club, a radical Wise Use group who are a direct pun on the Sierra Club, whom they call 'elitist snobs'. Their name is designed to cause maximum confusion. They actively promote 'dirty tricks' against activists and are masters of misinformation. Their logo is a hand strangling an Earth First! member. They are open in their hostility to environmentalists whom they call 'guilt ridden eco-arseholes', 'slime', 'maggot-infested hippies','eco-suckfaces', 'eco-freaks', 'jerkweeds', 'eco-pigs', 'snivelling, lying sayers of doom and gloom' and 'snob-bish scum', who spread 'panic and hysteria' and 'eco-gospel'.[164]

'Do anything and everything you can do to discredit the bastards,' advocate the Sahara Club in their Newsletter Number 24:

> When you attend some sort of public debate with eco-freaks, take a contingent of people with you, prepared to raise some hell . . . yell out long and hard and call them names. 'Liar! Land-grabber! Nature Nazi!' . . . When it gets to that point, find the opposing ring leaders and get right in their face. Spit in their face. Kick chairs over . . . Make the eco-freaks afraid to attend meetings. Plan deception. Create dissension. Send fake memos or letters from one eco-group to another. Drive them nuts . . . A small plastic squeeze bottle filled with red cayenne pepper will repel even the most obnoxious and violent eco-freak, if squirted in the face.[165]

Dirty techniques were used by the Sahara Club at the time of 'Redwood Summer' in 1990, where violence would be used against Earth First! and Judi Bari. Moreover, the Sahara Club routinely publish vehicle number plates and phone numbers and addresses of Earth First! activists, in order to intimidate them. Intimidation is left up to the Sahara Clubbers, whose message to EF! is simple: 'Our special division of Sahara Clubbers was simply gonna . . . do justice that the authorities wouldn't do.'[166]

The Sahara Club are receiving information on the environmental movement from both the John Birch Society and magazines associated with Lyndon LaRouche.[167] 'Fantastic source of information! It's called *21st Century Science and Technology*,' report the Sahara Club, 'Since we have been trading information and news with these folks they offered a special deal to Sahara Club members.'[168] They also repeat the accusation that environmentalists are terrorists.[169]

MONKEY-WRENCHING AND TREE-SPIKING

The advocacy of monkey-wrenching and tree-spiking was the reasoning and rationale behind labelling environmentalists as terrorists. 'Monkey-wrenching', the sabotage or 'ecotage' of property to stop ecological destruction, was originally born in the mind of Edward Abbey who wrote a fictional novel entitled *The Monkey Wrench Gang* published in 1975. The novel, which became a cult book, is about four characters who roam the west carrying out covert actions of ecotage. Dave Forman, a former lobbyist for the Wilderness Society and the founder of Earth First!, was to immortalise Abbey's writing in his book *Ecodefense: A Field Guide to Monkeywrenching*, which described various ways of monkey-wrenching. Whilst these tactics are considered sabotage by some, to others it is called terrorism. 'Sabotage is violence against inanimate objects: machinery and property. Terrorism is violence against human beings,' argued Edward Abbey.[170] Earth First! has always maintained a policy of non-violence towards people, although sabotage against property was advocated. But there is one tactic which was to prove contentious both inside and outside the direct action movement.

Earth First!'s most controversial tactic was its advocacy of tree-spiking, a practice that has now been stopped and denounced by the more progressive groups. If Earth First! had ever wanted to give a public relations coup to its opposition, tree-spiking was it. The theory behind tree-spiking was simple. If you wanted to protect old-growth forest from being logged, you banged a large spike or nail into a tree, which had the potential to damage either a blade or a saw when the tree was either being cut down or sawn in the mill. Because of the danger to equipment and personnel, it was believed that either the Forest Service or timber companies, once told a region had been spiked, would leave the trees alone. Of course they did not and felling continued.

Despite press reports to the contrary, only one injury has ever been attributed to tree-spiking, that of timberman George Alexander near Elk, California. In 1987, Alexander suffered facial injuries when a nail embedded in wood caused a massive saw blade to break.[171] As a result of the accident, Alexander's jaw was broken in five places and he suffered huge blood loss. He is lucky to be alive. Although it was the nail that caused the blade to break, it seems safety issues were coming to a head at the plant. Fresh saw blades had been delayed and the situation with his own saw had become so bad that Alexander had nearly not gone to work that day.[172]

The press quickly condemned the incident as 'Tree-spiking terrorism', and the company labelled it 'terrorism in the name of environmental goals'. Before he had fully recovered, Alexander was asked to go on a tour by a pro-industry group denouncing Earth First! and tree-spiking. He refused and returned to work. The company put forward a $20,000 for information leading to the conviction of whoever was responsible, although Alexander had to file a private lawsuit for personal injury damage. He received just $9,000 and was moved on to night shift. When L-P closed the mill down, he was laid off.[173] No proof has ever been offered that anyone from Earth First! was responsible for the incident, although the indifference of some EF! members in response to Alexander's injury angered Judi Bari and other activists. As a result, progressive EF! groups denounced tree-spiking and have signed a declaration of non-violence.

'ECO-TERRORISM WATCH'

Despite this change, groups like Earth First! and Greenpeace, who have a policy of non-violence, are still being labelled as 'terrorists'. In a move that singles the further merger between Wise Use activists and the LaRouche network, Barry Clausen, a Wise Use activist and private investigator, teamed up with Roger Maduro to produce a subscription journal entitled *Eco-Terrorism Watch*. The product is vintage LaRouche. *Eco-Terrorism Watch* has become another outlet for

LaRouche, with articles quoting from the *Executive Intelligence Review* and *21st Century Science and Technology* but also Ron Arnold and Alan Gottlieb, as well as Clausen's own book *Walking on the Edge: How I Infiltrated Earth First!*. Clausen's book is publicised on the Sahara Club Borderline Bulletin Board and promoted by *21st Century Science and Technology* as well as by the American Land Rights Association, Chuck Cushman's organisation. Alan Gottlieb's Merril Press distributed the book, although the Washington Contract Loggers Association published it. Ron Arnold designed the cover.[174]

Barry Clausen teamed up with Maduro because he is 'one of the best known environmental writers today. Maduro has done an excellent job debunking major environmental frauds such as ozone depletion and global warming.'[175] Clausen claims to have worked undercover in Earth First!, paid to infiltrate the group by timber, mining and ranching interests.[176] He is active in spreading a climate of fear in timber communities across the west and has appeared at meetings with former army officers and security agents, warning of a 'terrorist threat' from Earth First!. One such meeting in Potlatch, Idaho, was designed to induce fear and hatred towards Earth First!, according to people who attended.[177] Taylor has worked with Ron Arnold, too, appearing at 'anti-terrorism' workshops for the timber industry with him.

When once asked to classify his working definition of terrorism, Clausen responded that, 'I'd just as soon not answer that question.' When pushed by the interviewer, he responded, 'I bet that if you look it up in the dictionary, it would be spelled EARTH FIRST.' Clausen also admitted that, from his experience, only a very small proportion of people in Earth First! engage in these 'terrorist' activities.[178] He has also conceded that he had never seen any illegal activity being undertaken by EF!.[179] Still Clausen attempts to label the whole environmental community as terrorists.

Clausen's attempts to tar the environmental movement as terrorists capable of mass homicide took an ugly turn after the Oklahoma bombing, when he appeared on Vancouver television. 'Former Vancouver resident Barry Clausen warns the tragedy in Oklahoma could happen closer to home, and he says it may not be the work of radical right-wingers, but radical eco-terrorists,' warned BCTV Lynn Colliar. 'Many of these people are advocating eliminating people. They want the planet left for the trees and the animals. And they want us out of it,' said Clausen in scaremongering rhetoric.[180]

Despite there being no proof, the private investigator has attempted to link the notorious Unabomber to the environmental movement. The Unabomber had waged an 18-year periodic bombing spree that killed three people and injured twenty-three, before Theodore Kaczynski, a 53-year-old former Berkeley Maths teacher was arrested in April 1996, on suspicion of being the bomber.

The summer before Kaczynski's arrest Clausen had attempted to link EF! and the Unabomber via a publication called *Live Wild or Die* and the *Eco-fucker Hit List*, despite the fact that EF! had not even issued the publication. 'I think this list is where he drew some of his victims,' said Clausen, singling out two people on the list from the California Forest Association and Exxon.[181] Even though no-one from Exxon had ever been targeted by the Unabomber, ABC News, using information supplied to them by Clausen, reported that the 'Unabomber claimed responsibility for the death of a New Jersey advertising executive who worked for Exxon', and linked the bomber to EF!.[182]

In the immediate aftermath of Kaczynski's capture ABC repeated their story on their *World News Tonight* programme, again interviewing Clausen.[183] The right-wing press also linked the Unabomber to the environmental movement and Greenpeace.[184]

In what is obviously a concerted effort, Clausen and Wise Use activists are now attempting to label the whole environmental movement one big terrorist organisation, capable of perpetrating an Oklahoma outrage. Essentially, the campaign hopes to manipulate public opinion regarding the increase in political violence about resource issues. Clausen's message has been now taken up by *EnviroScan*, a newsletter produced by Public Relations Management Ltd, a Canadian PR outfit with close ties to the Wise Use movement. 'This [violence] is the darkness of the environmental movement. It's never far from the surface. And, no critic or opponent is immune,' warned *Enviroscan*.[185] Arnold and Gottlieb's CDFE has also established the 'Ecoterror Response Network' to 'compile the first comprehensive list of attacks against Wise Users – and to expose the environmentalist smear campaign to stigmatise the victims'.[186]

While no-one can condone violence, what is unjustifiable is when those who do counsel violence also accuse the victims of violence of being terrorists, which is what happened to Judi Bari and Darryl Cherney of Earth First! who were blown up by a car bomb (see next chapter). Furthermore, an often repeated allegation by LaRouche and Ron Arnold is that the Greenpeace activist Fernando Pereira who was killed when the French Secret Service blew up the *Rainbow Warrior* in Auckland Harbour in 1985 was a member of the violent terrorist organisation, the 2nd June Movement.[187] These allegations have been repeated in the mainstream press, such as in a *Forbes* article on 11 November 1991.

THE RESULTS OF THE PARADIGM PROCESS

The ultimate paradigm shift is that the non-violent people who are being labelled as violent are having violence used against them. Violence is not a by-product of the paradigm shift, it is the end result of a process designed to dehumanise. If people continue to scapegoat environmentalists as terrorists there is only one result – violence against the environmentalists.

Reframing the environmental movement as criminal subversives and terrorists has other effects. It tars them with the same brush as the extremists who planted the bomb in Oklahoma, killing 168 people. This outrage gave America a benchmark with which to define the meaning of 'terrorism'. After the bombing President Clinton talked about the 'climate of violence' being bred on right-wing radio stations. 'They leave the impression, by their very words, that violence is acceptable,' said Clinton.[188]

This said, Wise Use leaders have dismissed inflammatory rhetoric as potentially leading to violence. 'We don't believe that rhetoric is violence,' maintains Arnold.[189] Others disagree. 'Ultimately, some people persuaded by these scapegoating arguments conclude that the swiftest solution is to eliminate the scapegoat,' concludes Chip Berlet.[190] In the aftermath of the Oklahoma tragedy, the former President of the American Academy of Psychotherapists, Howard Halpern, wrote about language and violence:[191]

> Social psychologists and demagogues have long known that if ordinary citizens are to be provoked to violent actions against individuals or groups of fellow citizens, it is necessary to sever the emphatic bond with those to be attacked by painting them as different and despicable.
>
> We are unlikely to harm a friendly neighbour because she has strong views about equal rights for women, but if we call her a 'femi-Nazi' she becomes 'the other' – evil, dangerous, hated. We are unlikely to harm the couple down the block who are active on behalf of protecting endangered species, but if we call them 'environmentalist wackos' they become 'the other' – weirdos who must be vilified and suppressed as enemies to 'normal' Americans. When our shared humanity with those with whom we disagree is stripped away, it becomes acceptable to blow them up.[192]

This is exactly what is happening.

6

THE PRICE OF SILENCE

Surveillance, suppression, SLAPPs and violence

I used to love the silence. Now it haunts me.

Pat Costner, victim of arson[1]

THE BARI BOMBING

Judi Bari was one of the leaders of the new breed of Earth First! activists and a key organiser of 'Redwood Summer', an event planned to highlight the destruction of the redwoods in Northern California in 1990. A former union activist, Bari posed some serious problems both to the timber companies and to the federal authorities.

This charismatic mother of two was a mobiliser who could inspire people to undertake high-profile, non-violent direct action. This differed from tree-spiking which was predominantly low-profile and potentially violent. In April 1990, Bari led the renouncement of tree-spiking, saying that 'the real conflict is not between us and the timber workers, it is between the timber corporations and our entire community'.[2] By denouncing tree-spiking, and declaring their commitment to non-violence, EF! activists were potentially scuttling any effective counter-PR campaign. It would now be much harder to label these activists as mindless eco-terrorists.

Bari was concerned about worker rights and safety. She was organising a timber workers' coalition with both union members and Earth First! and had represented five Georgia Pacific employees who were contaminated by PCBs at work, after the company and union colluded in a cover-up.[3] At Bari's behest, Earth First! dropped its advocacy of tree-spiking fearing injury to loggers and millworkers.[4] Earth First! was undermining the paradigm process being undertaken by the timber companies and Wise Use groups to scapegoat the environmental movement by blaming them for job losses and violence.

Timber companies themselves were mainly to blame for the rapidly deteriorating job situation in the region, and had been overcutting for years and were planning on accelerating the practice.[5] Redwood Summer was going to highlight the unsustainable practices of the companies. Bari's message was simple: that environmentalists and loggers have the same goal to maintain forests for future generations, and current multinational corporate forest practices are unsustainable.

Increasingly, letters of intimidation were being sent to Earth First! members, *per se*. Several women received the following letter:

> It has come to our attention that you are an Earth First! lesbian . . . not only have we been watching you and have distributed your phone number to every organised hate group that could possibly have hostile tendencies towards you . . . we will specifically hunt down each and every member like the lesbians you really are.[6]

Meanwhile, EF! men received letters calling them 'fellatio experts [who] suck dicks in outhouses'.[7] But it was Bari who was really being targeted with intimidating letters, and increasingly these were death threats: 'After the first thirty, I stopped counting,' says Bari.[8]

A letter postmarked 10 April 1990 read 'Judi Bari: Get out and go back to where you came from. We know everything. You won't get a second chance.[9] The threat was on plain white paper; a later comparison revealed that it was typed on the same typewriter and matched the style of an anonymous letter sent to the Ukiah police by a police informant who had tried to set Bari up on drug charges eighteen months previously. This was unbeknown to Bari at the time and was to raise further questions after the bombing.[10]

Then came a letter to Darryl Cherney, Greg King and Judi Bari, signed the 'Stompers' – 'We are Humbolt County employees of the Forest Products Industry . . . we know who you are and where you live. If you want to be a martyr, we will be happy to oblige.'

At the beginning of May 1990, a picture of Bari playing the fiddle with a gun sight's cross hairs drawn on her head was pinned with a yellow ribbon, the symbol of logging industry solidarity, to the door of the Mendocino Environment Center. For added effect, faeces had been left on the doorstep.[11] Bari was so concerned she went to the Mendocino County Sheriff's office for help. 'We don't have the manpower to investigate. If you turn up dead, we'll investigate,' she was told.[12] Bari had reason to be worried. The year before, a logging truck had rammed her car from behind, in broad daylight and in the middle of the town of Philo. Miraculously the four children and three adults in the car were not permanently hurt, although badly shaken. The driver was the same person Bari had blockaded the day before in a logging protest. Two EF!ers had been injured with no police response. Mem Hill had had her nose broken in a logging demonstration; Greg King had been punched to the ground during a protest at a Louisiana-Pacific mill.[13]

The heads of the timber companies seemed to be openly endorsing the violence. 'As soon as we find the fellow who decked Greg King,' Pacific Lumber Public Affairs Manager Dave Galitz wrote to the company President, 'he has a dinner invitation at the Galitz residence.'[14] Galitz mailed copies of the vitriolic newsletter of the Sahara Club to the Timber Association of California, which falsely credited Earth First! with a death trap and talked about the Sahara Clubbers, less than a month before the bombing. 'It is so good, we had to share it,' wrote Galitz, 'I may join if only to enjoy the writing style.'[15]

Three weeks after the last death threat Judi Bari was to become a victim of a brutal assassination attempt, whilst driving through Oakland with fellow activist, Darryl Cherney. The smouldering violent rhetoric and the intimidating climate that had been building up over the last few months finally violently erupted. On 24 May 1990, whilst on a tour designed to drum up support for Redwood Summer, a bomb exploded under Bari's seat severely injuring her as well as hurting Darryl Cherney. Bari's pelvis and coccyx were shattered in the blast and her spine dislocated. Although incredibly lucky to be alive, she would be mentally scarred and physically maimed for life. For the moment, though, the pain was so intense she just wanted to die.[16]

Eleven days previously, on 13 May Redwood Summer organisers, with Judi Bari, had drawn up the 'Redwood Summer Code of Non-Violence,' which was to be signed by anyone participating in Redwood Summer. 'We will use no violence, verbal or physical, toward any person; we will not damage any property; we will not bring firearms or other weapons to any action, or base camp' read the declaration.[17]

What happened over the next few minutes, hours, days, weeks, and years after the bombing would send a chill wind blowing across the face of democratic America. The FBI tried to frame Judi Bari and Darryl Cherney. The victims would be made into the villains. In the immediate aftermath of the bombing, instead of being sympathetic, the media, fed by the titbits of misinformation from the authorities, embarked on a feeding frenzy of speculation and sensationalism, with the aim of blaming Bari and Cherney. Moreover, the bombing would be used as an unprecedented event to collate information and intelligence on environmentalists by the FBI.

The FBI misinformation started within minutes. One of the first people on the scene was Special Agent Frank Doyle, a 'certified bomb technician,' followed by fifteen agents from the FBI's Terrorist Squad. Bari has said that it is 'uncanny' how fast the FBI arrived.[18] Her attorney commented that 'it was as if the FBI agents were around the corner with their fingers in their ears'.

More uncanniness is revealed in documents surrendered under jurisdiction from the court which include an FBI memo dated the day of the bombing. It describes a bombing training exercise by the FBI on Louisiana-Pacific's land, taught by Frank Doyle less than a month before the bombing. Not only were cars blown up with pipe bombs, but two out of three of the bombs were placed under the seat, an unusual place for a bomb to be found. Bombs, said the bomb school instructor, were normally placed either underneath a vehicle or in the engine. 'This thing is strong enough to smell,' said Dennis Cunningham, Bari's lawyer. Videotape of the car bomb scene shows a laughing Special Agent Doyle, telling his student, 'This is it, don't you think? This was the final exam.'[19]

Agent Doyle, a veteran of twenty years experience in the FBI's International/domestic Terrorism squad, who had processed over 150 bombing crime scenes, concluded that the location of the bomb was 'immediately behind the driver's seat'.[20] His expertise was relied on by the local police department and the press.[21] The logical conclusion was that Bari and Cherney must have been able to see the bomb, they must have known it was there and therefore it must be theirs. It was this single statement that would give the authorities the justification and jurisdiction to hold Bari and Cherney. Furthermore, police records show that within half an hour of the bomb blast the duo were both misleadingly linked to a federal case of attempted destruction of nuclear power plants,[22] in a further effort to characterise them as 'terrorists'.[23]

FBI agent John Raikes told the Oakland Police the two 'qualified as terrorists', although the only thing Cherney had been arrested for was hanging a protest banner and Bari only for non-violent civil disobedience in a 1985 anti-war protest.[24] Even though Bari was barely conscious, her pelvis was shattered in ten places and her leg was in traction, the two were arrested and bail conditions were set at $100,000 as they posed such a 'threat to the community, and 'flight risk'. They were charged with 'illegal possession of explosives'. Moreover, the authorities held a press conference to tell the world they were 'no longer considering other suspects'.[25]

Plate 6.1 Photographs clearly show that the epicentre of the blast was below Bari's seat.
Source: Oakland Police Dept. This photograph was acquired through disclosure

For whatever reason, Doyle was wrong in his analysis. Police photographs, taken immediately after the bomb went off, show that the epicentre of the blast was not behind Bari's seat, but directly underneath it.[26] Furthermore, lab analysis by the FBI showed the bomb to be an anti-personnel bomb, found to contain a mechanism which would trigger an explosion if the vehicle in which it was placed was moved. A blue towel wrapped around the bomb hid it from view, which only the bomber and subsequently the FBI knew.[27]

The resulting media frenzy and spectacular federal witch-hunt against environmentalists give us clues as to why the FBI misrepresented the position of the bomb. The ensuing media message hitting national and international newsstands was simple. 'Earth First! Terrorist Blown up By Own Bomb'.[28] Untold damage was done to the effectiveness of Redwood Summer's anti-logging offensive, to Earth First! and the environmental movement as a whole. 'Even today,' wrote Bari in 1994, 'particularly in places where there is no active EF! movement, bombs and tree spikes are the only things many people know about us.'[29]

With the damage already done, on 17 July, two months after the bombing, the charges were dropped due to a complete lack of any incriminating evidence.[30] Judi Bari and Darryl Cherney felt they had no choice but to sue for wrongful arrest because they believed that the FBI had made a deliberate and malicious mistake. Just under a year after the bombing, on 24 May 1991, Bari and Cherney filed a $2 million suit against the FBI and Oakland Police Department for wrongful arrest and gross violations of their civil rights. The FBI has tried, unsuccessfully, three times to have the case thrown out of court. The case continues, with the first full hearing set for summer 1996.

Bari's first impulse, which she still believes, is that it was not someone working for the police or FBI who had actually tried to kill her, but that the enforcement agencies were more or less aware of which political forces might have committed the crime, and that they were ideologically sympathetic to them. They were therefore covering up for the true bombers by refusing to investigate the question of who had really planted the bomb in her car.[31]

After reviewing more than 5,000 pages of previously classified FBI documents relating to the case, Bari concluded:

> The FBI case file on the bombing shows that the FBI never conducted a legitimate search for the bomber. Instead, they presumed our guilt, they distorted or falsified the evidence in order to support that presumption. They also avoided investigating key leads that seem to implicate police and timber interests in the campaign of threats and harassment surrounding the bombing . . . These files raise many more questions than they answer, and indicate FBI misconduct on an even greater scale than we had previously suspected.[32]

FBI records show that the authorities were told about the death threats and physical harassment, and that this was an assassination attempt but they chose to ignore this evidence.[33] The authorities were given the originals too but failed to investigate them in any way. They were never even sent to the lab for fingerprint checking. The FBI had not only become obsessed with linking their only two suspects to the bombing but also to another bomb that had been defused a month previously outside a Louisiana-Pacific Corporation sawmill at Cloverdale. Their components were so similar that it was believed that the two bombs were made by the same people.[34] The search warrant for the Bari residence lists her and Cherney as 'members of a violent terrorist group involved in the manufacture and placing of explosive devices'.[35] Of the 111 items seized, nothing was found to be similar to anything in the bombs apart from two nails found at Bari's which were determined to have come from the same manufacturer. These nails were common along the north coast of California and sold in over 200 outlets.[36]

Many FBI documents which should have been part of the files handed over to Bari were missing. There are reports from Special Agent Doyle on collecting evidence, but none describing the scene or the physical damage. Although there were two extensive lab reports from Special Agent Williams, there was no report on his visit to Oakland to determine the bomb's position.[37] The documents also reveal that Bari and Cherney were under surveillance during the summer of 1990 and that FBI intelligence had found a 'hideout' where Bari was hiding. It was, in fact, Bari's home. Also missing from the files is a death threat against Bari which talks of Bari's 'hideout', much like the FBI's 'hideout', that was reported to them by the local police.[38]

Another suspicious event was a letter that was to become known after its signatory, 'The Lord's Avenger'. Originally sent to the media, the letter, written in heavy religious undertones, claimed responsibility for the bombing. It was not dismissed as a hoax because the writer of the letter gave intricate details of the make-up of both the Cloverdale and Oakland bombs, which led the authorities to conclude that the writer of the letter was somehow involved in the bombing. The Lord's Avenger letter and the unpublished FBI files were correct in their bomb description. The Avenger's letter also listed the bomb components. Bari concluded, however, that after it had failed to frame her and Cherney, the FBI itself might have concocted the note to create a diversion by 'providing a plausible lone assassin not connected to timber or the FBI'.[39]

Instead of investigating the whole myriad of suspects such as the people within the timber industry or those with extreme right-wing connections, the FBI used the Lord's Avenger letter to target north coast environmentalists. They contacted fifteen local newspapers asking for letters-to-the-editors to be sent to them from known environmentalists, to see if they matched the typeprint in the letter. None were found. The FBI then collated information from Wise Use groups and industry on environmentalists. Then finally, the FBI files show that the bureau compiled a list of 14 EF! associates' phone calls out-of-state. A print out of the 634 out-of-area phone calls was made and the FBI then researched each of these numbers, compiling information that included names, addresses, places of employment, physical descriptions, criminal records and in some cases, political associates of the people who received a call from EF!.

Richard Held, who was in charge of the FBI's case, quietly dropped the case in October 1992, not notifying the District Attorney's office for a further five months that the case was closed. Held had been a key figure in the FBI's counter-intelligence programme, or COINTELPRO, whose purpose was to 'expose, disrupt, misdirect, discredit, or otherwise neutralise' political dissidents. Held coincidentally resigned his post when photographs were released under a court order, showing that the bomb was underneath Bari's seat, not behind it.[40]

COINTELPRO AND OTHER FBI ANTICS

The FBI has a history of collaborating with right-wing groups to attack social justice and peace movements[41] as well as framing and discrediting them by labelling them as terrorists. The brainchild of FBI Director J. Edgar Hoover, COINTELPRO had started in the late 1950s and used classic misinformation, surveillance, psychological warfare, legal harassment, dirty tricks and infiltration techniques against political and social change activists. Most of its activity was illegal, but that was no deterrent to the FBI.

It framed Black Panther leaders in the 1960s, Indian leaders in the 1970s and CISPES (Committee in Solidarity with the People of El Salvador) members in the 1980s. Martin Luther King had been a favourite target of Hoover's, something that continued even after King was assassinated. 'Although the programme was allegedly terminated in 1971,' writes Ward Churchill, 'several former agents have charged that only the name was dropped.'[42] The Bari case raises serious concerns as to whether COINTELPRO does actually continue to this day.

In fact COINTELPRO had been active against the environmental movement since the very first Earth Day, way back in 1970.[43] Edward Abbey, activist and author of the *Monkey Wrench Gang* had been under surveillance for twenty years, considered by the FBI a threat to national security.[44] The peace movement had been under surveillance for decades.[45] It would be a logical move to extend surveillance to EF!. The FBI had been out to neutralise EF! since 1986, when power lines from a nuclear power plant were tampered with, although they had been keeping an eye on EF! since its inception in 1980. They were particularly obsessed with netting Earth First!'s big fish, its founder, Dave Foreman. An *agent provocateur*, FBI special agent Mike Fain, known to EF! as Mike Tate, infiltrated EF! and became friendly with key activists. In time, Fain began plotting an action to sabotage the transmission power lines of a nuclear power plant with three other activists. On a planning trip, the three unsuspecting activists were arrested in a sting operation by some fifty FBI agents who were lying in wait for them.[46] One of the activists charged said that Fain 'had not only encouraged the plan but had facilitated its implementation by renting the acetylene tanks, filling his truck with gasoline and driving the crew out to the desert where SWAT lay in wait'.[47]

The FBI did not manage to implicate Foreman directly, but he was arrested on conspiracy charges for handing some money to one of the activists. A tape recording which Fain did not know was being recorded at the time, gave an insight as to why the FBI so badly wanted Foreman. ''Cause this [Dave Foreman] isn't really the guy we need to pop – I mean in terms of actual perpetrator. This is the guy we need to pop to send a message'.[48] The FBI spent some $2 million getting its message across, a message designed to discredit a movement. 'They targeted the perceived leaders of the Earth First! movement,' said spokesperson Dale Turner, 'with the clear intent of both eliminating the masterminds within environmental circles and discrediting the whole environmental movement.'[49]

Other bombs have been used to discredit environmentalists, even though there has been no evidence to link the two. In April 1995 a small bomb exploded in the offices of the California Forestry Association, an industry lobbying group in Sacramento. Police and the FBI linked the bombing to the Unabomber, although an Association spokesperson said that, 'In my personal opinion, this is the work of extreme environmentalists.'[50] Barry Clausen, private investigator and Wise Use activist, has attempted to implicate EF! in an Oklahoma-style bombing, as was mentioned in the last chapter.[51] Environmentalists were incriminated in bomb attempts at MacMillan Bloedel's offices in Vancouver and London. John Paine, from the company's London office, linked the London bomb to Greenpeace's anti-clear-cut campaign. All the major environmental groups working on the issue have a history of non-violence and denounced the bombing and MacMillan Bloedel's subsequent tactics.[52] Bombs or bomb scares have also been used to discredit environmentalists in Australia (see chapter 9).

VIOLENCE BEGINS TO SPREAD

The violence and intimidation against American environmentalists continued during the summer of 1990. On 14 June 1990, Jim Blumenthal of Greenpeace

Action received a telephone threat asking if Greenpeace was aligned with EF!. When told that they were not, but were helping with Judi Bari's legal fees, Blumenthal was told 'well then, you will be next'. FBI files show that 'no further action will be taken' after the incident.[53] Later that summer, the Sahara Club, whose vitriolic rhetoric has never been investigated, held a 'dirty tricks' workshop with Candy Boak, of the Wise Use groups We Care and Mothers Watch. Much like the Sahara Club, Boak had been distributing false information about EF! and openly keeping EF! meetings under surveillance. By August, a Sahara Club member had been arrested, caught planting a fake bomb in an EF! centre.[54] The Sahara Club stepped up its personal hate campaign against Judi Bari and Darryl Cherney after the bombing, criticising the way they had 'handled their pet bomb' and calling them 'lice-infested grubs' 'Darryl "Spare Change?" Cherney and Judi "slugbody" Bari, or Judi the Junkie'.[55]

It soon became clear that the Oakland bombing was just the tip of the violence iceberg. By 1991 the Movement Support Network at the Center for Constitutional Rights had recorded over 300 suspicious incidents and 150 unexplained break-ins against social activist groups.[56] 'Some environmentalists,' wrote Eve Pell from the Center for Investigative Reporting (CIR) in 1991, 'now worry that the fight against them has escalated to include violence, arson and provocateurs'.[57] By 1992, Jonathan Franklin, who had worked on a four-month study of violence against environmentalists for the CIR, had 'discovered a pattern of death threats, fire-bombings, shootings and assaults targeting "green" activists across the nation'. The CIR had 'logged more than 100 reports of attacks and harassment that have taken place since 1988'.[58] By 1994, they had logged 124 and were still counting.[59] Sheila O'Donnell, a private investigator and the most experienced and respected person researching violence against activists, believes that there are thousands of individual acts of harassment.[60]

The same year saw Dave Helvarg's book, *The War Against the Greens*, devote two whole chapters to violence and harassment against environmentalists. Helvarg's book still remains the most detailed analysis of violence against US greens.[61] It is impossible to quantify the specific level of violence against environmental activists in America, because violence is designed to silence. There will be hundreds of acts of intimidation that will go unreported because they have succeeded in their aim, simple intimidation to 'chill' the person concerned.

What we can tell is that it is coming from companies, workers, the Right, the Wise Use movement, and increasingly the militia. The cross-over between these later two movements is increasingly becoming stronger. Moreover, the government are also implicated. 'Given what we know, I can't imagine that the federal intelligence agencies are not involved at some level,' says private investigator Sheila O'Donnell, who adds:

I think the interests that are threatened are corporate, and I see the federal government supporting those interests. It is hard for me to imagine that there is not collusion. There has to be collusion, there always has been, so I don't know why it would change now.[62]

What we also know is that it is mainly grassroots activists, miles from the safety of big cities who are suffering the most. The majority of these activists are women, who are involved in local environmental problems. Activists who live in remote areas or in blighted neighbourhoods are also singled out for attack. Furthermore, the support these 'frontline' activists are receiving from the mainstream environmental movement has been verging on non-existent. 'I think we isolate people when we don't speak out against violence and we make it safer to attack them,' says Sheila O'Donnell, 'that is why I am concerned that we have not heard more voices coming out of the environmental movement, talking about Wise Use and violence.'[63]

Asked why she thinks this is so, O'Donnell replies:

I think denial plays a very big part of it. I have to think that if Pat Costner [see below] was a white man, with an office in London, New York or Chicago or Washington DC, and her library had been torched, I think there would be an outcry. I think if an environmental organisation's office blew up in a city, everyone would jump right too, I think it would be quite clear that there was a major problem. It's the question that if a tree falls in a forest and no-one is there, does it make a sound.[64]

This process of denial and the rural/urban split 'are very important in why it is not being solved,' alleges O'Donnell:

I certainly do not think it is a bad heart or lack of interest, because if you ask any of the leadership of any of the major environmental organisations what they think about this, they would be horrified, but because it does not immediately threaten their self-interest, so to speak, they do not pay attention. I have to say that I think a lot of it has to do with ignorance as well. I am not sure, if you were actually to ask the heads of most environmental organisations, if they knew about this, what you would get for an answer. I am not sure if they would be able to cite the kind of cases that are going on.[65]

WHO'S TO BLAME?

People who are marginalising the environmental movement, scapegoating it, must take blame for the violence, because violence is the logical conclusion of such actions. The ultimate result of the paradigm shift is violence. 'I believe that people who promote the marginalisation of the environmental movement as green terrorists, as people who are costing jobs and creating hunger, I think people who say that are ultimately responsible for the violence against the environmental movement,' argues Chip Berlet.[66]

Berlet's concerns are borne out by other people. 'There is no question that the tactics Wise Use activists have used, in attempting to marginalise environmentalists, to falsely attribute all economic problems experienced by natural resource communities to environmentalists, are at least in part responsible for intimidation, harassment and violence,' says Tarso Ramos, who tracks violence against activists in the Western States.[67] 'The level of fear that these people can provoke is quite intense,' says Paul de Armond'.[68] 'It's not a common thing to hear violent language, certainly not in the press,' argues Sheila O'Donnell. She adds:

> I think the warlike mentality of the Wise Use movement is quite unique in environmental issues, and we see it in the vigilante movement and the militia and we see it with racialists and we see it with anti-abortion people and drug dealers.[69]

Lois Gibbs of the Citizens' Clearinghouse for Hazardous Waste, a veteran grassroots activist whose house was broken into three times during her Love Canal campaigning, sees the Wise Use movement as a bunch of bullies, who have 'changed the way people behave. They are in fear and submissive. Society needs to see what they have done,' says Gibbs.[70]

For example, Wise Use leaders maintain they are non-violent and recently they have been trying to tone down their public inflammatory rhetoric. Chuck Cushman will tell his audience that they should not go out and be violent and has circulated at least one declaration of non-violence at a Wise Use conference.[71] 'The violence issue is just something the preservationists use to try and get at us by implying we advocate or promote violence,' Cushman told journalist Dave Helvarg. 'I've never advocated or called for violence. Personally, we've always advocated non-violence like Martin Luther King or that guy from India, what's his name.'[72]

Arnold's CDFE has even published a declaration of non-violence whereby signatories:

> absolutely and unconditionally reject and denounce the use of weapons or personal violence against our opponents or vandalism against their property. We absolutely and unconditionally accept the power of unarmed non-violent moral conviction as the only standard of behavior in confrontations between our two movements.[73]

Although this initiative is a positive step forward, written non-violent language can come too late to stop violence that follows inflammatory scapegoating.

Chuck Cushman's inflammatory rhetoric brings him the most criticism. 'He is well known for his fiery, rhetorical style,' says Dan Barry from CLEAR, 'It seems that, in looking at the way he presents his arguments and in the way he presents himself, that he's treading very close to the line of instigating active intimidation.'[74] Tarso Ramos adds,

> Cushman is a very inflammatory speaker. His rhetorical strategy very much relies upon instilling fear, anger and hatred in the minds and hearts of his audience. So I think that Cushman is an example where it is legitimate to take him to task for the kind of harassment and intimidation that's followed in the wake of his organising drives.[75]

Moreover, there is 'no evidence that I've seen that the more mainstream elements of the Wise Use movement are distancing themselves at all from Rick Sieman [Sahara Club] and his kind of activism,' says Dan Barry, from CLEAR. 'In fact they seem to be embracing it in certain forms.'[76] William Perry Pendley has offered support to Sieman in a letter to *21st Century Science and Technology* and Putting People First have distributed at least one document through the Sahara Club.[77] Sieman's response to environmental activists being beaten up is simple. 'Good,' he replies.[78]

Judi Bari is in no doubt that the climate created by the Wise Use movement certainly contributed to the assassination attempt on her life:

> I don't know who put that bomb in my car on May 24, 1990, but I do know that the Wise Use Movement helped stir up a . . . mentality against Earth First! in our community right before the bombing. The Wise Use Movement says they just want to put people back in the environmental equation. But their actions show that they have a different agenda.[79]

Journalist Dave Helvarg came to a startling conclusion about the Wise Use movement's tactics. 'Unlike the Christian Right, with its millions of grassroots followers, Wise Use/Property Rights lacks the membership base to parlay the coverage into effective local action without resorting to confrontational tactics,' writes Helvarg.[80]

But Chip Berlet spreads the blame further than Wise Use and the militia to the PR companies: 'PR companies should take some blame,' argues Berlet:

> It's like someone who organises a hit and then claims no responsibility . . . When you hire a company that has no moral values whatsoever, and is willing to use any means whatsoever to scapegoat the attacks of the corporate image, you have a moral responsibility for the outcome, and in fact these PR firms are creating an environment of scapegoating. . . .
>
> There is a legitimate place in societies around the world for a debate over the balance between economic development and environmental concerns. But when you decide you are going to fight the battle by scapegoating an environmental activist in a society where there are economic tensions, and you start blaming the environmental activist for lack of food, lack of jobs, lack of economic development, you are basically saying that if you want to survive in the society you must go out and get the environmentalists. . . .
>
> That may not be what people sitting in a boardroom in Washington DC at Burson-Marsteller are saying to each other and their clients, but that is what is happening in economically strapped places around the world, where people are fighting environmentalists. People in the public relations sector and the corporate sector need to face what the outcome of those decisions are, and the outcome is harm to the people who are scapegoated.[81]

And people have been killed. They have had their houses torched, been shot at, had their dogs poisoned. They have been assaulted, harassed and received countless threats. They have also been victims of vandalism and theft.

THE VICTIMS OF VIOLENCE

Four children belonging to anti-herbicide activists, Carol and Steve Van Strum were killed in a fire in the early hours of the New Year 1978. 'It was commonly believed in Five Rivers,' wrote Lewis Regenstein in *America the Poisoned*,[82] 'that the house was deliberately set on fire because of the Van Strums' leading role in fighting against aerial spraying.' Carol Van Strum had forced the government to act on the safety of herbicides, and had been successful in having two herbicides banned.[83] The following year, water rights activist, Tina Manning Trudell, the pregnant wife of the American Indian Movement national Chairman, John Trudell, was killed with their three children and Tina's mother in an arson attack on the Duck Valley reservation in northern Nevada. Tina Trudell had been a leading organiser in the struggle to retain water rights in the area.[84]

In 1991, Lynn 'Bear' Hill, a former worker at a liquid waste disposal (LWD) hazardous waste incinerator in Calvert City, Kentucky, who had bulldozed barrels of waste into the ground for ten years was found dead 'in a very bloodied condition', half-sitting in his truck, with his truck keys in his hand. His kitchen was covered in blood, his abdomen was bruised and his nose had been pushed into his brain. Despite this, Hill's family were told by the coroner that he had died of pneumonia, although the medical examiner said Hill had a ruptured oesophagus and had bled to death.[85]

Hill had become concerned over the illegal dumping of 5,000 drums of toxic waste under concrete at the plant and had finally reported the matter to the editor of a local newspaper, *Tell It Like It Is*. 'I've been trying to talk to you for long time,' said Hill. 'Calvert City's LWD is a toxic time bomb just waiting to go off.' He feared for his life, if his name became known and asked for his identity to be disguised. When the editor, Bob Harrell, contacted the local Governor's office for verification he was asked not to print the story. 'You've got a real good little paper, but this will put you out of business,' was the response. Moreover, Harrell fears that by doing this, someone was able to figure that Lynn Hill was the source of the story and killed him. Harrell eventually published the story and subsequently had a dead bird, with its head cut off, left on his porch.[86]

Native American environmentalists in particular have suffered in the green death toll. On 9 October 1993, the decomposed body of Leroy Jackson, one of America's most prominent Native American environmentalists, was found under a heavy blanket inside his car on US Highway 64. Jackson, a 47-year-old Navajo activist, who was fighting logging operations on the Navajo Reservation, had been receiving death threats over the years. The summer before he died, he had been hung in effigy by angry loggers. Although Jackson neither drank, took drugs or smoked, his autopsy concluded he had died of intoxication by methadone, a synthetic substitute for heroin. His death came just days before he was due to fly to Washington for a meeting with the Bureau of Land Management to protest at logging. However, with unemployment high in the local area, Jackson's activism had angered not only timber officials, but tribal leaders and fellow Navajos.[87]

All the evidence points to suspicious circumstances around his death. Jackson disappeared on 1 October, and on the next day, 2 October, a hiker claimed to have seen his car parked beside the road. However, a tow truck driver looking for business did not see it there on the 4th and when his friend and doctor travelled the road on the 6th looking for him, they never saw the car either. Despite this, the state police claimed the death was accidental.[88]

Seven months later on 21 May, another tribal environmentalist, Fred Walking Badger, set out on a short trip in his car near Sacaton, Arizona. He never came back. His car was found burnt out in the desert three weeks later. Fred Walking Badger and his companion on the fateful day, Aaron Leland Rivers, are still missing. Walking Badger's wife believes that Rivers was in the wrong place at the wrong time and concludes the two men were killed and buried in the desert. She also believes that her husband was killed because he was a prominent anti-pesticide activist, who had mobilised opposition against pesticide use on the Gila River reservation. Badger had forced the tribal government to look into the issue of pesticide spraying and had also contacted environmental groups about the quality of the water due to pesticide contamination. 'Sometimes when the winds are blowing toward homes, plants and garden, vegetables are killed,' says Badger's wife, Marilyn, 'All the people living here are afraid now to eat the wildlife.'[89]

On the morning of Tuesday, 2 August 1994, Torres Martinez and Cahuilla Indians began a non-violent protest against illegal toxic facilities on their reservations. Sewage companies were illegally dumping as much as 1,000 tons of sludge in the reservation per day and the Indians were attempting to stop it with Greenpeace's assistance. Protestors had already been harassed by members of a security company, who had warned people on the protest that they would be 'run down'. The day before, people had been alerted that they should 'back down or else'.[90]

On the Thursday, fifty Torres Martinez Indians, Greenpeace, and Mexican farmworkers blockaded three illegal sewage facilities operating on the Torres reservation. However, on the Saturday, the 14-year-old nephew of one of the Cahuilla leaders was shot dead. The boy's father, standing over the dead child, was clubbed by the police and charged with interfering with a crime scene and murder. The charges were later dropped. The following night, gunmen again attacked the home of Marina Ortega, an Indian who had brought many of the Indians together to oppose the waste companies. Luckily, that time no-one was hurt. Ten days later, gunmen again toured the blockade threatening people with guns. The sheriff refused to investigate any of the attacks.[91]

Shooting is a common method of violence, which has been used for over a decade to intimidate environmental activists and officials. In March 1983, EPA officials investigating an illegal waste dump site were shot at and harassed in Atlanta and Philadelphia. In Seattle they were fire-bombed. The same year, when a citizens' group began a campaign to stop the Middlesboro Tanning Company from dumping toxics into a creek, they were fired at, had their dogs killed and their vehicles sabotaged. One shooting incident left an activist covered with broken glass, after his windscreen was hit by bullets.[92]

More recently, environmental activists have been fired at in Arizona, Colorado, Kentucky, Texas, and West Virginia. Pets are targeted as well. 'First thing they do to you in Tennessee is shoot your dog,' says Larry Wilson, an activist who has already lost two pets. The worst time, though, was when he found the skinned corpse of his dog on his lawn.[93] Lauri Maddy, an activist who has campaigned against the Vulcan chemical company has had one dog shot and one drowned. She has also been nearly run off the road twice. The company concerned showed videos of her to the workforce saying she was responsible for job losses. Diane Wilson, a shrimper from Texas who was protesting against a proposed plastic plant, had her dog shot by gunmen in a helicopter. Twice her shrimping boat was sabotaged. She was also under surveillance, has also been called a lesbian, a prostitute, a bad mother and a racist.[94]

In March 1993, a doctor who was assisting a couple in their lawsuit against a pharmaceutical corporation, had his dog stolen and its throat cut. The couple involved, the Winters, were filing suit against the company claiming that health problems and the death of their unborn baby were caused by contamination from Merck's plant. During that summer, there were eleven break-ins at the Winters' home, although nothing was ever stolen. They also had their vehicle sabotaged. In April 1993, another activist was shot at after he disturbed intruders at his home in Alaska. He also had his house set on fire.[95]

Anti-logging activists in New Hampshire, Jeff Elliott and Jamie Sayen and anti-pesticide activist Michael Vernon from Maine have all had their houses burnt to the ground. So too has former Wilderness Society president Stewart 'Brandy' Brandborg who lives in Montana. The Adirondack Park Agency building was set on fire in the late 1970s, and two of the Park Commissioner's buildings were torched along with her vehicle and boat. The Adirondacks have become a hotbed of anti-environmental violence targeted at government officials and activists. Adirondack officials have experienced a long violent campaign against them including death threats, vandalisation of property, and being fired at. Environmental activists have been punched and harassed. Activists have had their houses torched in Colorado, Florida, Montana, New Hampshire, New Mexico, New York, and Texas.[96]

Pat Costner, director of Greenpeace USA's science unit, was a victim of arson. On 2 March 1991, a month before she was due to publish a five-year investigation into toxic waste incineration, she returned to find her home of seventeen years and her extensive library reduced to ashes. Costner's report, ironically named *Playing with Fire*, was to name corporations who were targeting

Plate 6.2 The burnt-out remains of Costner's house, with the remains of her desk in the centre of the picture.
Source: Sheila O'Donnell, Ace Investigations.

poor rural areas in which to build incinerators, and to name companies that were to blame for illegal releases. Ironically, it was Costner herself who, by targeting such companies, was playing with fire. Subsequent investigations showed that the temperatures reached in the fire could only have been achieved if a flammable liquid had been used to ignite the blaze. The fire was a deliberate case of arson. Four months previously, two men who bragged at having been to a marine training academy, had been asking directions to Costner's house. Two weeks after the fire, the phone lines to her remote cabin were cut. When Costner phoned the National Toxics Campaign over assistance to replace lost files, she found out that their office had been broken into and their files stolen, too.[97]

Costner is not the only scientist and researcher who has suffered harassment for her work. After George Baggett successfully testified against a local hazardous waste incinerator he was accosted and threatened by two men, returning home from a victory party. Baggett also believed he was under surveillance and received obscene phone messages around the time he was testifying.[98] Another researcher opposing waste disposal was Sam Bishop, an incinerator expert. Days before he was to testify against the Brooklyn Naval Shipyard incinerator, his office was broken into, and all his files relating to the case were removed. New York City police detectives called it a 'typical corporate intelligence hit'.[99] A scientist from the Kentucky's Natural Resources Department was fired in 1989 for opposing liquid waste disposal (LWD) in Calvert City, Kentucky.[100]

But it is female activists who have suffered some of the most horrific attacks, such as Stephanie McGuire who was campaigning against Proctor and Gamble's cellulose mill in Florida. McGuire was a member of a community group called Help our Polluted Environment (HOPE) which was threatening to sue the company, forcing it to clean up its pollution. Both McGuire and her business partner, Linda Rowland, and the leader of HOPE, Joy Towles-Cummings had received numerous death threats and harassment. They also believed they were under surveillance, but they could not have believed what was to happen next. On 7 April 1992, five men dressed in camouflage gear, two of them wearing ski masks, attacked McGuire while she was alone.

Sheila O'Donnell, who investigated the attack, recalls what happened:

> The men beat her, kicked her in the head and body, and stomped on her hand injuring it severely; they tore her shirt off and burned her breast with a lighted cigar and also cut them with a straight razor. As they assaulted her, they repeatedly referred to Proctor & Gamble and to McGuire's opposition to the company. They threatened to attack another activist, Joy Towles-Cummings, the founder of HOPE. While two of the men held her on the ground the third cut her face and neck with a straight razor. 'I'm going to cut you real slow,' he said, 'to make it as painful as possible.' Then he took polluted river water and poured it into the cut. 'Now,' he said, ' you have something to sue us over.'[101]

Two of the men then raped her. In the words of one of the assailants: 'This is what you get for talking about P & G.'[102] No-one has ever been prosecuted for the incident. Wise Use and property rights fax networks widely circulated that McGuire had been beaten up by her 'girlfriend'. Phone harassment continued until the women moved, finally being driven out of the county. Even then the calls still persisted and their dog was killed by meat poisoned with anti-freeze.[103] Paula Siemers is another activist who has been forced to move after continuous harassment, arson and attempted murder. She too had her dog killed, poisoned by a contaminated hamburger. She was beaten unconscious by rock-throwing youths. She had her porch set on fire, twice. Siemers, who became known as the 'pollution bitch' was circulating a petition against Queen City Barrel, and campaigning against the excessive emissions from local industry that were causing severe respiratory problems. When she was stabbed from behind by two men, she had finally had enough and moved.[104]

Other activists have not moved, but live in fear. Like former logger Gene Lawthorn who has tried to offer solutions which would suit both environmentalists and loggers in the polarised community of Sutherlin, Oregon. For his troubles he was threatened with castration and arson.[105] Two environmentalists who also live in Oregon are Andy Kerr, from the Oregon Natural Resources Council and Rick Bailey from the Hell's Council Preservation Council. Both were hung in effigy and tarred and feathered just before the opening of a Wise Use conference in Joseph, Oregon, on 30 September 1994.

Speakers at the conference included Kathleen Marquardt, Perry Pendley and Ron Arnold. 'This burning in effigy is a symptom,' says Bailey:

> They have openly said they will run me out. They are trying to get me kicked out of my office space. This was clearly Wise Use . . . Ron Arnold was there. It was held on Friday afternoon but the tarring and feathering was clearly part of the conference.

Kerr calls the hanging 'a media event for the Wise Use conference'.[106]

Plate 6.3 The two tarred and feathered effigies of Rick Bailey and Andy Kerr and the Wise Use demonstration.
Source: Walawala Chieftain

Bailey had seen a real change in the area over the last six months. 'Something is inspiring people,' he says. Death threats are now common to both Kerr and Bailey. Bailey has discovered kewpie dolls, stuck with pins, drenched in red, hanging from his mailbox. 'Enviro-nazi beware' said one.[107] Hanging people in effigy is also becoming something of a Wise Use speciality, for example, in June 1992 Citizens for Property Rights hung twenty political leaders in effigy in Vermont.[108]

Dave Helvarg documents other incidents of intimidation and violence directly associated with Wise Use. 'I am going to blow your fucking brains out' is how a People for the West! member attempted to intimidate an activist from giving a slide show. EAGLE, a local Wise Use group in Alabama has threatened local activist, Lamar Marshall over a period of time. New Jersey lawyer, Craig Siegel, resigned after Wise Use flyers called him an eco-terrorist and anti-growth advocate which led to hate mail and physical threats.[109]

Three weeks after the Bailey and Kerr incident, Royal Lilly, a member of RECAP, Romulous Environmentalists Care About People, which is a grassroots movement trying to stop environmental injustice in the area, found a small box on his wife's car. The group had been campaigning against deep well injection in the area. Inside the box was the head of a bobcat and a note made of letters cut out of newspaper. 'RP Lilly. We are tired of you wasting our tax money so you can be a slum lord you are nothing more than a nigger loving hypocrite who will suck a politicians dick for a penny its time to retire for health reasons if you show up at another council or zoning board meeting next time we will deliver your wifes of one of your granchildrens heads' (*sic*), read the note, which was signed: KKK, or the Ku-Klux-Klan.[110]

Given the merger of various sections of the Right and the continuance of inflammatory rhetoric and polarising paradigm shift, violence seems set to maim the debate for the foreseeable future. 'The threats are continuing,' says Lois Gibbs of the Citizens' Clearinghouse for Hazardous Waste in May 1995, a veteran grassroots activist. 'They are becoming more overt. Congress and the militia are the perfect examples. The US community is now feeling more free to express themselves than ever before,' continues Gibbs, 'Bullets and violence are a new way.'[111]

In fact by the mid-1990s there seemed to be heightened violence directed at both the environmental movement and also federal employees who work on environmental or land issues. Some of this violence, at least, was coming from the militia. In October 1994, a bomb exploded at the office of the US Bureau of Land Management office in Reno.[112] The following month, an Audubon

Society activist, Ellen Gray, was approached by two men at a public hearing of the Snohomish County Critical Areas ordinance. 'This is a message for you,' said one as he placed a hangman's noose on a chair. The other told Gray simply, 'If we can't get you at the ballot box we'll get you with a bullet. We have a militia of 10,000.' The first man was distributing cards with a picture of a noose on it with 'Treason = Death' on one side and 'Eco fascists go home' on the other.[113] The man with the noose was not only with the militia, but also a property rights activist, Darryl Lord, of the Snohomish County Property Rights Alliance.[114] Although Gray reported the incident to the local police, they said they were too busy to look for the men.[115]

Attacks or threats by militia increased in 1995. 'The majority of militia-related incidents have involved people who, one way or the other, are associated with the environment,' wrote Dave Helvarg in May 1995.'[116] 'Attacks and harassment by right-wing extremists have centered around land use,' says Eric Ward, from the Northwest Coalition Against Malicious Harassment, who has monitored the links between Wise Use and the militia. 'They paint environmentalists as faceless monsters, so it's okay to harass them. It's very similar to racial violence.'[117]

But it is not just activists who are being targeted; intimidation and violence are occurring against federal officials, too. Three Fish and Wildlife Service Agents were forced to retreat off a rancher's property, after the local Sheriff claimed their warrant was invalid and he would return with an armed militia.[118] A bomb exploded at the US Forest Service office in Carson City in March. The very same day, it was discovered that an outhouse at the Humbolt National Forest had also been bombed. Other bombs went off at Lamoille Canyon, Nevada.[119] The day after the Oklahoma bombing, death threats and bomb scares were received by the Bureau of Land Management, the Nature Conservancy, the Audubon Society and the Native Forest Council.[120]

The violence and intimidation levied against the Forest Service became so bad that employees were told they did not have to wear an official uniform or travel in official vehicles for fear 'it would jeopardize the employee's safety'. In the Toiyabe National Forest in Nevada, the Forest Service has stopped routine maintenance.[121] In fact the situation had become so intolerable by the spring of 1995 that Rep. George Miller called for Hearings by the House Resources to investigate the violence.[122] 'The rage and hate are beginning to well up,' said Robert Marriot, from the National Park Service, at informal hearings into violence, 'We've always gotten threats against our employees. But now we hear, "Death", "You're going to be killed" or "You'll be shot". In the past, it was just a rancher is mad at us.[123]

Later that year, two environmentalists were once again hung in effigy in November. This time it was in New Mexico, when a gathering of loggers and ranchers, reportedly angry over restrictions on firewood cutting in the Carson National Forest, hanged an effigy representing Sam Hitt, Director of the local Forest Guardians, and John Talberth, the Director of the Forest Conservation Council. The latter was first hanged near Forest Guardians' office, then carried to the office of the Levinson Foundation, where Talberth's wife works. The violence in New Mexico continued into 1996. In January a US Forest Service office was fire-bombed causing $25,000 in damage, although no-one was hurt in the blast.[124]

SLAPPS

Another tactic used against environmental activists are SLAPPs, or Strategic Lawsuits Against Public Participation. Just as violence is designed to chill people into silence through physical intimidation, so SLAPPs are meant to chill through legal intimidation. Hardly ever designed to come to court, and if they do, they hardly ever win, but they do chill people into silence. 'It is part and parcel of the overall counter-attack by the polluter industry against the environmental community,' contends Al Meyerhoff, an attorney from the NRDC.[125]

'A new and very disturbing trend is happening in America, with grave consequences for politically active citizens and for our political system,' write George Pring and Penelope Canan, two professors from the University of Colorado, who first invented the term SLAPP, and have subsequently studied the SLAPP phenomenon. They now reckon thousands of people are being, or have been, SLAPPed for speaking out.[126]

The basic goal of the SLAPP, one designed to chill public debate, is illegal under America's First Amendment in its Constitution which guarantees citizens the right of public debate. Therefore SLAPPs come wrapped as libel, defamation, and claims for business damage. SLAPPS are also designed to remove the debate from the public arena, into the more controlled private arena of a courtroom, with the intention to 'dead-end opposition through the chilling effect of a lawsuit', says Pring.[127] 'Anyone who becomes politically active is at risk,' says Pring, 'because in this country anyone who pays the filing fee can file a lawsuit.[128]

SLAPPs are typically filed against activists or citizens by the rich and the powerful. They are designed for the polluters to keep polluting, the developers to keep developing, the loggers to keep logging, the miners to keep mining and the toxic landfillers to keep filling without any recourse from the public, who are affected by these decisions.

Thousands of people have now been sued. Some for purely attending a meeting, others for simply signing a petition, others for reporting pollution violations, writing to public officials, even writing to the editor of the local newspaper, or testifying at public hearings or taking part in boycotts.[129] Moreover, it is not normally the large environmental organisations that are targeted, it is small community-based groups that have been targeted for stopping waste sites, and development. George Pring personally knows of people who have vowed never to go to another public meeting because they fear being sued.[130]

People have a right to be worried. The average amount sought in damages is $9,000,000 and the lawsuits drag on for an average of thirty-six months.[131] Defendants risk facing financial ruin if they lose, being dispossessed of their home and sacked from their job. The financial and emotional burdens are therefore immense, and many people back down under such legal intimidation.

Therefore, although some 80 per cent of SLAPPs are lost or never come to court, the legal intimidation works in other ways. 'There is no lack of victims,' write Pring and Canan, 'In the last two decades, thousands have been sued into silence. For them, and for the estimated hundreds of thousands who know about them, SLAPPs have "worked" even when they lose.'[132] 'Increasingly,' says Tarso Ramos, 'various different kinds of tactics by Wise Use, including the threat of SLAPP suits have resulted in environmental activists re-evaluating the risks that they're willing to take in order to continue to participate in environmental activism, and SLAPP suits are part of that.'[133]

SLAPPs have already been vigorously used to silence environmental activists, and Professor Canan estimates that some 60 per cent of all SLAPPs are against environmentalists and people fighting unlimited development.[134] Irene Mansfield was SLAPPed for $5,000,000 in 1986 for opposing a hazardous waste landfill in Texas, and calling it a 'dump'. Mansfield's husband was also named in the suit, for failing to 'control his wife', even though he had not protested against the landfill, owned by Hill Sand company. The case was dropped after three years of bitter legal battles in 1989.[135] Dixie Sefchek, along with three others of the community organisation, Supporters to Oppose Pollution (STOP) were SLAPPed for opposing the unsafe practices of an Indiana-based landfill operator. The SLAPP claimed loss of income, defamation of character and libel. 'It's like a death threat to your organisation,' said Sefchek. Although the SLAPP was thrown out of court the following year, the organisation had already lost members, money and support from people frightened by the legal action.[136]

Well-known activists and groups are targeted too. In 1990 the Pacific Lumber company sued Earth First! activist Darryl Cherney for $25,000 for sitting in a redwood tree.[137] The Sierra Club were also SLAPPed for $40,000,000 by a timber company, because they had started to represent Native Alaskans who wanted to prevent logging on their traditional lands.[138]

The answer for people who have been SLAPPed is simple: to SLAPP-back, to sue the company that is doing the suing. SLAPP-backs have become extremely successful, and Canan does not know any that have actually failed. 'SLAPP-backs send a reverse chill,' adds Pring, 'when enough of these messages add up, SLAPPs will become an unpleasant part of history.'[139] In May 1991, a Missouri activist who had been SLAPPed in 1988 for criticising a medical-waste incinerator was awarded an $86.5 million judgment against the incinerator's owner for wrongfully SLAPPing her. A plumber's union attorney, Raymond Leonardini, won $5.2 million from Shell Oil in damages after he SLAPPed-back against the company. Shell had originally sued him, after he had raised fears about the carcinogenity of their plastic pipes.[140]

SLAPPs are becoming more common internationally as well, with the tactic appearing in Canada, Europe, South East Asia and Australia. For example, in Canada the Ogden Martin company threatened to sue fifty-two doctors who endorsed a letter rejecting their proposal for a 3,000 tonne-per-day waste incinerator in Ontario. The doctors, concerned over the health impact of the proposed scheme, were told by Scott Mackin, Ogden Martin's general counsel that, unless they dropped their support they would be sued for defaming the company.[141] In 1993, MacMillan Bloedel, dropped a SLAPP against the Galiano Conservancy Association, who opposed MacMillan selling company land for residential development.[142]

In total, therefore, what we find is that violence and intimidation are on the increase around the world against environmentalists.

7

TO CUT OR NOT TO CLEAR-CUT

A question of trees, truth and treason?

I believe they [Greenpeace] have thrown truth out of the window on this one. They simply seem unwilling to listen to the science on their whole forestry campaign.[1]

Patrick Moore, Director of the BC Forest Alliance

ARNOLD SETS THE SCENE

Due to the geographical proximity between the British Columbian forests on the west coast of Canada and those in the western USA, there are many similarities in the backlash being suffered by environmentalists working to protect those forests. Given that Ron Arnold, a leader of the Wise Use movement, lives close to the Canadian border and has been a regular visitor to the country for the last fifteen years, it is easy to see where much of the polarisation process that pervades the forestry debate has come from.

Back in 1980, Arnold warned the delegates attending a BC Professional Foresters Conference that there was a 'monotonous thread of totalitarianism running through the public utterances of environmental leaders'. He told his audience of a 'hidden agenda of environmental leaders to stop production where possible'.[2] The following year, whilst addressing the Ontario Agricultural Conference on the 'Politics of Environmentalism', Arnold warned of 'terrorism' by anti-pesticide activists, likening their campaigning literature to Hitler's *Mein Kampf*.[3] In 1984, he told his audiences that environmentalists were Marxist-Leninists, and that the Soviet Union was funding the German Green party in order to destabilise the western economy.[4] He also advised industry to set up citizens activists groups. 'It takes a movement to fight a movement,' was his message to the Educational Seminar of the Atlantic Vegetation Management Association, a pesticide trade association.[5]

Just weeks after being appointed Executive Director of the Center for the Defense of Free Enterprise, Arnold was back in Canada in Ontario in 1985, this time talking to the Canadian Pulp and Paper Association.[6] When he addressed the Alberta Forest Products Association a year later, he had refined his theme into a simple one: the environmental movement is a communist conspiracy led by marijuana smokers.[7] After a year's break, he was back talking to the Ontario Forest Industries Association in February 1988. His speech, 'Loggerheads over landuse' was reprinted in *The Logging and Sawmill Journal*. Arnold warned of the 'unfinished agenda' of environmentalists, who were engaged in 'genuine psychological warfare' to create 'true believers who will fight and die for the cause'.[8] He accused environmentalists of being eco-terrorists, when he addressed the recently formed Temagami Forest Products Association in Ontario, the same year.[9]

Arnold was busy in Canada in the late 1980s, speaking at the BC Professional Foresters Conference in February 1989. The same month he also spoke on 'multiple land use – cultivating citizen supporters' at the 57th annual convention of the Prospectors and Developers Association of Canada. Four months later he addressed the Alberta Registered Professional Foresters' Association inaugural meeting.[10] When MacMillan Bloedel hired him as an advisor, Arnold told it to 'give them [the coalitions] the money. You stop defending yourselves, let them do it, and you get the hell out of the way. Because citizens' groups have credibility and industries don't'.[11]

'Arnold,' wrote *Vancouver Sun* journalist, Mark Hume, 'has done more than influence the rhetoric being used in the resource debate in BC. He's offered a blueprint for the future struggle.'[12] Just weeks after Arnold had talked to BC Foresters, Gerry Furney, the mayor in the logging town of Port MacNeill, set up his own group, calling it the BC Environmental Information Institute.[13] The Institute claimed to 'encourage responsible, integrated and sustainable resource development which meets present needs without foreclosing those of future generations' and to occupy the 'middle ground to counter the "emotional" argument put forward by environmentalists'.[14] Western Forest Products and MacMillan Bloedel were quick to donate money to the Institute.[15]

NORTHCARE AND SHARE

In fact, pro-industry citizens groups had begun appearing in the mid-1980s in Canada, although they mushroomed towards the end of the decade and in the beginning of the 1990s. Most of these groups formed around the issue of clear-cutting Canada's old-growth temperate forests in Ontario and British Columbia. In Ontario, the movement is called NorthCare, in British Columbia, the 'Share' movement. Much like the terminology of its sister Wise Use movement in the US, the Share and NorthCare movements use the same soft ambiguous language to promote themselves and 'multiple and Wise Use'.

In Ontario, NorthCare formed in 1987, about the same time Arnold paid his visit to the region. Despite the different name, for NorthCare and its sister organisation in British Columbia the rhetoric is much the same. For example, NorthCare, which stands for Northern Community Advocates for Resource Equity, believes in: 'Sharing the resources and beauty of Northern Ontario; strong, vibrant communities based on the Wise Use of all Crown land resources; and conserving the land of Northern Ontario as a viable economic entity for the benefit of future generations.'[16]

With slogans such as 'We care' and 'sharing our resources for enjoyment and employment', Northcare membership included three local Chambers of Commerce, the Temagami Forest Products Council, steel workers, trappers, hunters, lumber and sawmill workers and some seventy timber town municipalities.[17] It also received money from multinationals such as Falconbridge, Red Path, and E. B. Eddy.[18] NorthCare is not the only anti-environmental organisation to be formed in Ontario. AG Care or Agricultural Groups Concerned About Resources and the Environment, is a coalition of pro-pesticide farming organisations.[19]

But it is the forestry debate in British Columbia that has spawned the majority of anti-environmental 'Share groups'. Fifty per cent of all the logging in Canada takes place in British Columbia. Ninety-five per cent of these forests are owned by the provincial governments and in general, as in North America, the conflict is greatest in areas of forest which are publicly owned. Exacerbating the problem, timber rights in British Columbia are concentrated in the hands of a few transnational companies with over two-thirds of timber assets owned outside the state.[20] Large-scale clear-cutting of old-growth forest has accelerated since the 1960s, when transnationals took over the small logging companies. Since then, there has been conflict between environmentalists, Indians and certain politicians and populations willing to preserve the old-growth forest, and loggers wanting to preserve their jobs and companies wishing to preserve profit maximisation through clear-cutting.

TO CUT OR NOT TO CLEAR-CUT

In a nutshell, the conflict is one of how to exploit the forests. Despite the claims of anti-environmentalists to the contrary, no ecological group actually proposes preserving all forest *per se*. They all agree there should be some logging – it's just the nature and intensity of logging that are the bone of contention. Communities want logging to be done in a sustainable way, but large timber companies, such as multinational MacMillan Bloedel, believe the only option is to clear-cut, a practice where all timber in an area is indiscriminately felled. This technique uses least labour and yields greatest profit for the companies – it also causes the maximum ecological damage. Traditionally, the companies have always been given what they wanted, which has led to an arrogance that any interference in operations by outsiders is misguided or malignant.[21]

Although the forest industry proclaims that, through its reseeding programmes, it can recreate old-growth forest, in reality this does not happen. A canopy of trees is necessary for seedlings to grow properly. If clear-cutting takes place there is no canopy and the resultant second growth are low-growing trees with inferior quality wood. Award-winning journalist Joyce Nelson writes of the 'industry's dirty little secret: coastal rain-forests don't grow back in clearcuts. Trees grow there, but they will never be the giants that stand there now.'[22] Observers highlight how there is no comparison between liquidation of a complex ancient forest eco-system that has evolved over thousands of years and planting a man-made forest where the trees are for public fibre.

Environmentalists and first nations contend that clear-cutting is utterly unsustainable and causes widespread, irreversible, ecological destruction. They believe that eco-forestry practices should go beyond clear-cutting in the setting up of forestry reserves. Clear-cutting is extremely unsightly and is therefore opposed by many in Canada's booming tourist industry. A scarred moonscape mountainside does not have the makings of a picturesque postcard. There are also serious concerns regarding losses of biodiversity, of habitat and species, and fragmentation of the forest with thousands of clear-cuts and access roads.

Opponents to clear-cutting propose a more sustainable, selective cutting technique, which still provides a timber yield, but is far more labour-intensive and therefore is not so profitable. It does, however, protect both long-term jobs and the old-growth forest. However, this option is unacceptable to the timber multinationals, whose ultimate responsibility is profit maximisation for their shareholders and not damage minimisation to the forest for future generations. Economics not ecology is the ultimate driving force behind large-scale clear-cutting.

Many loggers are caught in a Catch-22 position. They know that current forestry practices by the multinationals are wasteful, ecologically harmful and unsustainable, but, being dependent on the companies for employment, they are afraid to express their concerns. They are deemed the 'silent majority' in the forestry debate, caught in the cross-fire of the clear-cut wars, but condemned never to speak out.[23] But in a country where trees are more popular than politicians, people do voice their anger at the continuing clear-cutting of 250-feet high, 1,000-year-old trees.[24] Therefore, wherever the forest industry proposes widescale clear-cutting, there is confrontation.

SOUTH MORESBY

One of the first and most controversial areas was the Queen Charlotte Islands, off the west coast of Canada, once known as 'Canada's Galapagos'.[25] In 1985, after fourteen years of debating the future of forestry, logging was banned on Lyell, one of the smaller islands in the group, but home to Windy Bay, an area of ecological significance in the area. Western Forest Products, the transnational that intended to log the South Moresby Wilderness Proposal organised its workers to form the Moresby Island Concerned Citizens (MICC) group. Almost all of the employment on Lyell were temporary jobs for workers from Vancouver and elsewhere.

The debate in South Moresby was underpinned by the forest industry attempts to generate a climate of hate against environmentalists, which was achieved through the newspaper, the *Red Neck News*, sent free to loggers in South Moresby. Written by R. L. Smith, a former logger and employee of Beban Logging, a company working in the Queen Charlottes, *Red Neck News* vilified environmentalists, calling them 'hippies', 'dope-smokers', 'leeches', 'welfare bums, cheats, frauds, draft-dodgers and criminals'. The themes running through *Red Neck News* were similar to ones Arnold would reiterate, mainly that city-based dope-smoking terrorist greens would knowingly deceive to shut down the timber industry. Environmentalists were 'the lowest form of life there is'. *Red Neck News* targeted specific environmentalists, namely the Valhalla Society and Colleen McCrory. Such scapegoating and scaremongering led to violence when a Valhalla Society member was beaten up, and their office vandalised. The campaign of hate resulted in a successful private prosecution against the assailants and a 1,100-page report to the Ministry of Justice documenting the hate-campaign, but the damage was already done: community-life in many towns in BC had been sabotaged by the forest industry.[26]

Red Neck News ceased production when the owner of Beban Logging died of a heart attack. One of Smith's protégés, Patrick Armstrong, took Smith's crude style and turned it into a far more sophisticated PR campaign. Armstrong also became Chairman of the MICC.[27] However, the MICC could not prevent what the corporations feared most: in July 1987, the Governments of British Columbia and Canada signed a Memorandum of Understanding leading to the creation of the 1,450-square-kilometre South Moresby National Park Reserve, making the majority of the largest of the Charlotte Islands off-limits to logging. 'The BC government has made a serious mistake,' said Armstrong at the time.[28]

It was a devastating blow to the industry, which saw itself having not only lost the public relations war but fighting growing public support for wilderness preservation throughout the province. 'South Moresby really shook the industry,' said Peter Pearse, a forest economist at the University of British Columbia, 'They realised they had no friends left.'[29] British Columbia's Council of Forest Industries (COFI) attempted to remake friends by ploughing C$1.5 million into a major three-year, multi-media public relations campaign, called *Forests Forever*.[30] 'The Forests of British Columbia are forever' read one of the adverts:

> They are our heritage and our soul. They give us peace. Tranquillity. And refuge. We live, play and work in their magnificent presence. But forests are also our economic lifeline. They create jobs and provide us with lifestyles unequalled throughout most of the world.[31]

It was classic greenwashing. The aim was to convince the public how responsible the forest companies were and how important the forest sector was to jobs throughout Canada.[32] In different disguises it is a message that the industry and its myriad of citizen action groups have been attempting to cement in the minds of the policy-makers and the public ever since. It is a message that the forest industry has had to build across continents as criticism has spread internationally.

But South Moresby also acted as catalyst both to the logging communities and the forest industry to form the pro-industry citizens groups that Arnold had advised. The 'MICC . . . presented an organisational template for the subsequent proliferation of Share groups in the province,' wrote Dr Mike Mason, in his study of Share groups.[33] The forest industry, with Armstrong in tow, moved to the next hotspot, the Stein Valley. In September 1987 the Share the Stein was formed to promote the logging of the Stein Valley. Equipped with C$200,000 for one year's funding from the Council of Forest Industries, BC Forest Products Ltd, and the Caribou Lumber Manufacturers' Association, Armstrong was employed as the head of a Forestry Industry Task Force for the Stein.[34]

SHARE THE WORLD

Armstrong began to pick up on the Wise Use tactics of paradigm shift. Vilify the environmentalists, but verify yourself. Calling the Western Canada Wilderness Committee's publications on the valley 'doomsday literature', he promoted 'multiple use' and 'sharing' for sustained development of the local communities (rather than sustainable development).[35] Moreover, the forest industry public relations strategy to log the Stein went into full swing with the distribution of a slick, 150,000-copy, eight-page tabloid called *Share the Stein*; tours of saw-mills for journalists, helicopter flights and poolside barbecues.[36]

Armstrong and the forest companies were also instrumental in initiating other Share groups. Share the Stein was the first of many Share groups formed to fight local forestry battles: Share our Resources (preserve special places and protect the working land), Share our Forests, North Island Citizens for Shared Resources, North Caribou Share our Resources Society, Slocan Share Group, Kootenay West Share Society (the people who care about people), Share the South Okanagan Society, and Share the Clayoquot, to name the most prominent ones.

Many of the individual Share group members are workers in rural communities with real and legitimate worries over resource-dependent jobs who are fed fear tactics by the timber companies, and in this sense, Share's message is identical to industry's and their rhetoric is indistinguishable from Wise Use.[37] 'The clear indication from the companies,' said BC Member of

Parliament Bob Skelly, 'was that environmentalists are trying to take away your jobs, and therefore if you attack the environmental groups, your jobs will be protected.'[38]

Industry was reacting in its age-old way of polarising the debate into jobs versus environment. The Share groups started using this Wise Use jargon, advocating 'multiple use', and 'sustained yield', calling loggers the real 'endangered species'. Publications promoted their's as the voice of reason, balance and sincerity against extreme 'preservationist' propaganda whose agenda was 'the destruction of all resource related jobs' with no more logging in BC. Preservationists were 'masters of deception, who use half-truths and outright lies about forestry'.[39]

SHARE AND WISE USE

Links between leading Share advocates and Canadian businessmen were extended when forty people attended Arnold and Gottlieb's first Multiple Use Conference at Reno in August 1988. MacMillan Bloedel even flew a private company jet down to the meeting. Signatories of the Wise Use Agenda included the following from Canada: the Caribou Lumber Manufacturers Association; Council of Forest Industries; Furney Distributing, Ltd; MacMillan Bloedel Ltd; Mining Association of British Columbia; Jack Mitchell, Alderman, City of Port Alberni; Northcare; Share Our Forest Society; Share the Stein Committee; Truck Loggers Association; Western Forest Products, Ltd; and the Temagami Forest Products Association.[40]

By the time of the Reno conference, the forest debate had become a heated issue in the province. Further spice was added to the boiling pot of the debate by journalist Mark Hume's article in *The Vancouver Sun* in 1989, 'Resource-use conference had links to Moonies cult', which detailed Arnold's links to the Moon-linked CAUSA and the AFC.[41] As the result of this article two Canadian MPs, Jim Fulton and Robert Skelly, commissioned a federal parliamentary report into Share's links with right-wing groups, which was published by the Canadian Library of Parliament at the end of 1991. The report makes some extremely valid observations and highlighted definite parallels between Share and Wise Use. It also caused an outcry from the logging industry and the Share groups. A barrage of condemnation was levied against the two MPs from the Share groups who vehemently denied being linked to the 'Moonies'.

But the parallels between the two movements is valid. The results of the tactics employed by Wise Use and Share have been the same, an effective paradigm shift against the environmental movement. Just as the Wise Use movement has polarised the debate in the USA, so Share's 'apparent objective has been to pit labour against environmentalists and environmentally-orientated persons,' said the Share report. 'Their effect has been to divide communities and create animosity in the very places where honest communication and consensus should be encouraged.'[42]

The report continued: 'The rhetoric and vocabulary of BC Share groups are identical to language used by Ron Arnold and The Wise Use Agenda, particularly phrases such as the "unfinished agenda of environmentalists", "wise use", "multiple use", "sharing", "preservationists"'. It also analysed how, just like its American counterpart, 'typically a Share group presents itself to the public as reasonable, objective, conciliatory, neutral and "middle of the road". Spokespersons go to considerable lengths to emphasise that their activities come from community participation – that they are "grassroots" organisations.'[43]

Due to the barrage of adverse publicity that rocked British Columbia over Share being linked to Wise Use and the Moonies, Share leaders were quick to distance themselves from Wise Use and Arnold. Mike Morton, Executive Director of Share B.C., was adamant that there never has been, nor will there ever be, a link between Share groups and Arnold.[44] Despite this, two Share group contacts, Lou Lepine, from the North Island Citizens for Shared Resources and Ray Deschambeau from Share the Stein are listed in William Perry Pendley's *Hero Network*, published by Alan Gottlieb's Merril Press. Other Canadians listed are Patrick Armstrong, Moresby Consulting and Liza Furney, from the Truck Loggers Association.[45] Both Armstrong and a representative from the Truck Loggers Association were panellists at the first Share B.C. conference in Chilliwack.[46]

Armstrong has reiterated that there is 'not a single shred of evidence' to link the Share and Wise Use movements.[47] However, he had been a Director of Our Land Society, a Wise Use group from Idaho which had itself received support from the Moon-affiliated, American Freedom Coalition.[48] Darryl Harris, the President of Our Land Society, almost paraphrased Alan Gottlieb and Ron Arnold when he wrote that 'Wise Use is the "new environmentalism"' and that the 'goal of old environmentalism is to destroy or at least cripple industrialized civilisation'.[49] During the late 1980s and early 1990s Armstrong's wife was also a Board Member of Canadian Women in Timber (CWIT), an anti-environmental group, modelled on sister Wise Use organisations in the USA, such as Women in Timber, which encourages the 'wise use of forest resources'.[50]

Armstrong also advised the first Share network, the Citizens Coalition for Wise Land Use, which was formed in February 1989 and which was the forerunner to Share B.C., the coalition that formed ten months later. He had also forged a loose network of the industry's major trade associations in the province.[51] The Citizens' Coalition for Sustainable Development is its full name, or Share B.C. for short, was formed as the umbrella organisation for the individual Share groups, much like the Alliance for America is for many Wise Use organisations. It is an 'Alliance of citizen groups from throughout BC which supports environmental protection and economic prosperity'.[52]

Although the Share groups constantly reiterate their grassroots nature, it receives corporate logistical and financial support, and there are key links between Share B.C. and MacMillan Bloedel. Its first chairman, John Bassingthwaite, was a long-term employee of MacMillan Bloedel and most other key Share organisers have had links with the forest multinational. MacMillan Bloedel has also reportedly given funds, office space and corporate encouragement to the Share groups.[53] The forest industry is funding the Share movement, but figures are hard to come by. In 1991, Share B.C. admitted some 60 per cent of its annual budget was paid by forest companies.[54] The following year, however, Mike Morton, Share B.C.'s new director, maintained that none of the 1992 budget was from corporations. By 1993, however, Share B.C. was receiving corporate money again.[55] Moreover, individual Share groups, such as Share Our Resources, have received financial support from MacMillan Bloedel and the Ministry of Forests and the Forest Alliance.[56] Others have acquired corporate back-up: in October 1989, Fletcher Challenge Canada admitted it had sent Share the Stein literature in unmarked envelopes.[57]

There are on-going links between key Wise Use and Share activists: Patrick Armstrong, BC Forest Alliance Chairman Jack Munro, and Wise Use activist, Bruce Vincent from the executive Committee for the Alliance for America, addressed the third Share B.C. Conference.[58] This was not the first time that Vincent had been to Canada. In 1990, speaking to the Kootenay Livestock Association, Vincent told the audience that Earth First! was using terrorist equipment and that it advocated using explosives against equipment and promoted the shooting of cows. He urged citizens to get involved in the political process.[59]

Three years later Vincent was back, this time touring BC at the invitation of Share groups and the Interior Lumber Manufacturers Association (ILMA).

Once again he was urging people to participate in the political process, whilst advising people not to accept the 'big lie' – that only by conserving resources can you preserve them. *The B.C. Environmental Report* described how Vincent's 'speeches in BC fall into line with one of the Wise Use strategies; to give grossly oversimplified and distorted claims'.[60] One such speech was sponsored by Share Our Resources. A few weeks after his Canadian trip, Vincent travelled to Australia to promote the Wise Use message down under.[61]

Given continued contact between anti-environmentalists in America and Canada, the rhetoric between Share and Wise Use continues to be similar. One of polarisation rather than conciliation. Howard Goldenthal, writing in *Now Magazine* in 1989, accused Arnold of stirring a 'simmering cauldron of fear and violence'.[62] By 1990, Colleen McCrory, a long-time environmental activist and winner of two prestigious environmental awards, the Goldman Award and the UN's Global 500 Award, was accusing the industry, government and Share groups of 'fomenting hatred towards environmentalists. Ultimately hatred turns to violence. There's a well organised hate campaign happening in British Columbia.' McCrory also blames the anti-environmental movement for scapegoating the environmentalists for the real problems in the forest industry, which are mismanagement and automation.[63] 'To this day,' wrote award-winning journalist, Kim Goldberg in 1992, 'Share groups continue to employ precisely the same language and tactics of the burgeoning Wise Use movement in the US, suggesting the link has not been broken.'[64]

In 1996, showing signs of on-going co-operation, Al Biex, the President of Share B.C., attended the Alliance for America's Fly-In For Freedom.

ANTI-GREEN SENTIMENT SPREADS

In essence the Canadian environmental movement has gone through the same paradigm shift as its American counterpart, with anti-environmentalists repeatedly attempting to demonise environmentalists. For example, at the first inaugural Share conference, environmentalists were called 'fly by night, long-haired, dope-smoking', 'preservationists', and 'zealots' who 'do not care', who 'manipulate' people with their 'blatantly deceptive' 'lies' and 'propaganda'.[65]

However, this vehement anti-environmental verbosity has seeped through other sections of industry and the press, especially the industry press. In August 1988, the Assistant Chief Forester of MacMillan Bloedel, in a speech to

Vancouver's Rotary Club, gave a speech mirrored on Arnold's rhetoric, entitled, 'The Unfinished Agenda'. In it he warned of 'preservationists' 'manipulating the media' and being 'long on archetypal symbolism and short on rational argument'.[66]

In December 1988, *The Truck Logger* labelled environmentalists 'zealots', a year later it called environmentalism 'the pseudo-religion of our times'.[67] 'Save a logger, Bugger a Hugger', 'Save a Logger: Spike a Preservationist' shirts appeared at loggers' rallies in May 1989.[68] Ken Williams, the Chief Forester at Macmillan Bloedel, alerted his suppliers and customers about 'preservationists' in August.[69] Two months later, the *Globe and Mail's Report on Business Magazine* labelled environmentalists as 'something approaching an environmental terrorist movement' and 'tree-saving zealots engaged in environmental guerrilla tactics'.[70] A month afterwards, Adam Zimmermann, the Chairman of Noranda Forest Products, which owns MacMillan Bloedel, called environmentalists 'extremists' and 'terrorists'.[71] The Provincial Forests Minister Dave Parker said in August 1989 that 'one of the best ways to get economic chaos in North America is to stymie development. One of the best ways to stymie development is to get an environmental preservationist movement going.'[72] Parker also commented that the environmentalists' long-term agenda was to create economic instability.[73]

By 1990, there was said to be 'deep fear approaching paranoia' that environmentalists 'harbour a hidden agenda' at the Second Share Conference.[74] Canadian Pacific Forest Products Limited warned of 'preservationists' 'polarising' the forestry debate, whilst advocating 'sustained development'.[75] A year later Thomas Buell, President and Chief Executive of Weldwood of Canada Limited, a BC forest company, also cautioned his employees of 'extreme preservationists whose hidden agenda is an end to the forest industry in BC'.[76] In 1991, Judy Lindsay, writing in *The Vancouver Sun*, called boycotts against BC forest products 'environmental terrorism'.[77] The Canada Pulp and Paper Association called greens 'extremist environmental theologians' and 'dogmatists'.[78] The head of Quebec Manufacturers Association labelled environmentalists 'eco-fascists' and 'eco-terrorists'.[79] By the third annual Share conference, in 1992, the message was simple: 'Don't let your love of wilderness blind you to the needs of your fellow man',[80] whilst environmentalists were called 'eco-terrorists', 'fruits, nuts and flakes' who were placing loggers in the 'bull's-eye for extinction'.[81]

Such language targeted at the environmental movement angered Stephen Hume from *The Vancouver Sun*, who wrote in one article, 'I have met the environmental terrorists and they are us!' In another Hume remarked that environmentalists, I'm told, are terrorists who threaten worker safety, although the cold record of statistics clearly indicates that the biggest threat to the lives of loggers comes from conventional practices on the work site . . . I'm told again and again that environmentalists threaten jobs in logging. The clear evidence, however, is that the biggest threat to jobs has been political resource-use policy in Victoria, economic decisions in air-conditioned board-rooms by distant executives determined to rationalize costs, and the introduction of new technology.[82]

Hume's arguments seem to backed by the figures. According to journalist Kim Goldberg, some 27,000 forestry jobs disappeared in BC between 1981 and 1991. Of these, 2 per cent were sacrificed to parks, the remainder to mechanisation. In 1991, BC timber companies exported the equivalent of 2,700 jobs in raw logs. Half the trees logged in all of BC's history have been cut since 1972, and Environment Canada says all of BC's most valuable timber will be gone in just sixteen years.[83]

It is this concern over BC's forests, that has made environmentalists consider a whole range of tactics to raise awareness of the issue. The stakes were increased the following year, however, when the rumours of a European boycott of forest products began to permeate through the province. Environmentalists found themselves not only called 'terrorists', they now became 'traitors', 'guilty of treason'. 'It was obviously a well orchestrated and intentional use of those words,' says Greenpeace's Tamara Stark, 'because they appeared over and over and over again.' It was the perfect tactics of paradigm shift once more being rallied against environmentalists. Meanwhile the forest industry and the government were interested in 'protecting real jobs for real people in real communities'.[84]

Jack Munro, the then President of the International Woodworkers of America (IWA), initially stated that environmentalists who advocated boycotts of BC products were guilty of treason.[85] Munro and Frank Oberle, the Federal Forest Minister also called people who had appeared in a German documentary, which was critical of current Canadian forestry practices, traitors. David Suzuki, a nationally respected television presenter, who had himself been interviewed in the programme, countered by saying that

the use of a word like treason is a cheap ploy to avoid discussing the serious issue of the non-sustainability of current forestry practices. It shouldn't be dignified with a response, but in the furore over the future of B.C.'s forests, this rhetoric has a real potential to escalate into violence.[86]

THE BC FOREST ALLIANCE

Munro made his 'treason' remarks at the April 1991 launch of the newest player in the PR wars. Munro had just been appointed the head of the BC Forest Alliance, the forest industry's latest weapon, having quit his job as the Chairperson of the International Woodworkers of America Union. The Alliance was designed as the industry's stealth bomber, packed with influential people and fuelled with industry dollars, to blast holes in the environmentalists' forestry campaigns. However, it came packaged as a dove of concern and conciliation. 'We should make it clear we are not a lobby group,' said Munro, 'This is a group of concerned British Columbians that can see the well-being of our economy suffering because not enough people understand the middle point of view.'[87]

The Alliance was the brainchild of prominent anti-green PR company Burson-Marsteller, whose employee, Gary Ley, was appointed the executive director. 'Our key commitment is to tell the truth of what goes on,' said Ley.[88] B-M's involvement in setting up the Alliance made independent observers extremely sceptical though and the public relations firm's involvement gives insight into the reasons behind the Alliance's inception: pure public relations and forest industry propaganda rather than the promotion of change within the industry. Stephen Hume from *The Vancouver Sun*, called the Forest Alliance the 'son of Forests Forever, second cousin of Share the Resources, and love-child and bedmate of Burson-Marsteller'.[89] Journalist Kim Goldberg, labelled it 'a pseudo-populist, industry funded pressure group'.[90]

Munro was head of the Citizens' Advisory Board, whose other members included the Chairman of the Board of MacMillan Bloedel, a Professor of Forest Policy at the University of British Columbia, the Director of Canadian Women in Timber, the Executive Director of Western Wood Products Forum, the Mayor of Vancouver, a retired forester and a Professor of Forest Ecology. Many of these people were hand-picked by B-M.[91] Another key director was to follow in May. Dr Patrick Moore, one of the founders of Greenpeace and member of the BC Round Table on the Economy and the Environment.[92] A year later Munro became the full-time Chairman of the Alliance, with Munro and Moore the Alliance's two main vocal spokespeople. Moore was acting as both a consultant and Director.

Moore's appointment was to cause problems within the environmental movement *per se*, but in particular for Greenpeace. Greenpeace's forest campaigner, Tamara Stark says:

> The fact that they chose Patrick Moore as a former Greenpeace spokesman was a really dangerous sign for us in Canada as it created a confusing message for the public, in particular because Patrick was not shy at all in repeatedly billing himself as a Greenpeace founder . . . but you have to admit it was a brilliant strategy.[93]

The Alliance language was a mixture of corporate greenwash and Wise Use rhetoric. Formed to 'find common ground' and promote 'Forests for all', it was 'a voice of reason and information between the hard rhetoric of the preservationists and the tough talk of the forest industry. The Alliance will monitor the environmental practices of the industry and report frankly to the people of B.C.'[94] The very same companies the Alliance promised to be so frank about were the ones paying the bills. The Alliance's $1 million first year budget came from thirteen forest product companies including MacMillan Bloedel, Fletcher Challenge Canada, Weyerhaeuser of Canada, West Fraser Mills and Enso Forest Products.[95] One of the companies, Weldwood, is owned by US forest giant Champion International, on whose board, Alan Gotlieb, B-M's Canadian Chairman, sits.[96] Despite the corporate funding and close personal connections, the Alliance portrays itself as a 'citizen-based organisation' that promises it is 'dedicated to achieving a balance between a healthy forest industry, with the accompanying jobs and economic benefits, and a clean environment'.[97]

B-M gave the same advice to the industry as Arnold had years before: the companies could not be relied on to solve their public relations problems, and that a so-called independent citizens' organisation was a panacea for their predicament. 'The most effective spokesperson for a subject is one with no obvious interest in it,' Wayne Pines, B-M's head in Washington advised his clients, 'In public crises, seek out third-party support and use it with the public.'[98] It was widely believed that the 'Forests Forever' campaign run by the Council of Forest Industries had failed because it lacked credibility with the public. Moreover, just months before the Alliance's launch, a Vancouver-based PR consultant had told the industry that 'no amount of old-fashioned PR or television advertising can reverse the growing credibility gap now facing the industry'.[99]

The Forest Alliance hoped to reclaim the lost ground in the integrity stakes.[100] In the words of MacMillan Bloedel chairman: 'It was created to provide a credible source of information and a better understanding of an emotional issue facing the industry today.'[101] The counter-credibility campaign was quick into action with the Alliance producing some seven half-hour television programmes in 1991. 'The forest and the people' series was created by B-M. The producer of the series, Ken Reitz, was exposed by *The Vancouver Sun*'s reporter Ben Parfitt as being involved in Richard Nixon's scandal-ridden re-election campaign of 1972. The programmes were also criticised for merging news and public relations propaganda.[102]

Parfitt, who had been the paper's forest reporter since 1989, paid dearly for his important reporting on B-M's role in the Forest Alliance and his critical analysis of the forest industry, by being pulled off his beat. In 1991 the paper had five reporters covering forestry, logging, environmental and energy issues and native affairs. By 1993, only one remained: the environment reporter, who was told to concentrate on Greater Vancouver and the lower mainland. 'An area,' writes Kim Goldberg, 'conveniently free of large tracts of old growth forest.'[103] Mark Hume, who with his brother Stephen, has been invaluable in covering the forestry and environmental debates for *The Vancouver Sun*, was highly critical of his own paper, complaining that, due to the forest industry, 'an awful lot of stuff does not get written anymore'. '*The Vancouver Sun* has pretty much turned a blind eye to the biggest environmental story in North America, which is happening right here in its backyard,' said Hume, 'and that's the termination of the temperate rainforest.'[104]

Meanwhile the Alliance's PR effort went into full swing. An economic survey was undertaken to assess how valuable the forest sector was to the economy.[105] A twelve-day 'fact finding' trip to Sweden, Finland, Austria and Germany took place in the summer of 1991.[106] A New Code of Sustainable Development was issued. Stephen Hume dismissed the code as 'an exercise in hypocrisy: Trotting out a list of principles which are supposed to guide logging practices – a list devised while those very same principles were being systematically violated across the province – seems less than an adequate response'.[107]

'A healthy economy and a healthy environment. You can help' began appearing across BC's transit buses.[108] The Alliance even attempted to join the BC Environmental Network, but it was turned down as no other environmental group would second its membership.[109] But the Forest Alliance has been successful in one respect: over the last two years it has replaced Patrick Armstrong and the Share movement as the premier and most quoted voice of the BC forest industry.[110]

CLAYOQUOT SOUND

The nucleus of the resource conflict became centred around logging activity in Clayoquot Sound, a 350,000-hectare area of ancient forest, coastal estuaries,

Plate 7.1 Fragmentation of old growth forest by clear-cutting on Vancouver Island.
Source: E. P. L./Dylan Garcia

alpine tundra, and estuaries, on the west coast of Vancouver Island. It is home to three of the last six old-growth watersheds on the island and one of the last remaining fragments of ancient temperate forest anywhere.[111] Clayoquot was a traditional homeland and hunting ground belonging to First Nations. It was a picture postcard that the companies wanted to moonscape.

After a four-year consensus-building exercise between interested parties failed, the government announced a new land-use decision, the Clayoquot Compromise, which allowed logging in about half of the Sound in April 1993. Environmentalists and many affected First Nations were outraged at the decision and Clayoquot Sound has been the scene of conflict ever since, as people attempt to stop it being clear-cut.

Loggers and industry were pleased, though, and some 5,000 industry supporters turned out in Ucluelet. 'We're here to support the decision of the government because it is right and proper. It represents a compromise, it represents democracy and it represents a balance,' Munro told the crowd, some wearing their 'Bugger a Hugger' and other openly violent anti-environmental T-shirts.[112]

The two companies who benefited from the decision and who held the Tree Farm Licences (TFLs) to log Clayoquot Sound were MacMillan Bloedel and Interfor. The TFLs had been issued to the companies by the provincial government of Canada, despite the fact that First Nations, specifically the Nuu-Chah-Nulth people opposed the logging and had never ceded their territorial right to the Sound to the government.[113] Environmentalists further cried foul, because just two weeks before the decision was made to open up Clayoquot, the provincial government bought $50 million worth of stock in MacMillan Bloedel.[114]

Another bone of contention is the forest industry's environmental record, which many environmentalists and journalists have criticised. Stephen Hume, from *The Vancouver Sun* remarked that:

> while Munro and Moore ran the dog and pony show up front, their industry bosses were out back. They trashed watersheds, caused a massive landslide, built shamefully sub-standard logging roads, cut trees without proper reforestation permits and logged in prohibited areas, all the while seeking to have the public barred from Crown land and protestors jailed.[115]

Protestors were getting arrested and people were being jailed. In their hundreds. MacMillan Bloedel had succeeded in obtaining an injunction against the blocking of its logging roads in 1991, and two years later in the summer of 1993, the police started to enforce it. An estimated 800 people were arrested for continuing to blockade the roads. In some cases people were severely punished, having to spend forty-five days in jail. People from all walks of life, and at all ages of life, were prepared to get arrested which was an indication of the strength of feeling surrounding the issue.[116]

'Forests are integral to the Canadian identity and Clayoquot Sound has come to symbolise the last Canadian stands of old-growth forest,' said Karen Mahon, a Greenpeace forestry campaigner.[117] Only one person, Tzeporah Berman, was charged with aiding and abetting the protests, although she never actually joined the blockades. 'I think it's quite clear that the charges are political and it's an attempt by the government to quell public dissent,' said Berman, 'It's an attempt to intimidate and silence organisers and it sets a very dangerous precedent.'[118]

By October, the government was looking for a resolution to the Clayoquot conflict which had grown out of its 'Clayoquot Compromise' earlier in the year. It announced a scientific panel would study forest practices to come up with a blueprint for sustainable forestry for the Sound. Forests Minister Andrew Petter said he wanted the panel to 'translate the results of their scientific investigations into a practical prescription for sustainable forest practices in the Sound'.[119] It would be eighteen months before they would deliver their dynamic findings, and in the meantime the forestry debate raged on.

IS BC THE BRAZIL OF THE NORTH?

Due to the controversy raging at home, it was inevitable that, in time, the forestry issue would eventually generate international attention. As foreign condemnation began to grow, the Alliance became increasingly preoccupied with polishing the industry's image both at home and abroad. With the Share groups fighting environmentalists in the communities and at the local level, the Forest Alliance, with its heavy financial backing and prominent heavyweight anti-environmentalists, was ideally suited to this task. The forest industry was unbelievably concerned about adverse criticism, especially from abroad. The clear-cut issue, as symbolised by the conflict of Clayoquot Sound, had become extremely politically embarrassing. Industry was worried whether it was going to become economically damaging, too.

When journalists and environmental activists started labelling British Columbia the 'Brazil of the North,' drawing on the similarities in the way the two countries were destroying their rain-forests, the Alliance responded vehemently. Although Canada's forests are temperate and the main method of logging is clear-cutting, and Brazil's tropical forests are felled and then burnt, many similarities can, indeed, be made. 'Brazil is losing one acre of forest every nine seconds. We're losing one acre every twelve seconds,' said Colleen McCrory, who originally coined the 'Brazil of the North' phrase.[120] MP Jim Fulton outlined his analogy of Canada to Brazil in a letter to Jack Munro. Reasons for the comparison included similar figures for the size of the countries; the size of the forests; the amount of timber felled annually; and the per centage of forest 'not sufficiently restocked' or destroyed. Distinct dissimilarities exist in Canada's favour, though, in the amount of forest burnt per year and the amount of regeneration of forest that occurs.[121]

Seven Alliance staff, including Jack Munro, were dispatched on an eleven-day tour of the Latin American country. 'B.C. is NOT the Brazil of the North,' declared Munro on his return. He called environmentalists 'masochistic mouthpieces' who were threatening jobs. Colleen McCrory stood her ground though, calling the trip an 'absolutely ridiculous' public relations exercise.

'I am not backtracking . . . The comparison stands. Just as Brazil was in a crisis, B.C. is in a crisis.' Moreover, McCrory went further, 'I owe Brazil an apology,' she said, 'their logging has improved considerably and they have protected a lot more of their forests than we have in British Columbia.'[122] When the World Fund For Nature published a report on the globe's vanishing temperate forests, of which a couple of pages were on clear-cutting in Canada, the Alliance took unprecedented steps to criticise the report. Patrick Moore even flew to London to have a head-to-head meeting with the author, ecologist, Nigel Dudley.[123]

Moore has increasingly become a prominent pro-industry figure in the Alliance and the international debate over Canada's forests. What makes this quite surprising is that he was a founder member of Greenpeace, a fact that he exploits to the full. Although he left the organisation over ten years ago, he still uses the Greenpeace name to gain credibility in the environmental debate. For this, he is lambasted by his critics as an arch eco-villain. 'I think Moore is a sell-out. I think he is a traitor to his own people,' says Joe Foy of the Western Canada Wilderness Committee.[124] 'I think a lot of people question his credibility,' says Tamara Stark from Greenpeace, 'you still have to look at the fact that he did leave the organisation, not necessarily by his own choice . . . Patrick behaves often as if he has been hurt by the organisation.'[125]

Industry executives, however, see Moore as the voice of eco-reason, against his old colleagues who personify eco-extremism. He has been busy spreading his anti-green message: speaking to Canadian Pulp and Paper Association (CPPA), Wood Pulp Section in 1993 and the Western Forest Industries Association in 1994. Whilst speaking to the CPPA, Moore argued that environmentalists were engaging in 'adversarial, alienating rhetoric' and 'inflammatory, purposely misinformed invective'. Sadly the same can be said about him, too.[126]

INFLAMMATORY RHETORIC SPREADS

By 1994, Moore too was beginning to talk of 'eco-extremism'. An article that appeared in *The Vancouver Sun*, written by Moore, had uncanny similarities to Ron Arnold's rhetoric. In Arnold's book on James Watt in 1982, he wrote that environmentalism stood for a new religion, which was anti-humanity, anti-civilisation, and anti-technology.[127] In his article, Moore wrote about Greenpeace and then went on to warn that eco-extremists are anti-human, anti-technology, anti-science, anti-organisation, anti-trade, anti-free enterprise, anti-democratic and anti-civilisation. Arnold has long said that environmental organisations are riddled with communists who want to destroy industrial civilisation. Moore repeated that one too.[128] He also reiterated the same message to the Canadian Lumbermen's Association and in the *Timber Trades Journal*.[129]

Jim Fulton, one of the MPs who had commissioned the parliamentary investigation into the Share groups, and the newly appointed Executive Director of the David Suzuki Foundation, was quick to criticise Moore. Calling the piece 'a nasty piece of propaganda', Fulton wrote that it 'reveals Mr Moore as the real zero-tolerance extremist' who uses a 'classic form of innuendo and guilt by association'. Fulton continues: 'I am not a member of Greenpeace, but I will not sit idly by while such unsubstantiated attacks are made on anyone.'[130]

As Moore's anti-environmental writings became more fervent, the Alliance was strengthening its ties with other anti-greens, welcoming Mike Morton, the executive Director of Share B.C., and Linda McMullan, the Chair of Canadian Women in Timber, to the Board.[131] However, anti-green inflammatory language was not only coming from the Forest Alliance. The Save Our Jobs Committee, formed in 1994 to fight the 'worldwide misinformation campaign' being sponsored by 'international preservationist groups' helped organise a meeting of forestry workers where Harvey Arcand, the IWA Vice President, called 'zealous' environmentalists tyrants and likened them to Hitler and Idi Armin. Arcand was also telling loggers to 'get in the face of these lying sleazebags . . . stay in their goddam face. Give 'em shit, keep 'em honest all the time and don't ever give up.'[132] The 'comments such as those made by Harvey Arcand are heard around the dinner tables of many forest workers, myself included', wrote one logger to the newspapers, after reading Arcand's remarks.[133]

In true paradigm shift fashion, Dennis Fitzgerald from MacMillan Bloedel was calling himself and his company 'true environmentalists' whilst lambasting environmentalists as 'preservationists'.[134] Brian Kieran, writing in the *Vancouver Province*, was picking up on Wise Use language, describing Friends of Clayoquot Sound as 'zealots' 'driven by their loathing for the works and mistakes of man'. European environmentalists, meanwhile were 'Greenpeace-nurtured eco-loonies' who 'worship ancient trees'. Kieran's suggestion to the debate was 'I if were Premier Mike Harcourt, I'd pave the Clayoquot.'[135] If he advocates policies like these, he never will be.

GOVERNMENT INVOLVEMENT INTENSIFIES

In early 1994, as Premier Harcourt went to Europe to defend BC's forest practices he was dogged by Greenpeace and other environmental organisations. He was also preceded by Moore, who was sent to Europe to counter Colleen McCrory who had gone to Brussels ahead of Harcourt to tell the 'Brazil of the North' story.[136] Back home, the Environment Minister accused Greenpeace of 'an intolerable level of hypocrisy'.[137]

It was not the first time the government had intervened in the forestry debate, either. It has a history of back-room deals to prop up the industry's PR efforts, especially abroad. Back in 1991, Thames TV, one of Britain's independent television companies ran a series of five-minute public information bulletins on the timber industry, which included one on BC. The bulletin remarked that 1,000-year-old trees were being felled faster than in Brazil. Thames TV were warned by the Canadian High Commission 'to be careful about the use of any further information on Canadian forestry obtained from your current sources' which included Greenpeace and the Women's Environmental Network.[138]

In May of that year, the Federal Ministry of Industry, Science and Technology gave C$15.6 million and the provincial BC government C$14.8 million to a PR initiative to expand the forest markets abroad. Moreover, in 1993, on the same day that the 'Clayoquot Compromise' was announced, the provincial and federal governments donated another C$6 million to a PR campaign purely aimed at the European market. As the blockades at Clayoquot that summer became politically embarrassing, the Canadian Pulp and Paper Association (another B-M client) was promised another C$4.5 million in federal money to set up an office in Brussels to protect foreign markets and counter environmentalists' claims.[139]

Journalist Joyce Nelson pondered the rationale behind the government's stance, calling it 'illogical and financially and environmentally bankrupt'. Nelson pointed out that the provincial government was losing about C$1 billion a year from the sector, with the industry raking in some C$3 billion in federal and provincial subsidies.[140] Despite this, the government decided once again to bail out the forest industry's PR campaign, and started to fund the Forest Alliance.[141] Vicky Husband, the Chair of the Sierra Club, called the decision outrageous and unacceptable.[142]

It employed a new Senior Advisor, Eric Denhoff, to fight the 'misinformation campaign against B.C.'s forest practices'. Within days Denhoff was in Europe with the Alliance placing adverts in British and German newspapers 'to counter Greenpeace misinformation' and 'to educate Europeans about B.C. forest practices'.[143] The advert in the German newspaper was called 'deliberately misleading' by the Sierra Club.[144] As Greenpeace and Friends of Clayoquot Sound toured European cities with 'Stumpy', a 400-year BC tree stump, to highlight the practice of clear-cutting, the Alliance launched its 'Dog the log' tour with Jack Munro in tow, harassing the organisations whenever the stump went. Munro and six other officials even turned up to slide shows to heckle the speakers.[145] Industry was left using tactics originally associated with environmentalists.

CUSTOMERS START TO CANCEL CONTRACTS

Tensions rose even further between Greenpeace, MacMillan Bloedel, the Forestry Alliance and the Canadian government when two major BC customers, Kimberly Clark, the makers of Kleenex, and Scott Paper cancelled their respective C$2.7 million and C$5 million contracts with the timber company in response to Greenpeace's protests. An outraged MacMillan Bloedel took out a full-page advert in *The Vancouver Sun* and *Globe and Mail*, accusing Greenpeace of threatening its customers and using incorrect information, 'strong arm tactics', 'intimidation' and 'unreasonable attack'.[146] 'Its demands are not based on sound science or practical experience,' claimed the company.[147]

Jack Munro warned that the 'potential loss of a couple of contracts has sure as hell shaken or awakened a lot of people that this is a hell of a lot more serious than we thought'. Meanwhile, the government accused Greenpeace of 'blackmail' and said it should be held responsible for job losses as a result of its actions.[148] Potentially at stake were the C$2 billion of BC forest products exported to Europe annually, half of which goes to Britain. But other European markets were important too: in July, German telephone directory publishers announced they too would not be buying any more pulp and paper from MacMillan Bloedel.[149]

The Alliance response to the cancellations was to spend more money on national and international public relations. At home, adverts commissioned by the Alliance, refuted to cost C$200,000, were shown on BCTV.[150] Internationally, a further advert was placed in the British *Times* newspaper. When the £37,000 advert did not appear on time, Moore commented that

'I see this as further evidence that blackmail is at the centre of its campaign.' However, a spokesperson for News International, the owners of *The Times*, denied that Greenpeace was pressurising the newspaper.[151]

THE BATTLE CONTINUES AT HOME AND ABROAD

Meanwhile, as the stories of boycotts still ricocheted around the forest industry, the loggers and Share were fighting their own battle closer to home, protesting against the government's Commission on Land Use and the Environment, or CORE, proposals. These included increasing the protected area of Vancouver Island from 10.3 per cent to 13 per cent, but only a mere 6 per cent of forests, something loggers were fundamentally opposed to. The backlash against CORE culminated when some 15,000 to 20,000 loggers, fuelled by a fear of losing their jobs, descended on the legislature lawn in Victoria on 21 March 1994 for a mass rally. Yellow ribbons, the sign of logging industry solidarity, were flying in force.[152]

The technique of industry scaring its workforce over heightened job losses and then mobilising them to become its grassroots soldiers fighting the industry cause is another technique learnt in the American timber wars.[153] However, the tactic worked. Due to forest industry opposition Premier Mike Harcourt's plans were watered down; whilst maintaining 13 per cent as parkland, the percentage off limits to logging fell.[154] Vicky Husband of the Sierra Club accused the forest industry of using 'fear and coercion' against CORE. Husband also reiterated that it was mechanisation and overcutting that were leading to job losses, not the CORE proposals.[155]

Meanwhile the Alliance were still busy trying to burst Greenpeace's credibility. It launched the *Top 10 Greenpeace Lies*, which was subsequently expanded into a booklet *What Greenpeace Isn't Telling Europeans: A Response to Greenpeace's European Campaign of Misinformation About British Columbia's Forests*, which reported fifty-one pieces of 'misinformation' from Greenpeace. Tamara Stark from Greenpeace dismisses the book, saying that 'they put in quotes, most of which are not actually from Greenpeace'.[156] Stephen Hume from *The Vancouver Sun* said of the booklet: 'many of these "facts" are really factoids – that is, carefully worded nuggets removed from their context and presented as unbiased truth. As with all PR, green or anti-green, the intent is persuasion, not education; image, not actuality.'[157]

By September 1994, the BC Forest Minister, Andrew Petter, was doing his own European tour, offering the province's new forest code as a model of respectability and sustainability. It was meant to keep Europe buying BC forest products.[158] But Colleen McCrory was back in Europe in March 1995 along with the Western Canadian Wilderness Committee (WCWC) at the climate negotiations in Berlin, once again calling Canada the 'Brazil of the North'. 'Canadian governments are repeating the problem in the fragile boreal forests of the north – allowing levels of harvesting that cannot possibly be sustainable,' said McCrory. The worst offender in the country was the BC forest industry, she alleged. 'The BC forest industry is so unsustainable that it's reaching out to consume forests in Alaska, the Yukon, Alberta, Manitoba and Saskatchewan, because there's not enough wood to feed the mills in BC,' added McCrory.[159]

But it was not only customers in Europe that the Canadian government and the Forest Alliance were worried about. In April 1995, environmentalists targeted Hollywood asking them not to film in BC until clear-cutting was stopped, forcing the Alliance on another damage limitation tour. Delegates included Patrick Moore and Mike Morton, the latter claiming that Clayoquot Sound was a model of progressive forest practices.[160] Internationally acclaimed film-director, Oliver Stone sided with the environmentalists, though. 'Clear-cutting must stop and the rainforests must be preserved,' said Stone.[161] Whilst in California, the Forest Alliance tried to portray itself as a 'Citizens' Environmental Coalition' rather than a 'corporate front-group'. Patrick Moore, in debate with Randy Hayes from the Rainforest Action Network, made much of their 7,000 members who paid C$10 a year annual subscription, but reluctantly had to concede that the remainder, some C$1,930,000, of their now C$2,000,000 annual budget, came from the forest industry.[162] The Alliance now has some 170 corporate sponsors.[163]

The Alliance was also running new ads against Greenpeace in which its founder and now arch-foe, Patrick Moore, appeared. The soft-spoken Moore has recently become the main spokesperson used by the Alliance in meetings and in adverts, replacing the more rough and ready, but less polished, Jack Munro. In the 30-second advert Moore said:

> I think the hardest choice for environmentalists today is to decide when to talk and when to fight. If people would just inform themselves about the facts of forestry and the art and science of growing trees they would learn that in many cases clear-cutting is the best way to go about regenerating a new forest.
>
> There is simply no doubt of that, and silviculturists worldwide agree. Even the chief conservation biologist with the World Wide Fund for Nature in Europe, which is the

leading mainstream environmental group, agrees that clear-cutting is necessary in some cases and is the best method in some cases . . . Clear-cut free is not a silviculture programme. Clear-cut free is an advertising slogan that is being used for fund-raising purposes in Europe and it has no basis whatsoever in science.

. . . I believe they have thrown truth out of the window on this one. They simply seem unwilling to listen to the science on their whole forestry campaign.[164]

Unfortunately for Dr Moore there were a couple of problems with the advert. WWF were 'very disappointed to see that WWF's name is being used in support of BC forest policy'. Moore was also charged with 'grossly misrepresenting' WWF's position, something WWF 'deplored'. In addition, according to WWF, this was not the first time the Forest Alliance had 'selectively quoted, distorted and misrepresented statements by representatives from WWF'. Moreover, the 'chief conservation biologist for WWF in Europe' does not even exist.[165]

CLEAR-CUTTING ISN'T SOUND

Worse was to come for the industry, though. The potential death-knell of clear-cutting was sounded in May 1995, when the government's Clayoquot Sound Scientific Panel recommended an end to the practice. The panel, made up of 'blue-ribbon' scientists and 'First Nations' representatives rejected clear-cutting in favour of a more selective way of logging called 'variable retention silvicultural system'. The experts recommend that the size of cut areas should be reduced and that between 15 and 70 per cent of forest should be retained in cut blocks, which would secure the existing structure and function of the original old growth eco-systems. This essentially means that, based on science, no existing clear-cutting practices should be applied. Under the new system the size of an area that could be cleared would be no greater than about 200 metres in width. This would be a fraction of the current size of clear-cuts which now average about 40 hectares.[166]

The report was a victory for the anti-clear-cut campaign, turning traditional forest practices upside down, despite the millions of dollars paid by the industry to fund forest industry front groups, and to fly people the world over to fudge the real story, and who proclaimed they had science on their side. In the light of the panel's recommendations, the *New York Times* announced it was reviewing whether it would continue buying newsprint from MacMillan Bloedel. This was a slight change of direction by the *Times*, who four months earlier had met with Dow Jones and Co. and Knight-Ridder at the Canadian Pulp and Paper Association's annual meeting to blunt environmental protests against forestry practices in the USA and Canada. Six months later the *Times* announced it was not renewing its contract with Macmillan Bloedel.[167]

In July, the BC government announced that it had accepted the scientific panel's recommendations, which signalled the end for conventional clear-cutting in Clayoquot Sound. 'The government recognises the unique values present in Clayoquot Sound and is committed to implementing the world's best forest practices there,' said Forests Minister, Andrew Petter, announcing the decision.[168] So the so-called traitors, guilty of so-called treason and eco-terrorism, who were supposedly guilty of manipulating the truth, who had been treated to intimidation and violence, who had been tried in the law courts, had finally been vindicated.

Meanwhile in May 1995, WWF accused the Canadian Standards Association of pressing the International Standards Organisation (ISO) to take short cuts on the labelling of international forest products.[169] In June, proposals to develop a joint Canadian-Australian Environmental Management System for 'sustainable forest management' for the ISO, were dropped after pressure by WWF and other NGOs, who feared that no minimum environmental standards for forestry would be established.[170] By November, though, environmentalists were accusing the government and timber companies of breaking promises made when the scientific panel's report was published over Clayoquot Sound and called for a moratorium on logging in the region. The government rejected the claims.[171]

So the clear-cut campaign is not over. Only 10 per cent of Clayoquot Sound has been clear-cut, and the debate of whether to clear-cut or selectively cut continues unabated. The forest companies maintain that if their profit margin is threatened they will move south to Brazil, where the fight for the forest rages unabated.[172] The inflammatory and polarising language looks set to continue. As in America, where you have this type of incendiary rhetoric, violence follows. The resource debate in British Columbia has already been marred by violence, with environmentalists being attacked and intimidated. However, the violence in Central and Latin America is much worse.

8

THE FIGHT FOR THE FORESTS OF CENTRAL AND LATIN AMERICA

I don't want flowers at my funeral because I know that they would be taken from the forest.

Chico Mendes[1]

THE RUBBER TAPPERS

In the Amazon Basin the conflict between development and destruction, on the one hand, versus conservation and sustainable utilisation on the other, is at its most intense. This is nowhere more apparent than in Brazil, where the struggles for land and power are indivisibly interwoven. The fight for agrarian reform, and to protect the Amazon, has taken a heavy toll on activists. It is where the battle is bloodiest.

Brazil is a land of inequality, between the rich élite and the desolate poor. Some 45 per cent of land is owned by just 1 per cent of the richest landowners. The poorest 53 per cent own just 3 per cent of the land. Annually during the 1980s some 40,000,000 acres were taken by large landowners, whilst 30,000,000 rural people had no land at all. By the late 1980s there were thought to be some 8,000,000 landless migrant families. Some 93,000,000 cattle are owned by less than 2,000 ranchers. In 1989, the wealthiest 20 per cent of households earned 63 per cent of income, the bottom 20 per cent, just 2 per cent.[2]

Exacerbating this inequality, during the 1970s and 1980s there was a political push to open up the Amazon for development, induced by some $3 billion of subsidies and tax-breaks for the élite to buy up land and 'develop' it.

Plate 8.1 A tapper collecting rubber in the Amazon.
Source: E. P. L./Alois Indrich

The military junta wanted to 'flood the Amazon with civilisation'.[3] Leading the call for the conservation of the rain forest during the 1980s were the rubber tappers (*seringueiros*), workers who for centuries had walked trails in the forest collecting latex from rubber trees. Many tappers who traditionally were illiterate and innumerate spent their lives in debt bondage (*aviamento*), never being able to pay off the debt that they inherited to the rubber barons (*seringalistas*). *Aviamento* was their real concern rather than one of *desmatamento*, or deforestation of the Amazon. In time, the tappers would be forced to find a new vocabulary and cause, one of *ecologica*, in order to receive outside assistance and international attention.

Inevitably there was increasing conflict between rubber tappers and these new landowners, whose motive was profit maximisation, not concern for the people who already lived on their recently acquired land. The most effective method used by the landowners to reinforce a land-claim was to set fire to the forest. Moreover, many new owners were ranchers who would burn the forest to make way for cattle, which would also irreparably destroy the tappers' way of life.[4] The *seringueiros* were left with two options: to fight for survival and for their right to a sustainable future or to resign themselves to becoming slum dwellers, consigned to life on the edge of humanity.

There are similarities between the environmental movement in the North and the rubber tappers in the South, but one has to be careful when drawing comparisons. Both were trying to save forests from extinction and destruction. Many northern environmentalists were working to save the Amazon without the knowledge of the tappers. Another similarity was that certain sections of both groups used direct action to achieve their goals. Rubber tappers staged *empates*, whereby chain-saw gangs were confronted and asked to leave the land that they were clearing. Often the tappers would dismantle the makeshift shacks belonging to the chain-saw gangs, forcing them to move away.

But there were significant differences between the two groups. Northern environmentalists were primarily concerned about the fate of the forest and were slow to realise that people lived in the Amazon who were also an integral part of the eco-system. Around 86 per cent of the protected areas in the region are inhabited.[5] Traditionally it had been up to groups such as Survival International, Cultural Survival and many individual anthropologists and film-makers to champion the cause of the forest dweller in the northern hemisphere. Thankfully, some environmentalists have started adopting a more realistic, holistic approach. Rubber tappers, on the other hand, were initially interested in land reform and social justice, and only later recognised the importance of the forest ecology.

This said, some of the policies that northern environmentalists have advocated have been totally ignorant of forest dwellers and have done more harm than good. There are, for example, still problems where northern environmentalists have advocated the setting up of national parks where the traditional forest dwellers have been excluded and expelled from their land. Whilst causing untold personal harm by displacing the forest dwellers, the policies have also actually increased deforestation as ranchers and industry have invaded areas previously protected by the evicted inhabitants.

Protecting the Amazon and the people who live in it is another case of where environmentalists have to find the right solution to the balance of biodiversity preservation with sustainable utilisation of resources. Traditional forest dwellers and indigenous peoples cannot be excluded in the name of conservation or environmental protection of the region's biodiversity, because they are part of that biodiversity. In fact, there are many ways in which the forest can be sustainably utilised by harvesting fruits, nuts and rubber.

CHICO MENDES

Land conflicts in Latin America have almost always been accompanied by a high level of violence. On the night of 22 December 1988, Francisco Alves Mendes Fiho, known to everyone as Chico Mendes, a trade union leader who had recently found a new vocabulary in *ecologica*, was assassinated by two gunmen, as he walked out of his back door to take a wash. Mendes knew he was going to die. He had become too powerful, too successful and had taken on a local violent landowner and had won. 'I'm not going to make it to Christmas,' he told a friend, days before he was shot.[6] He even knew who would order his killing.

The Brazilians have a particularly perverse system of assassination. Death threats come with an *anúncio*, described as a 'matter of fact' which 'is meant to prolong the victim's torment as he waits for the inevitable'.[7] For Chico Mendes, the inevitable had been a long time in coming. Over the previous decade he had been forced into hiding to avoid *pistoleiros*, to evade the police who called him a communist and who were spying on the tappers. Mendes had survived at least three attempts on his life, numerous death threats and several *anúncios*.[8]

Mendes was the fifth rural union president to be murdered that year and just a week after his death another union leader was blasted in the face by a shotgun. In all, some forty-eight activists and workers were killed in 1988. In the 1980s over 1,000 people were murdered in rural Brazil in land disputes. Fewer than ten killers were convicted and none of the masterminds, or *mandantes*, have ever been tried.[9] The 1980s were a bloody decade: ten years of violence against the unions and rubber tappers had started in July 1980, when Wilson de Souza Pinheiro, a union activist and rubber tapper was shot dead by hired *pistoleiros*. Pinheiro's crime had been to organise a demonstration which had stopped the chain-saw gangs cutting down the forest. It was the first real assassination of the land wars in the 1980s and it set a gruesome pattern for the decade.

Chico Mendes was so threatening to the authorities and ranchers because of his uncanny ability, like Earth First! activist Judi Bari and Ogoni leader Ken Saro-Wiwa, to organise people to participate in protests. Furthermore, he could force one region of rubber tappers to support another, something that was unheard of before Mendes and Pinheiro. Mendes was also instrumental in organising a National Conference of Rubber Tappers and in forming a national union. Moreover, the rubber tappers, under Mendes' leadership, joined forces with their traditional enemies and formed the Alliance of the Peoples of the Forest.

Mendes and an anthropologist, Mary Helena Allegretti, had been the driving force behind educating the tappers and empowering them to break out of the vicious circle of debt bondage and set up cooperatives to look after their own affairs. Mendes was also brilliant at initiating *empates* to stop the clearing and burning of the forest. Until his death, Mendes preached non-violence, and although *empates* were confrontational and often ended with houses being destroyed, they were rarely ever violent against people, despite the aggression being used against the *seringueiros*. Moreover, *empates* worked as a tactic: before Chico Mendes' murder, some 3 million acres of the Amazon had been saved.[10]

But success comes with a high price, and, as the *empates* became more successful, violence and intimidation increased against the tappers, both from the authorities and from ranchers. The police routinely beat the demonstrators. Ranchers increasingly hired *pistoleiros*, or gunmen, to kill whoever was in the way. There have always been plenty of *pistoleiros* in the Amazon. Tappers were frequently rounded up at gunpoint, intimidated, beaten, or shot. The first President of the National Council of Rubber Tappers had his house set on fire and he survived several death threats. The authorities did little, if anything, to stop the violence, even after they recognised the organised nature of it. For example, 40 per cent of all murders carried out in Brazil in 1986 were by *pistoleiros*.[11]

Another reason for the increased burning of the Amazon and the heightened level of violence sweeping the region were the agrarian reforms. As these were pushed through in the mid-1980s, the large landowners felt increasingly threatened. In 1985, with the dual threat of agrarian reform and the success of the rubber tappers movement, the ranchers responded and formed their own union, the *União Democrática Ruralista* (Rural Democratic Union) or the UDR. Its intended function was to promote 'Tradition, Family, and Property'. 'Beneath the surface, there were persistent signs that the UDR was a front for organised violence against the leaders of the agrarian reform movement, and it soon became the bane of the rubber tappers, small farmers and landless peasants throughout rural Brazil,' wrote Andrew Revkin in his book *The Burning Season*.[12]

In some ways the UDR is similar to a Wise Use group. It is an anti-environmental organisation. Its name suggests that it is something it is not – democratic – and its backers are wealthy ranching interests. Both profess to have the backing of the family and tradition. The UDR's inauguration event was even said to raise money for public relations and lobbying. In Brazil that means that 1,636 firearms were distributed to all members.[13]

It was well known that members of the UDR were actively plotting Mendes' death and that Mendes, along with other leaders, appeared on a UDR assassination death list.[14] One UDR representative, who bought up a local newspaper, warned of the threat of the 'internationalisation' of the Amazon and the threat to national security, that this brought with it. This is similar in vein to the right-wing/Wise Use who warn that national parks are the first step to a totalitarian regime, and international control under the UN. The UDR also advocated destroying 95 per cent of the forest and replacing it with nuclear reactors.[15]

George Monbiot, Visiting Fellow at Green College, Oxford, also became aware of the language used to vilify foreigners on his travels to the Amazon. 'Land vested interests see every threat to them as subversion, communism and internationalisation', recalls Monbiot. 'This is the great word you hear.' Moreover, he adds the UDR would put out the message that:

> the only reason that anyone in Brazil would be interested in the Amazon, the environment or the Indians is because they had been subverted by foreign agents. The reason for this is that foreigners wanted the great wealth of the Amazon for themselves and they wanted to steal it from Brazil.[16]

Mendes' biggest triumphs came both home and abroad where he was championing his cause under the umbrella of 'environmentalism'. He had enjoyed a rapid rise to both national and international fame and Chico Mendes had become the human face of the fight to save the rain forest. With Mendes' involvement with the rubber tappers, the authorities could no longer dismiss efforts to save the rain forest as foreigners interfering in Brazilian affairs. But they would still call his companion, Mary Helena Allegretti, an American agent and label Mendes as anti-Brazilian, anti-progress and a CIA puppet.[17] Mendes was worrying for the authorities, because wherever he went abroad, be it Washington or London, he generated publicity for his cause. Mendes also won the prestigious 1987 Global 500 Award from the UN. This antagonised the authorities even more.

Mendes' success at home was not just stopping destruction of the forest with *empates* but banning it altogether with the initiation of 'extractive reserves'. These were areas where only sustainable traditional practices such as rubber tapping and forest harvesting of nuts, fruits, cocoa and oils would be allowed. Extractive reserves offered an alternative form of development for the region: one of preservation rather than desolation.[18] They meant sustainable development for the tapper and the Amazon, not systematic destruction of it by the ranchers. Extractive reserves were areas of sanctuary that would guarantee survival of the rubber tappers and of the forest on which they depended. They would also ensure Chico Mendes' death.[19]

In the summer of 1988, there was heightened tension between tappers and a local villain and rancher, Darlí Alves da Silva, who had a string of killings behind him. Da Silva had bought some land at Cachoeira, a rubber estate where Mendes had been brought up. By a mixture of threats and bribes da Silva was attempting to force the current rubber tappers off the land. In retaliation, by using *empates*, Mendes and the tappers were preventing da Silva from cutting down the forest. Mendes saw the attempt to burn down Cachoeira as a direct threat against him personally and the rubber tappers' movement as a whole.

In order to defuse the situation, the authorities bought out da Silva and declared Cachoeira an extractive reserve, which meant that Mendes and the tappers had won. It also signalled the beginning of the end for Chico. Furthermore, he asked the authorities to arrest da Silva for a murder the rancher had been charged with, but never arrested for, in a different state. When the authorities refused to act, his death seemed inevitable.[20]

Although Darlí Alves da Silva ordered the killing of Chico Mendes, the authorities, through their inactivity to protect him, must take some of the blame for his murder. 'I only want my assassination to serve to put an end to the immunity of the gunmen,' Mendes wrote just before his death.[21] The international outrage that Mendes' killing provoked forced the Brazilian authorities to find the killers. Darlí Alves da Silva and his son Darci were arrested, Darlí for ordering the killing, his son for carrying out his father's orders. Darci went so far as to actually confess to the murder. Even though these two were arrested, there is an enormous amount of evidence to suggest that many more people knew about and planned Mendes' death. But once again the authorities have refused to act.

The legal process surrounding the trial dragged on for two years, but finally in December 1990, the two were each sentenced to nineteen years in prison. However, an appeal court in 1992 overturned the father's conviction, although the Mendes family appealed.[22] The killers subsequently escaped from prison and remain uncaptured. In July 1995, the pair were reported to be living in a Bolivian town close to the Brazilian border. Brazilian police were reported as having asked Interpol to assist in the search for the missing killers.[23]

CONTINUING CARNAGE

But the killing did not end with Chico Mendes' death and the bloody battle for land reform in Brazil continues. Both bodyguards who had been with Mendes on the night he was murdered were also shot dead. In 1989 and 1990, Amnesty International expressed concern for Chico Mendes' relatives and rubber tapper leaders after relatives of the killers arrived in the region for the trial. Amnesty's concerns were borne out when Mendes' widow was assaulted and his brother threatened.

Violence against witnesses continues, as it does to rubber tappers generally. Júlio Barbosa, Mendes' successor as President of the Xapuri Rural Workers' Union, has been repeatedly intimidated. So too has Osmarino Amâncio Rodrigues, who also succeeded Mendes as a tapper leader. One assassination attempt against him was foiled. He has been denied police protection, despite the fact that the death threats against him and three other rubber tapper leaders continued through 1989 and 1990.[24]

Official indifference to the murders persists and the deaths of rural leaders continue unabated. In 1990, four agrarian reform activists were murdered.[25] Another activist, Manoel Pereira da Silva began receiving death threats after he started highlighting illegal logging in the Figueira Extractive Reserve.[26] The following year, Jose Helio da Silva, and Expedito Ribeiro de Souza were both assassinated. Other leaders received death threats and were fired at.[27] In the three-year period after Mendes' death, some 185 more people were murdered in land disputes.[28] Amongst the continuing carnage there has been one concession. In 1990, some 6,553 square miles of forest in the regions of Acre, Rondônia and Amapá were declared extractive reserves. They were named after Chico Mendes.[29]

On 13 March 1991, activists announced 'A Day Against Violence', which culminated in a rally where over 1,000 people protested at the ongoing bloodshed.[30] In December 1992, the President of the Rural Workers Trade Union of Imperatriz, Valdinar Pereira Barros, who had supported the setting up of the Mata Grande and Siriaco ecological reserves was wounded in an assassination attempt but survived. He had been receiving death threats for some time. Following the attack, the government's own environmental agency, The Brazilian Institute for the Environment and for Renewable Resources (IBAMA), expressed concern over the level of violence against 'those who seek to live integrated with their environment'.[31] Three years later, a gunman was arrested who claimed that his next victim was to have been Father Ricardo

Rezende, one of the organisers of the 1991 rally and a land reform activist. Father Rezende's name was on a list of about forty people who have been threatened with death because of their support of the rights of peasant farmers and for agrarian reform. Six of those people have already been killed in Xinguara, and two others have suffered assassination attempts.[32] The repression continues. Macedo, a leader of rubber tappers and Indians of Acre was arrested in 1995, part of the ongoing silencing of activists.[33]

Given the violent nature of the country it is hardly surprising that other environmentalists working in Brazil have been murdered. Indian environmental activists have also been threatened and killed: Davi Kopenawa, a Yanamani Indian who was awarded the UN's 1988 Global 500 Prize for defending the environment, has also received death threats.[34] The same year, fourteen Indians were killed and twenty injured whilst discussing ways of protecting their forest from loggers. They were murdered by *pistoleiros* hired by timber barons and seven years later still no-one has been charged with their murder.[35] In 1993, forty-two Indians were killed according to the Brazilian Indian Missionary Council.[36]

Furthermore, in the space of a few days in the spring of 1993, two prominent environmentalists were assassinated. Paulo Cesar Vinha was found brutally murdered on the Barra do Jucu sandbanks, which he had been working to protect from excavation, as well as highlighting the activities of Aracruz Cellulose, one of the largest global pulp producers. He had assisted the Tupiniquim and Guarani people to recover their land from the paper company. He was killed, however, because of his involvement to protect the sandbanks from illegal excavation.[37]

Brazil's environmental community were shocked three days later when another activist, Arnaldo Delcidio Ferreira, the leader of the Rural Workers Union of Eldorado Delcidio de Carajas, was found shot dead. The father of nine had worked with Greenpeace highlighting the deforestation of the Amazon and the violence being directed at those who were attempting to stop it. Six months before his death he had been with rubber tappers, activists and environmentalists protesting against predatory logging by the Maginco Company on the Rio Maria.[38]

Ferreira had much in common with Chico Mendes. Apart from also being a member of the National Union of Rubber Tappers, he promoted environmental conservation of the Amazon along with sustainable social and economic development. He had also fought for the end of rural violence, the end of deforestation and for agrarian reform in Brazil, policies that had cost Mendes his life, too. Ferreira had already survived three attempts on his life, and received countless death threats. One attempt in 1985 had killed a nun he happened to be talking to.[39] It is believed that Ferreira was killed by gunmen hired by ranchers.[40] Other people have also been killed or harassed for speaking out in other parts of the Amazon.

A CRUDE BACKLASH[41]

Oil company devastation of the Ecuadorian Amazon would never have reached the attention of the international community had it not been for the perseverance and courage of many people who have faced incredible personal danger. The American lawyer, Judy Kimerling, was so shocked by what she found whilst travelling in the Amazon in 1989 that, since then, she has fought for the peoples of the region and their environment. Being a visitor to a foreign country and taking on a government which receives half of its revenue from oil, Kimerling's work has not been without considerable risk. Exposing the environmental destruction caused by some of the most powerful multinationals whose life blood is oil has only compounded that risk.

'I went to Ecuador in February 1989, and I went there without really knowing at all that I would be working on issues related to petroleum development,' recalls Kimerling, six years after her first acquaintances with the Amazon. 'Frankly I thought that I was going to be looking at birds and butterflies, and I was really surprised and appalled to find toxic waste pits in the rain forest.' What Kimerling found was a region riddled with thousands of unregulated and unlined waste pits, spilling and seeping their toxic waste into the ground and water. The pipelines were leaky. Gases were being flared regardless of people living near by. Horrified by what she saw, Kimerling began to investigate the oil operations that were clearly having a devastating effect on the local people and environment. Practices like these would be illegal back home, she thought.

'I discovered this in June or July 1989, and that was my first trip out to the field and I went with the local indigenous organisation, and did a fifteen-day trip and visited a number of sites,' continues Kimerling. 'The second trip I took out to the Amazon was in August of that year and I was arrested during that trip.' Although she was never formally charged, Kimerling was arrested and detained for twenty-four hours. Initially imprisoned by the military she was then handed over to the police. Most unnerving to Kimerling was that the authorities seemed to know who she was.

One of the people arrested with Kimerling had accompanied her on her previous trip into the Amazon. He was heavily interrogated by the police as to why he was travelling with Kimerling and told that all foreigners were terrorists who were solely intent on giving guns to the Indians. Kimerling, meanwhile, was being labelled a subversive. 'My crime was talking about contamination and pollution by Texaco and CEPE, Petroecuador's predecessor, that was the national oil company at the time,' she recalls. 'I think they wanted to scare me, I don't think they intended to hurt me.'

Kimerling was seen as threatening because she had visited the Amazon to see what was going on. 'That was a big step and a fairly radical method of work because before then the literature that was being published by environmental groups basically said that there was no harm from the oil industry,' Kimerling says. Environmental groups were maintaining that the only problem related to oil development in tropical rain forests was due to road-building, which opens up previously inaccessible areas for colonisation, exploitation and deforestation. In Ecuador, as elsewhere, the settlement patterns of colonists have followed the oil roads – 1 million hectares of rain forest have now been colonised in the Oriente. Moreover, with the new colonists have come serious epidemics of new diseases for indigenous peoples.[42] Once alien diseases, such as flu, chronic coughs, skin diseases, ulcers, malaria, pneumonia and gastrointestinal diseases are now widespread.[43]

Kimerling recalls:

> But there was absolutely no talk about contamination, and in fact the literature published by the environmental groups was misleading, because it took the industry line and repeated it without questioning it . . . I found it hard to believe that there was no problem in oil development in itself.

Why, thought Kimerling, if roads were the only problem associated with oil development, were people protesting about oil and gas activity in other parts of the world, where colonisation of an area was not a problem?

But there was a problem. By venturing into the Amazon, Kimerling uncovered a rather oily and unpleasant can of worms hidden from the unsuspecting world by the remoteness of the region. Kimerling's research, published in the book *Amazon Crude* has been called *The Silent Spring* of ecological awareness of the Amazon. The international attention that followed the publication of the book was to bear down on the Ecuadorian government and oil companies such as Texaco.

TEXACO LEADS THE WAY

Texaco, the company that pioneered oil development in the region, first acquired an interest in the Oriente in March 1964. In 1967, the company had found oil near Lago Agrio in sufficient quantities to build a 1,000-barrel-a-day refinery, prefabricated in the USA. But Texaco needed to get its oil to market, so it built the 500-km Trans-Ecuadorian pipeline in the summer of 1972, stretching from the Oriente into Peru.[44] It signed a twenty-year production contract with the government which expired in 1992, when Texaco handed over operations to Petroecuador, which now controls 100 per cent of production.[45]

Other companies such as Du Pont's Conoco, British Gas, Occidental, ARCO, Petro-Canada, and Maxus have all subsequently been involved in exploration and/or production in the country. Although Texaco's operations in the country have now ceased, its legacy continues. 'Texaco has been the dominant force in oil development in the region and has set precedents and standards for oil operations,' wrote Kimerling in *Amazon Crude*. 'Texaco's actions have thus contributed directly and indirectly to the current widespread contamination and deforestation in the Oriente'.[46]

Ecuador's 13 million hectares of rain forest, known as the Oriente, is an internationally important area from the point of view of biodiversity, containing at least 10,000 species of plants. It is regarded as one of, if not *the* most, biologically diverse areas in the world, with high levels of endemism and endangered species. The region is also home to eight groups of Indians, an estimated 90,000 to 250,000 people and has the highest density of indigenous peoples in the Amazon.[47] This biological and culturally rich region has become a hive of oil activity, with some 90 per cent of the Oriente being consigned to oil activity, with 400 active wells.[48] Over 1,000,000 hectares of the Oriente are used for oil production activities covering over one tenth of the Ecuadorian Amazon.[49]

The waste from these operations was staggering. Kimerling calculated that some 5,000,000 gallons of untreated toxic waste were discharged every year from producing wells, with flowline spills accounting for up to a further 500,000 gallons of oil spilled annually. Nearly 20 billion gallons of toxic waste had been dumped since 1972, threatening soil, surface and groundwaters. Despite this, Kimerling found no clean-up equipment in the Oriente.[50] The Trans-Ecuadorian pipeline had also been an ecological disaster. Built in an earthquake zone, at least 16.8 million gallons have been spilt in over thirty reported spills, with one spill reaching the Atlantic, in Peru and Brazil.

Moreover, there is significant air pollution in the region. An estimated 53,000,000 cubic feet of gas is burned daily despite the fact that Ecuador imports gas for domestic use.[51]

The pollution has caused immeasurable health, environmental, cultural and social problems in the Amazon. Pollution, coupled with loss of land and infestation of new diseases by colonists, have directly or indirectly devastated populations of local Indians. In one area, the number of Cofan Indians had been reduced from 70,000 to 3,000 in twenty years. Decimated by oil exploration and other factors, they are on the edge of extinction. Another tribe, the Siona Secoya, is also struggling for survival, as a result of Texaco's operations. The Quichua Indians have also been adversely affected by oil activities, and oil companies have actually encouraged settlers into the Oriente, which exacerbates the problems being suffered by the Indians.[52]

THE BACKLASH BEGINS

When the first results of *Amazon Crude* became known, there was inevitably going to be backlash. 'At the time I was told by people against speaking out, that my name should not appear in the papers, that this was a very dangerous thing,' recalls Kimerling. Advice was coming from both friends and officials. 'The words from officials was "as friendly advice", but it certainly shook me up a little bit, because that was the first time that I was publicly identified with this work,' says Kimerling. After her initial reservations Kimerling felt 'that in order to be more protected I needed there to be some recognition of my work, so that if something did happen, there would hopefully be some public interest and some questions as to who was responsible'.

An official from Petroecuador's environmental unit said to her, 'If you go out to the Amazon at this time, the military will probably arrest you, because just by going there it would be seen as a provocation.' This kind of warning would set a pattern for the future. Kimerling also received a series of phone calls from people who had been at meetings with officials from the US oil companies, from the Ecuadorian national oil company, and from the military. Kimerling had been discussed and it was believed that she was going to be deported, ostensibly for not paying taxes.

The situation changed slightly after Kimerling met the US Ambassador in Ecuador to discuss her book. 'The Embassy let members of the oil industry know that if they tried to deport me that it would really be a big mistake.' The Embassy's stance, coupled with a heightened awareness in Ecuador of human rights abuses, made Kimerling feel safer, although she still felt concerned for her safety. She began making plans to be sure that when she was actually in the Amazon people knew she was coming and knew where she was, a practice she has followed until this day.

But the backlash did not just come from the authorities. It also came from other Ecuadorian groups. 'I also was given a very hard time by Fundación Natura, when I was arrested,' continues Kimerling:

> They were concerned about their image. Rather than saying sorry to me that I was treated in this way, I was actually asked what are you doing running around in the jungle with these Indians, and told that I was going to hurt their image.

There was also trouble from missionaries in the area. Rachel Saint pioneered the evangelisation of the Huaorani, living with them for forty years, until her death in 1994. Kimerling met her before she died. 'After the meeting she got on the radio, and told the Huaorani that I am a very dangerous woman and that they shouldn't have anything to do with me,' remembers Kimerling. Saint also remarked that if the Huaorani took Kimerling in the field that 'it is going to make the oil companies and the military very angry and they would go to jail'. Saint also told NBC News, a visiting American news channel, who were in Ecuador to make a film about the Huaorani, that Kimerling was a communist. Kimerling maintains that, according to a Huaorani who became evangelised, Saint had an 'open check-book from Texaco, to relocate the Huaorani out of the oil producing area'.[53]

TEXACO RESPONDS

The response from the oil industry to Kimerling's book was hardly favourable, as one would expect. Texaco, which came under an immense amount of corporate criticism as a result of *Amazon Crude*, responded in various ways. Whilst calling the report anecdotal, Texaco reacted to pictures of waste pits that had appeared in *Amazon Crude*, and which now began appearing in magazines all over the globe, prompting other researchers to flock to the Oriente. First, Texaco denied that the pits constituted a serious ecological problem, but second, it began bulldozing over the pits, making them far less photogenic for the photographers out to horrify their readers back home. But the company was not just making aesthetic improvements. Kimerling also recalls that roughly at the same time as *Amazon Crude* was published, Texaco also began to give a lot of money to the Missouri Botanical Gardens in New York, and the CEO at the time became a Director of the Gardens. The company also heavily sponsors Earth Day events and tree-planting initiatives in the area.

Also in response to *Amazon Crude* and mounting general criticism of its environmental record in the Amazon, which by this stage was coming from many indigenous and environmental organisations, Texaco carried out two audits of its operations. 'They say [the audits] prove that there is no irreversible impact,' says Kimerling. 'They say, well, our operations are not very pretty, but they are not dangerous, and are not a threat to the environment.' Despite this, Texaco told the American lawyer that one of the reasons it undertook the audit was its concern over its international image. Another company that has faced heavy criticism for its operations abroad is Shell in Nigeria. It too has funded an 'independent' survey to attempt to distract criticism away from its operations.

Many people were critical of Texaco's audit which was undertaken by the Canadian consulting firm HBT-Agra-Calgas, and estimated to cost $400,000.[54] 'Indications are that the Canadian Company HBT Agra, is severely limiting its inquiry, following threats from Texaco that they would cooperate only if narrow parameters were used for investigating their environmental record,' warned the Rainforest Action Network in 1993.[55] A Statement of International Delegates, who attended a 'Texaco' week in Ecuador in July 1993, declared that, 'We do not believe that the environmental audit that is currently under way by the Canadian consultancy HBT Agra, in conjunction with Texaco and Petroecuador, will be adequate to address our concerns, because it is neither independent nor transparent.' The delegates concluded that:

> None of the residents or organisations we spoke with have been informed, much less consulted with, by the auditors. In addition, we witnessed waste pits at wells in the audit area that have been covered with dirt (without first removing the wastes) or emptied into nearby streams. These are not acceptable environmental practices in countries we come from, and appear to be an effort to hide the contamination caused by Texaco, instead of investigating and cleaning it up.[56]

Furthermore, of the six people on the oversight committee who were in charge of the audit, four were from Petroecuador and its subsidiaries and two from

Texaco. Ecuadorian environmental and human rights groups were refused participation, despite numerous requests. Critical analysis was not undertaken either: river sediments were not included in the evaluation of water quality. The audit was only concerned with environmental issues. Social, cultural and economic impacts were not studied.[57] A coalition of Ecuadorian and international groups condemned the audit outright.[58]

OTHER COMPANIES RESPOND

Other oil companies distanced themselves from Texaco, portraying it as a bad company within the oil industry, whilst at the same time funding aggressive PR campaigns about their own operations. The underlying message was that these newer companies in the Amazon could extract oil from the rain forest without serious harm to the land or the people. This massive effort by the companies to portray themselves as green has had serious implications in the shift that has slowly taken place in Ecuador. As has happened elsewhere, there has been a slow marginalisation of the environmental movement.

Kimerling says:

> I do feel there is a backlash at this time, because initially when *Amazon Crude* first came out, after the initial denials by the government, there was a general feeling that the situation was 'Oh it's horrible and outrageous and we are going to do better and we can do better.'

Kimerling adds that:

> [the] post-Texaco generation of companies still had this 'trust us' attitude. They really made a major effort to convince people to trust them and it seems that they have convinced some people in Ecuador, especially in the government and the press, that they are in fact the good guys and they can be trusted. There is now more criticism of environmentalists.

Where there used to be respect now there is just ridicule. 'Let's laugh at the tree huggers,' can be heard ringing in the bars of Quito.

Petroecuador, the government-owned oil company, was beginning to respond to Kimerling's work too, calling it outrageous lies and reiterating that there was not a problem in the Amazon. In time, their attitude changed and they admitted that there was, after all, a problem. Petroecuador shifted the debate away from oil pollution to one of sustainable development and the right to development, and the fact that Ecuador was a poor country. 'The next line after that was that "well we are making great steps forward",' recollects Kimerling. She says:

Now the line you see is either 'well, this is a long-term problem, it is not something that can be corrected overnight, but we are very serious and we are taking big steps forward', which is one analysis. The other one which I hear a lot now is 'well, everything is OK and we have these great new clean-up technologies that are approved by the US EPA'. It is what I call an environmental faddiness.

Kimerling is worried by what she sees as a new trend: the aesthetic greening of the industry, although underneath there is little evidence of real change. 'That is part of the propaganda initiative that I see now and it is obviously very troubling, and it is the kind of thing that you would not really see in the US or Europe because it is just too simplistic,' she says. 'But in Ecuador, there is so little experience with this and so little understanding that people say these things and a lot of people really believe them.' This is despite the fact that neither the industry nor the government has brought in new environmentalists in high level positions. Furthermore, everyone within the industry is suddenly calling themselves an environmentalist.

Another group of people who are still fighting to get their voice heard in the debate about oil and the Amazon are the ones who should have been invited first to the negotiating table: the Indians of the Amazon. Moreover, the Indians working on the oil issue have been subject to a lot of pressure by both the government and the military. 'I think the line they use,' says Kimerling, 'is that you are anti-Ecuador, you are hurting our image, how can you do this and why do you have so many foreign friends because the foreigners are bad.' This is coupled with the underlying philosophy of the country that the development of the Amazon is in the economic interest of Ecuador. All the Amazonian countries see that the majority of the population have development needs which can be provided by oil, so the rights and concerns of the indigenous populations cannot overrule the needs of the majority.

It is Kimerling's impression that the Huaorani who speak out have received a lot of pressure for doing so, which actually has affected their actions. 'The word is that he is a dead man,' Joe Kane had been told about the Huaorani Indian he had been travelling with on a trip to the USA to highlight the cultural, environmental and social devastation of oil drilling in the Oriente. The Ecuadorian government were annoyed that the Huaorani had travelled to Washington, and he had been picked up for 'interrogation'. Kane had been working for some months with the Huaorani who would, in all probability, be wiped out by oil exploration, due to pollution, disease and the opening up of land for development.[59] The Huaorani individual concerned had been lucky to leave Ecuador in the first place, as the American Embassy had originally denied him a visa. He had effectively been accused of treason for opposing oil development by the Ecuadorian Ambassador, and now he was being punished by the authorities for raising the issue with people in Washington.[60]

THE FUTURE

Asked if she felt hopeful for the future, Kimerling replies that:

> I haven't lost hope, but I think in order to have serious meaningful change, both in environmental protection and respect for fundamental human rights of residents of the Amazon, for that to occur, there is going to have to be a tremendous amount of pressure on an international level. There will also have to be a lot of work on a national level and at the grass-roots. But the jury is still out.
>
> I feel like there has been a significant change in the rhetoric and the dialogue, but in the actions in the field, I don't see the change yet. I see some changes, but not necessary changes that are yielding a net environmental benefit.

To Kimerling the changes seem cosmetic public relations exercises rather than serious changes in operation practices. She says:

> You could say that covering a waste pit is better because the cows do not fall in and of course that is better for the people who own the cows, but if one is looking at the net environmental benefit, I don't really think that is a significant benefit for the environment.

The oil companies have changed their operations in other ways, too. Maxus used directional drilling on one occasion, highlighting how this was going to decrease deforestation. Kimerling says:

> Well if they are able to keep colonists out, which is still a big if, that may limit deforestation, and that is definitely an aesthetic improvement. But by doing directional drilling rather than direct drilling, they are generating larger quantities of waste with toxic constituents, and I do not think these wastes are being seriously handled. There is a question in my mind whether there is, in fact, a net environmental benefit.

Kimerling is similarly dismissive of other so-called environmental improvements that the industry have undertaken. Maxus once again made a large public relations announcement about their 'reforestation' project, bringing in various high-level consultants to work on the project. Kimerling says:

What they called a reforestation project is in fact a revegetation project, it is not reforestation. They have planted a grass that is native to Africa. All the public relations and even their environmental masterplan give the impression that this is reforestation with native species. Is that an environmental benefit as opposed to just a cosmetic benefit?

To make matters worse, there is no independent monitoring of the oil company operations. 'I really do not see a mechanism in place yet to control these oil companies or even to seriously evaluate their own initiatives,' she continues:

I don't think we can have a success story of the industry without that. Although it takes a long time to make changes, unfortunately the pace of destruction is much greater than the pace of positive change. I really think the jury is still out . . . But there really is no logic to the destruction of the Amazon apart from opportunism and short-term greed.

But the people living in a region devoid of logic have resorted to litigation for compensation. In November 1993, in an unprecedented move, Ecuadorian Indians issued a class action lawsuit against Texaco in New York, seeking damages of $1.5 billion. The lawsuit on behalf of some 30,000 Indians in twenty tribes, charges that Texaco has contaminated their waters and land, and caused cancer in the area. It also seeks for Texaco to clean up its mess. The suit also charges that Texaco deliberately dumped oil wastes into leaking pits.[61] In the suit, the plaintiffs allege that:

Texaco did not use reasonable industry standards of oil extraction in the Oriente, or comply with accepted American, local or international standards of environmental safety and protection. Rather, purely for its own economic gain, Texaco deliberately ignored reasonable and safe practices and treated the pristine Amazon rain forests of the Oriente and its people as a toxic waste dump.[62]

In December 1994, a group of Peruvian Indians filed a $1 billion class-action lawsuit against Texaco, charging the company dumped oil from its Ecuador operations into a river that flows into Peru.[63] Both cases continue.

OTHER LATIN AMERICAN HARASSMENT

Elsewhere in the region people are also being harassed for working on ecological issues. In 1992, ten members of the Peruvian organisation, the Comite de Defensa de los Bosques de San Ignacio, who were opposing local logging, were arrested and charged with terrorism. Amnesty International believed that 'they may have been charged for no other reason than opposition to the commercial exploitation of the woods'.[64] A year later, the Greenpeace Argentinean office received a phone call 'If you keep on working against nuclear power you'll be blown into a thousand pieces. Stop it. Last warning.' Greenpeace Argentina had been campaigning against a local nuclear power plant.[65] Staff received further death threats, four months later.[66]

The following year, in September 1994, two Peruvian environmental activists, María Elena Foronda and Oscar Díaz Barboza were detained by the authorities 'apparently' on suspicion of having links with the 'armed opposition', according to Amnesty International. Amnesty did not 'believe they have any such involvement', and feared that they would be tortured and ill-treated.[67] The coordinator of the Uruguayan Environmental Network, Jose Luis Cogorno, received three death threats against him and his family in December 1994. He had been campaigning against the import of toxic waste from America.[68] Just under a year later in November 1995, Luis Erasmo Arenas Hurtado, an environmental community leader, working to preserve the region's rain forest, was shot dead by a hired killer. Ten days before his death, Hurtado had written to the local mayor saying he was being threatened.[69]

CENTRAL AMERICA

In Central America, where there has been a spate of murders against environmentalists, the violence could be worse than in Latin America. In Mexico, in 1988, a fire destroyed the Xochicalli ecological house owned by Arias Chavaz. Two days previously, Chavaz, who was a prominent anti-nuclear protestor, had taken part in a confrontation with government officials. Police investigating the incident told him that the fire had been started deliberately by well-trained arsonists.[70] Six years later, the environmental and human rights organisation, Global Response, highlighted how drug and timber barons are destroying the Sierra Fired forest. 'Environmentalists working to protect the indigenous population and preserve the region's remaining 2 per cent of old-growth forests have been threatened, kidnapped, shot and tortured,' said Global Response.[71]

In Costa Rica, in 1989, Antonio Zuniga, an activist working against the illegal hunting of animals on indigenous lands in the country was murdered. The killing was never solved.[72] Three years later, the *campesino* leader Gerardo Quiros, who was working against deforestation in the country, was also murdered.[73] In December 1994, three members of the Costa Rican Environmental Organisation AECO, an organisation affiliated to Friends of the Earth, died when their house was set on fire. The three were all prominent activists in the country.[74] The official investigation established that the causes of the fire were accidental, but that it was not a criminal attack. However, environmental activists remained extremely sceptical and suspicious. A few days before the fire, the leaders of AECO had completed a successful campaign against the installation of a paper pulp complex which would destroy an important forest reserve in the south of the country.[75]

Moreover, the AECO activists had been receiving threatening phone calls and one had already been injured in a car accident after someone had tampered with his vehicle.[76] Five months after the arson attack, two grassroots environmental activists, Wilfredo Rojas and Elizabeth Gonzales, working against a proposed toxic waste dump in their neighbourhood, had their houses burnt to the ground. The two had been subject to death threats and harassment. Environmentalists in the country report that the climate of fear has become so intense that people are giving up working on ecological issues, and that the government seem to do nothing to prevent the rising level of violence.[77]

In Guatemala, meanwhile, violence over protecting the country's forests has worsened. Much of the logging is illegal, undertaken by people who have the support of the military authorities and who are protected by armed guards. In 1993, Omar Cano, a journalist from the prestigious newspaper *Siglo XXI*, who was researching deforestation in the Peten region, received death threats. The threats were so bad that he requested political asylum for him and his family in Canada. Officers of the government's National Commission of Protected Areas (CONAP), the institution responsible for the protection of the Maya Biosphere Reserve, were also threatened and attacked by armed bandits working for the loggers.[78] Two years later, two Greenpeace researchers working on deforestation in the Peten region heard that their lives were in danger from armed gangs and quickly had to leave the area. The level of violence and intimidation has become so bad that activists believe no environmental work will be carried out in the Peten region in the foreseeable future.[79]

Violence against environmentalists also seems to be spreading to Honduras. An environmental organisation of traditional fishermen in the south of the country, CODDEFFAGOLF, has been subject to horrific levels of violence, after starting a campaign to preserve mangrove forests in the country. Since the late 1980s, five members of the organisation have been killed by guards working for the owners of private shrimp farms in the Gulf of Fonseca. Shrimp farming threatens the mangrove forests where the fishermen catch their prey.[80]

In February 1995, one of the country's leading environmental activists, Blanca Jeannette Kawas Fernandez was hit in the neck by a bullet. She was killed instantly. Two days previously, Kawas had led a successful protest march against a proposed plan to develop an oil palm plantation in one of the country's national parks, Punta Sal. She was the President of an influential environmental group, Prolansate, which had been working against development of national parks in the country by timber companies, ranchers and oil palm developers. It is believed that one of these interests hired the contract killers to assassinate her. Two well-dressed and extremely fit young men were seen running from the scene of the crime. 'All factors indicate that Kawas was murdered by powerful groups with whom she disagreed on the exploitation of the Punta Sal nature reserve,' said Elias Romero of the Honduran Association of Environmentalist Journalists, after the killing.[81]

The Honduran police soon announced they were close to catching the killers, but six months later still no-one had been arrested. Moreover it is unlikely that anyone will ever be apprehended, given the corruption endemic in the country and the likely powerful vested interests behind the murder. Moreover, environmentalists believe that Kawas' killer is well connected with the government and that only foreign pressure brought to bear on the country will resolve the killing. Loggers, ranchers and migratory farmers have now started destroying thousands of acres of another reserve, the Sico Paulaya, a United Nations designated World Heritage Site. Environmentalists working to protect this are said to be in fear of their lives.[82]

In April 1996, the Mexican environmentalist Edwin Bustillos, who has survived three attacks on his life due to his work protecting the forests in the Sierra Madre Mountains of Northern Mexico, as well as Chico Mendes' colleague, Marina Silva, who became the first rubber tapper to be elected to the Brazilian Senate, were awarded the Goldman Environmental Award. Both activists were fearful of the future, due to the impact that free trade was having on the region. Silva highlighted how 'the new patent law in Brazil has opened the biodiversity in our forests up to foreign companies'.[83] Only time will tell if the people and the forests of Latin America can survive their bio-diversity being plundered for the profits of the First World multinationals.

9

DIRTY TRICKS DOWN UNDER

The military sometimes act like cretins and this was an example.

President Francois Mitterand, on the bombing of the *Rainbow Warrior*

Further evidence of an emerging international green backlash template comes from an unusual geographical source. Despite the fact that Australia and New Zealand are thousands of miles from the hotbed of anti-environmentalism of North America, the green backlash is flourishing 'down under'. All the trademarks of anti-green activism are on the increase: corporate front groups, the demonising of environmentalists as terrorists, the use of dirty tricks, physical and legal harassment and the use of violence against environmentalists. Networking is also occurring between key North American anti-environmentalists and their Australian counterparts.

Before looking at the current situation, it is worth remembering that, just over a decade ago, New Zealand was the scene of one of the most vicious attacks by a government against a non-violent environmental organisation.

YOU CAN'T SINK A RAINBOW

On 10 July 1985 two explosions ripped through the hull of Greenpeace's flagship vessel, the *Rainbow Warrior*, which was in dock in Auckland harbour. The blast killed the cameraman on board, Fernando Pereira, who went to recover his camera equipment after the first explosion and was caught in his cabin by the second blast. Pereira was the first person to be killed in New Zealand by international terrorism. The perpetrators of the crime were the French government and agents acting on its behalf.[1]

Plate 9.1 The *Rainbow Warrior* in Auckland Harbour after being bombed.
Source: Greenpeace/Miller

It was lucky there were not more casualties. Had the bombs gone off one hour later, nine people would have been killed; one hour earlier and fourteen people would have died.[2] This said, it does seem that the original intention was to scuttle the ship, not kill the crew. However, no warning was given before the blast, a tactic even used by the IRA at the height of their terror campaign. Both the US and British governments refused to condemn this act of state terrorism against a non-violent organisation.[3]

The *Rainbow Warrior* was in Auckland on her way to lead the 1985 Pacific Peace Voyage, a flotilla of non-violent protest vessels, to demonstrate against continued French nuclear tests in French Polynesia, especially at Moruroa Atoll. The French authorities were determined to put a stop to Greenpeace's action, using violence if necessary. It was a well-planned operation. Greenpeace's New Zealand offices were infiltrated by an agent working for the French Secret Service (DGSE), months before the *Rainbow Warrior* arrived. The agent, Christine Huguette Cabon, introduced herself as Frédérique Bonlieu, a scientific consultant. Cabon was one of ten agents despatched to New Zealand, although she was long gone before the bombing in July.

The bombing itself involved three independent teams in the country with one person in overall control, Lieutenant Colonel Louis-Pierrer Dillais. Back in Paris, the authorisation for the bombing came from Charles Hernu, the Minister for Defence and Admiral Pierre Lacoste, the Head of the DGSE. Four members of the DGSE came in by yacht, the *Ouvéa*, bringing with them explosives and the dinghy that were used in the bombing. The two other teams involved arrived by plane: including a supposedly married couple, Sophie and Alain Turenge, alias agents Major Alain Mafart and Captain Dominique Prieur who acted in a support role during the evening of the bombing and who were later convicted of manslaughter. But there were three more agents in the country, one of whom, François Verlet, walked on to the *Rainbow Warrior* the night it was bombed. The other two, Alain Tonel and Jacques Camurier, are the duo that investigative journalist Michael King, who wrote a book on the affair, believes actually placed the bombs that were to kill Pereira and would sink the *Rainbow Warrior*.[4]

Only the 'Turenges' would be caught by the New Zealand authorities, after what can only be described as a 'bungled' operation on the night of the bombing. The rest, some eight people, would escape from New Zealand and would never be prosecuted for the atrocity in which they had all played an integral part.[5] The French government, who refused to cooperate with the investigating authorities, used classic disinformation techniques to upset the whole investigation. Pereira was subsequently called a communist by the French. He was said to have links to the KGB and to be a member of the Baader-Meinhoff gang.[6] It was a vicious lie that was intended to discredit him and somehow justify the bombing. This smear is still repeated by anti-environmentalists ten years later.

The initial reaction from the French authorities was one of complete denial. 'In no way was France involved,' the French Embassy declared a day after the bombing. 'The French government does not deal with its opponents in such ways. France is not worried by the anti-Moruroa campaign planned by Greenpeace.'[7] Despite numerous attempts to whitewash the affair, eventually the truth emerged, forcing the resignation of Charles Hernu as Defence Minister and the dismissal of Admiral Lacoste from the DGSE. Laurent Fabius, France's Prime Minister admitted that it was Hernu who had given the order to 'neutralise' Greenpeace's vessel. Six months later, as if being rewarded for his actions, Hernu received one of France's most prestigious awards, the 'Legion of Honour'.[8] Ten years afterwards, France conceded that it had considered using biological weapons to stop the crew of a second Greenpeace ship sailing to the South Pacific, months after the *Warrior* was bombed.[9]

The agents in jail were treated as heroes, rather than villains. Jacques Chirac, who became President of France in 1995, and who resumed nuclear testing at Moruroa, said 'that the French Army had every reason to be proud of the two officers involved in the sinking of the *Rainbow Warrior*'.[10] The new Defence Minister, Paul Quilès, told the defendants on the day of their trial that the French people were proud of what they had done. The authorities made sure their agents were not stranded abroad, and although the pair were originally sentenced to ten years in prison, they did not serve their full time. The duo were released just over a year later and relocated to a French military base on the Pacific atoll of Hao after delicate negotiations between the Paris and Wellington authorities. The New Zealand government had been warned that their French butter and mutton sales might suffer if a deal was not struck. Even then, Prieur and Mafart did not stay on their paradise prison island for long. By 1987, Major Mafart had been returned to France on 'humanitarian and medical grounds' and by May 1988 Captain Prieur was to follow when she became pregnant. Mafart soon won entry into France's élite School of War.[11]

Speaking ten years after the incident, Captain Dominique Prieur called herself and Alain Mafart scapegoats for French 'political powers'. 'We paid the price in a game in which we were just the pawns,' said Prieur. Asked under whose ultimate authority she was operating, she responded: 'Frankly, I don't have a clue. It nevertheless seems unthinkable to me that such a decision could have been taken simply at the level of the DSGE.'[12] Moreover, Prieur confirmed that there was a third team involved in placing the explosives. She accused the Secret Service heads of mishandling the operation and of being responsible for the photographer's death, by demanding that two sets of explosives be used, when one would have been sufficient to scuttle the ship.[13]

This was not the first or last time France has used violence in an attempt to silence Greenpeace. It would not be the first time either that the French had lied to the world to protect their nuclear testing. Back in 1972, David McTaggart, now the honorary chair of Greenpeace International, and fellow crewman Nigel Ingram were badly beaten up by French commandos, after Greenpeace had sailed into the testing zone. McTaggart was beaten unconscious and has never fully regained the sight in one eye. The French said McTaggart had slipped, thinking they had destroyed all photographic evidence by throwing all their equipment into the sea. Smuggled photographs showed the world otherwise.[14]

Ten years later, the Greenpeace ship, *Rainbow Warrior II*, the original ship's

Plate 9.2 French commandos storm the new *Rainbow Warrior*.
Source: Greenpeace/Morgan

successor, arrived at the Moruroa Atoll on 10 July 1995, to celebrate the tenth anniversary of Fernando Pereira's death, and once again to highlight the renewed threat of French testing. They were once again met by force, as the French rammed the ship and commandos used tear gas to attack the protesters. Outrage at the assault spread through the South Pacific and around the world. The French Ambassador in Washington was summoned to explain his country's actions by the New Zealand Foreign Minister. 'In this country, tear gas is considered a weapon. On the French side, it is not,' said Ambassador Jacques Le Blanc.[15] The New Zealand Prime Minister called the action, 'totally over the top'. 'It does seem hard to justify why 150 French commandos would need to use tear gas to overpower 30 peace protesters,' added Senator Bob McMullen of Australia.[16]

Thousands of miles away, Marelle Pereira, Fernando Pereira's daughter, laid a wreath outside the French Embassy, which read 'You can't sink a rainbow'. Marelle lamented that her father's death had changed nothing, now that France was set to resume testing.[17] France has never apologised to Marelle or any other members of the Pereira family for her father's murder. The true murderers, those who ordered the attack and those who planted the explosives, have never been brought to justice. Nor will they ever be.

France also defiantly resumed nuclear testing at Moruroa Atoll in September 1995. Greenpeace, which led the campaign against the new tests, came under concerted criticism from the New Zealand press for supposedly manipulating the media and politicians. Although this may seem surprising in a country that has traditionally been vehemently anti-nuclear, it is probably understandable, given the anti-green sentiment that has been growing in the region for years.

BOB'S ANTI-BACKLASH CAMPAIGN

The fact that we know that the green backlash is increasing in Australia and New Zealand is very much due to Bob Burton from the Wilderness Society, an activist and researcher who has spent seventeen years working on forestry, mining and energy issues. Burton lives in Tasmania which has long been at the forefront of environmental campaigns or new approaches to ecological issues in Australia. In part, this is because it is home to some of the most spectacular forests and wilderness anywhere in the world. In that sense, many of the resource conflicts that have occurred in Australia, over logging, energy and mining, have happened in Burton's backyard. Having started researching the backlash in Tasmania, Burton has spent four years documenting anti-environmentalism in both Australia and New Zealand.

It was while travelling in the USA in 1992, that, confronted by what he saw of the emerging anti-environmental movement there, Burton began putting the pieces of the jigsaw together and the subsequent picture became clearer in Australia. 'The pattern was so similar, some of the instances were just so similar and on totally different continents,' says Burton.[18] What was interesting to Burton was how similar the situation was in Australia and America, despite the vast geographical separation. Burton says:

> I think there is very little I see that is different from the North American model. My reading of it is that Australia is probably a few years behind what has been happening in North America, and NZ is probably a few years behind what is happening in Australia, but the trends are all in the same direction.[19]

Moreover, just as in America, the problem seemed to be getting worse, and there seemed to be an increasing level of sophistication by industry, its PR gurus and the political Right in their attacks on the environmental movement. Although Burton believes that neither the scale or the intensity of the environmental debate have changed much in the last fifteen years, there are a couple of reasons why the backlash has suddenly blossomed. First, since the early 1980s, whilst the environmental movement has enjoyed public support, the political area was in a state of uncertainty, with a succession of different progressive and conservative governments. In certain cases, the majority of these governments was so small that green members of parliament actually held the balance of power. This was a worrying trend for industry, as it gave environmentalists previously unknown political power. Second, in the late 1980s, environmental issues suddenly rocketed in popularity, generating large amounts of media attention. It was then that industry realised that it had to become involved in the debate, and counter some of the environmental gains.[20]

THE FOREST PROTECTION SOCIETY

As in North America, much of the green backlash had been centred around the forestry debate. By the mid-1980s, the timber industry, more so than any other, had become politicised and had suffered years of sustained campaigning against it. As part of these campaigns, direct action tactics, such as physically obstructing machinery, were commonly employed by environmental activists. The debate spawned the first green-sounding 'corporate front group', the Forest Protection Society (FPS), in 1987.

The forestry industry has been actively involved in lobbying against conservation measures in Australia for the last twenty years and the 'front group', the FPS, is merely a manifestation of this. This said, the environmental movement won a potentially historic decision in 1995, when the Australian government agreed to a Commonwealth Position on National Reserve Criteria, which meant that a broad benchmark of 15 per cent of the pre-1750 forest area was to be protected within designated conservation reserves.[21]

The FPS was not the first organisation formed to work for the interests of the industry, there had been others, such as the Forest and Forest Products Industry Council (FAFPIC). The way the FPS differed from its predecessors was that it sounded like an environmental group, rather than an industry lobby group. FAFPIC was a tripartite council established by the then new Labour Government, whose members were from government, industry and unions.

Serviced by public officials, its charter was to develop a strategic approach to the development of the industry.[22] Within the year, FAFPIC had assisted in the formation of the Forest Industry Campaign Association (FICA) and was embarking on a public relations campaign on behalf of the industry, supported by some $1.7 million of tax-deductible industry money and staffed by a secretariat of public sector (government) employees.[23]

In September 1986 the Forest Industry's campaign was released in Melbourne by industry and union representatives but it soon hit controversy. One of its advertisements showed an environmentalist supposedly chained to a tree screaming 'I don't care'. The adverts caused an outcry, but there was other dissent growing too. Public service employees being used as public relations practitioners for a private industry caused a political spat and government support was suspended in February 1987.[24]

Following federal government withdrawal from FICA, the formation of the National Association of Forest Industries (NAFI) was announced in March 1987. Its founding president was Mr Dick Darnoc, the managing director of Weyerhaeuser Australia, the subsidiary of the US timber conglomerate.[25] This group has since become a powerful lobbying force with a hefty $5.6 million budget in 1993 and has ensured that environmental protection legislation has been watered down, especially the Commonwealth Endangered Species Legislation.[26] They have also sponsored overseas visitors who have promoted the industry's position.[27]

Eight months after NAFI was formed, the Australian Council of Trade Unions (ACTU) and FICA announced that they were setting up a grassroots voice for the timber industry, the misleadingly named Forest Protection Society. Claiming that it could have a potential membership of 100,000, Bob Richardson from the ACTU, said, 'We recognise that we are ten to fifteen years behind conservationists in this sort of action so today's resolutions, including the membership drive, is about catching up in the people sense.' Membership of the fledgling group would cost $10, but forest industry money would help it with start-up costs.[28]

In Wise Use-type rhetoric, the Forest Protection Society said it promoted 'balanced' conservation solutions and was 'promoting the wise use of Tasmania's natural resources today and for future generations.'[29] It is a message the FPS repeats in different states around Australia. The forest groups also started using new industry techniques to relay their message to politicians. They became the grassroots, third-party voice of the industry. FICA started advocating joining in the 'necessary evil' of 'industry demonstrations' to counter 'protesting environmentalists'. The FPS urged people to write letters to interfere in a film on the logging industry by complaining to the producers, by writing letters to the editor, and to phone and fax people showing solidarity against the film. They held media training exercises. They were determined to counteract 'misinformation about the forest industry'.[30] The FPS also picketed the Wilderness Society's shops in January 1995. Rather than adversely affecting the shops, sales in the following week jumped by over 20 per cent.

But charges of misinformation could be laid at the door of the FPS. They were listed as an 'Environmental Protection Organisation' in the 1994 Directory of Australian Associations, promoting a 'balanced use of Australia's forests'.[31] The FPS, whilst supporting the logging of rain-forests, proclaim in their fact-sheets 'one of the best ways to ensure that the rain-forests are not destroyed is to harvest the wood and sell it'.[32] Leaked minutes from Network Communications Limited, a company that shared the same address as the FPS, to the Tasmanian Women's Timber Support Group, outlined 'taking over local environmentalist meetings, with the result that they become distracted from their ongoing campaigns'. Furthermore, the minutes also referred to misinformation kits being distributed.[33] 'Everybody in the community ought to be concerned about this sort of thing going on,' said Labour Senator John Deveraux, commenting about the minutes that, 'there is no doubt that it is a threat to democracy'.[34] The FPS is no longer a client of Network Communications. Its current PR company should come as no surprise to people tracking the global green backlash: Burson-Marsteller.

The FPS has intimate links with the forest industry. Contacts for the FPS have included the public regional officer for the APPM – the Associated Pulp and Paper Mills, Australia's largest woodchip company and the Executive Director of NAFI. The state-based industry umbrella organisations that are affiliated with NAFI provide support and assistance to the FPS.[35] Some 80 per cent of its $588,942 budget for 1991–92 came from the Forest Industries Campaign Association. The following year NAFI gave $637,914, or 78 per cent of their total $819,845 budget.[36]

PR STRATEGIES

The forest groups have outlined their PR message to the public. 'Just locking away areas as National Parks and World Heritage areas does not protect them,' says Forest Industries Association of Tasmania's (FIAT) Public Affairs Manager, Kevin Broadribb. He continues:

> Without proper management they will be subject to ongoing destruction through the spread of exotic weeds and elimination of native fauna by feral animals. The scenic and aesthetic values of well managed regrowth areas are comparable to much of the existing native forests within the region. It is difficult to tell the difference between old regrowth and unlogged forest.

This line of thinking is one that resonates from different sections of industry and the Right in Australia, one of privatisation. Effectively, all public resources should be sold off to the private sector. The Forest Industries Association has circulated faxed letters entitled 'The Fight Against the Big Lie', which underlines how the industry is attempting to demonise the environmental movement:

> [To] avoid confusion care should be taken when referring to protesters, politicians and lobbyists who oppose forestry. Generally they should not be referred to as conservationists, environmentalists or greens. There are two problems with using general terms such as conservationist and environmentalist. The first is simply that most rational conservationists support forestry, the second is that foresters, and other people working within forestry, are excellent conservationists and generally have a good record in terms of protecting the environment . . .
>
> It is only the radical elements within the conservation movement who oppose forestry and the words used in statements and press releases should reflect this fact. Some of the following may be of use when putting thoughts together; the environmental lobby, radical conservationist, conservation extremists, environmental fanatics. Other words it may be useful to include; excessive demands, unreasonable, drastic claims, zealots, uninformed, rebellious, ultra radical, hysterical, bigots, biased, narrow-minded and uncompromising . . . the following terminology will often be appropriate: the anti-forestry lobby, anti-forestry extremists, the anti-industry movement, the anti-development lobby.[37]

INTERNATIONAL LINKS

In 1992 the FPS began to build links overseas with other anti-environmental organisations, with the editor of the FPS newsletter commenting that, 'I have heard and read of good things being done by some interesting organisations with goals very similar to our own. . . .Canadian Women in Timber, Forestry

Alliance of BC and Share Canada.'[38] In fact, the group they made direct contact with was the Western States Public Land Coalition in the USA, People for the West!'s parent organisation. 'We are beginning to discover a worldwide network of natural resource producers and rural communities who are fighting the same fight as we are,' People for the West!'s Communications Director, Joe Snyder wrote after the meeting.[39]

Two years later, the links between the two groups were expanded when the FPS's Tasmanian Coordinator, Barry Chipman and the Society's National Director, Robyn Loydel, attended People for the West!'s 1994 annual meeting, in Glorieta, New Mexico. 'Obviously we share a lot of common ground with the US organisation,' said Chipman. 'The conference is a great opportunity to exchange ideas and reinforce with the politicians present that the movement against excessive government pandering to minority groups is international and growing.'[40] By August 1994 the National Coalition of PFW! and 'its multiple use partners in Australia, the FPS' had 'taken their first step in linking an international cooperation agreement between the organisations'. The proposal called for 'international cooperation between the two groups to "promote the development and use of the world's natural resources to the benefit of mankind"'. Hopefully, said PFW!'s then Executive Director Bill Grannell, 'this will spark an international conference of like-minded groups to counter the growing shrillness of the green movement'.[41] Unsubstantiated reports are that, when Sydney hosts the Olympic Games in 2000, there will be the first global Wise Use conference in the city.

This is not the only networking between Wise Use and anti-environmental activists and the forestry industry in Australia. In 1993, Bruce Vincent, from the Alliance for America, addressed the NSW Logging Association annual conference. Vincent's trip was being organised by the Australian Logging Council, 'so that Bruce can present his important messages throughout Australia'.[42] At the invitation of NAFI, the BC Forest Alliance's Patrick Moore made a trip to Australia in July 1995, speaking at the national press club in Canberra.[43] Moore was back again for a month's tour of the country in March 1996. His increasingly vitriolic language was picked up by his sponsors who were once again NAFI: 'Dr Moore believes that many environment groups are so extreme that they pose a greater threat to the global environment than mainstream society,' declared NAFI in its media release. 'He believes environmentalists should resist attempts by left-wing extremists to hijack green politics. He says environmental extremists tend to be anti-human, anti-organisation, anti-trade, anti-free enterprise, anti-democratic and anti-civilisation.'[44]

Other prominent anti-environmentalists and anti-green organisations have also made the trip down under to Australia and New Zealand. As far back as 1986, Ron Arnold made the trip to New Zealand warning that 'groups such as Greenpeace are bent on world reform that would eliminate capitalism and private property'. He also equated the use of direct action with an upsurge in 'eco-terrorism' and accused environmentalists of blowing up power stations and shooting forest officials.[45] Arnold's comments were later proven to be totally unfounded in a court case brought by anti-pesticide activists in the USA who were claiming harassment by the federal authorities.[46] In 1993, another prominent anti-environmentalist, and arch-Greenpeace critic, Magnus Gudmundsson, visited New Zealand at the invitation of the New Zealand Fishing Industry Association (see chapter 13).

There are other links between international anti-environmentalists through Australian and New Zealand right-wing think-tanks. The most prominent is the Institute of Public Affairs (IPA), whose remit is to advance the cause of free business enterprise in Australia and which has had a substantial influence on environmental issues since 1989. They have direct and frequent ties to the principal US right-wing think tank, the Heritage Foundation, but also network with the Cato Institute and the American Enterprise Institute.[47] The IPA has links to some heavyweight businessmen. Rupert Murdoch sat on its Council in the 1980s as did the Chairman of the Western Mining Corporation. It has a current budget of some $1.5 million with some 1,000 corporate members and 3,500 subscribers.[48]

Another important Australian think-tank is the Centre for Independent Studies (CIS) which is primarily interested in academic right-wing ideas. It has links with American and British academics and the Hoover Institute in America and the Institute of Economic Affairs and Centre for Policy Studies in England.[49] Fred Smith, from the Competitive Enterprise Institute in the USA, has also toured Australia. Both the IPA and CIS have sponsored tours by two of the best-known climate sceptics, Fred Singer and Richard Lindzen. Singer delivered a seminar to the CIS in 1990 and then returned under the auspices of the IPA and Tasman Institute in 1991 to Australia and the Coal Research Association Conference in New Zealand.[50] The same year, the IPA published a collection of essays outlining the views of Fred Singer and Edward Krug from the Committee for a Constructive Tomorrow, another leading right-wing think-tank and member of Earth Day Alternatives.[51] In 1992, the IPA followed this up when they produced a background paper on global warming written by Richard Lindzen.[52]

Lindzen was to visit again three years later, in the run-up to the 1995 Berlin Climate Conference, when he was invited by the CIS and the New Zealand Business Round Table. During his visit Lindzen declined to debate climate change with Kirsty Hamilton from Greenpeace, stating that Greenpeace was an organisation that 'acts like Goebbels'.[53] The Science and Research Minister for the New Zealand Government, Simon Upton, wrote to the Business Round Table 'frankly surprised that you have chosen to sponsor a tour by a scientist who has to date failed to convince his peers in normal scholarly exchanges'.[54] Lindzen's highly sceptical line has been adopted by such organisations as the Business Council for Australia who highlight the uncertainty of the science of global warming, claiming that no proof has yet been furnished concerning the link between greenhouse gases and climate change.[55]

Furthermore, articles in the CIS's magazine *Policy* have reiterated the old right-wing rhetoric about the threat of environmental 'zealots' and the 'environmental priesthood'. 'Environmental extremism has become the principal means by which many collectivists hope to achieve their dream of a thoroughly regulated, controlled and planned economy,' writes one author warning that, 'The Red Star is burned out, but the Green Star is rising.'[56] The IPA also warns readers of 'green hysteria', 'the greens and beliefs in sorcery and witchcraft' and the 'religion of environmentalism'.[57]

One person who has long regarded the environmental movement as religious fanatics is political extremist Lyndon LaRouche, whose followers have been busy in Australia. They have flown key activists over, such as the Citizens Electoral Councils (CEC), paying for one of LaRouche's chief spokespeople, James Bevel, to visit in 1994.[58] They have also forged links with at least two politicians, set up intelligence networks and tapped into Australia's growing gun lobby. One of LaRouche's supporters in the country has called on people to start forming armed militia, just as is happening in the USA. Moreover, the LaRouchians have also established a substantial funding network.[59] Their agenda is simple. Uwe Friesecke, from the Schiller Institute, outlined the LaRouche blueprint for a nationalistic recovery in Australia in 1993: pro-development, pro-technology, pro-nuclear power and anti-environmental.[60]

Another anti-green group called the New South Wales Public Land Users Alliance, started in September 1993. Its spokesperson, National Party MP Peter Cochrane, has scapegoated environmentalists for causing the bush fires which raged through the region in 1994 and for supposed acts of 'ecoterrorism'. Following the fires, R. J. Smith from the American Competitive Enterprise Institute, gave the Alliance and other groups a speech entitled 'Endangered Rats,

Fire and the Federal Bureaucracy'.[61] In 1996, the Public Land Users Alliance, announced that it would be organising a national Wise Use rally 'in conjunction with our friends from the mining, forestry and agricultural industries', to which Ron Arnold would be invited.[62]

In the near future, it is in all likelihood that the direct links between individuals and groups propagating the anti-environmental message will not only continue but widen.

OTHER ANTI-GREEN ORGANISATIONS

The Forest Protection Society has also worked with other anti-green organisations in Australia, such as the Tasmanian Traditional Recreational Land Users Federation (TTRLUF) on issues such as the proposed closure of the Raglan Range four-wheel drive track in the Western Tasmania World Heritage Area.[63] The TTRLUF was formed in 1990 to oppose the designation of parts of Tasmania as World Heritage Sites. TTRLUF are the only community organisation to get front-page treatment on the Chamber of Mines newsletter advertising their meetings and have consistently refused to oppose mining in National Parks.[64] In a tone similar to his North American counterparts, TTRLUF's spokesperson Simon Cubit, who is also a senior Forestry Commission employee, warns people against Tasmanian environmentalists who are 'dangerous fanatics intent on locking up public land'.[65]

Other anti-green and animal rights groups exist too in Australia but some have misleading names which just confuse: the Conserve Our Residential Environment (CORE), which is a pro-freeway organisation that opposed Greenpeace on the building of the M2 freeway.[66] The Australian Federation for the Welfare of Animals 'is the largest national association representing people who are associated with animals in work and leisure and wish to put common sense back in animal welfare'. The group, which seems very similar to the Wise Use group, Putting People First, lists the advantages of animals for medical research.[67]

MOTHERS AGAINST POLLUTION

Another area of topical environmental debate that has sprouted industry groups is the packaging industry, especially over drinks cartons. Much like the B-M-instigated Alliance for Beverage Cartons and the Environment in Europe, there is the Association of Liquid Paper Carton Manufacturers (ALC) in Australia. Eight of ALC's ten sponsors also sponsor their European equivalent.[68] ALC is a lobbying organisation and spends most of its time defending the ecological impact of drink cartons made by its sponsors. However, it has also threatened Friends of the Earth with legal action for questioning the ecological impact of the carton. These SLAPP-style actions are becoming increasingly common in Australia, as explained later on in the chapter. ALC has also threatened the Queensland Conservation Council (QCC) with legal action after the QCC allegedly associated it with an organisation called Mothers Against Pollution or MOP.[69]

MOP, an 'Environmental, Educational and Information' organisation entered the packaging debate in the spring of 1993. The mainstay of its argument was whether cartons were more environmentally or nutritionally sound than plastic bottles. MOP's membership was initially free and it called itself a 'non-political and non-denominational' Queensland Environmental Lobby Group. Supposedly formed by concerned mothers, its aim was to 'provide families and individuals with simple, practical ideas and information on how to help protect and preserve the environment'.[70]

When MOP announced that it wanted 2-litre carton milk containers stocked in supermarkets, as compared to plastic bottles, people became concerned. MOP organiser Mrs Alana Maloney said her group was concerned the plastic bottles were not being recycled. 'We believe cartons are more environmentally friendly,' said Maloney, as, 'they are combustible, biodegradable and recyclable'.[71] MOP's next campaign was to offer free tree seedlings, as part of a state-wide seedling planting project. 'Milk cartons are great for growing seedlings,' said Maloney.[72]

It was MOP's preoccupation with the promotion of milk cartons that led to allegations that it could be nothing more than a 'corporate front group' for the carton manufacturers. Such fears led to a critical article in the *City Farm News*, warning of a bogus green group.[73] Subsequently, though, the City Farm Association received a letter from MOP requesting an apology and a retraction or the organisation would face legal action. In their April 1994 edition *City Farm News* issued an apology and retraction. This legal threat stirred them to further investigative action on MOP.[74] In their offending article, City Farm had revealed that MOP's address was an Australian Post Office, and the telephone number was an impenetrable answering service. They could not find MOP's sole representative, Alana Maloney, who appeared to be on none of the relevant electoral rolls. 'All in all, we are now doubtful Alana Maloney exists,' they concluded.[75] MOP's other two offices were found to be Post Office boxes and their phone numbers connected to an answering and pager service located in Melbourne.[76]

In February 1995, the truth about MOP finally began to emerge. City Farm's fears had been realised. 'Alana Maloney' did not exist. She was really Janet Rundle from J.R. and Associates, a public relations firm. Janet Rundle was also co-director of a family trust with Trevor Munnery of another public relations firm, Unlimited Public Relations. Mr Munnery was a consultant to the Association of Liquid Paperboard Carton Manufacturers (ALC). When contacted by a journalist, ALC denied any connection with MOP and threatened to sue the newspaper concerned if any links were made. Mr Munnery stated that he knew nothing about Ms Maloney or MOP but hung up when asked about Janet Rundle.[77] For months MOP disappeared from view but in August 1995, they emerged once more writing to members of the South Australian Parliament warning about damage to vitamins from milk exposed to light in plastic milk bottles. 'Alana Maloney', however, had disappeared, with the letter signed by Janet Rundle.[78]

SLAPPS

The use of SLAPPs, which have already been documented in North America, is growing. Bob Burton has chronicled over thirty-five cases as industry uses the threat of legal action in an attempt to silence environmentalists. Many of these have been in the forest debate in the last five years, although the earliest SLAPP dates back to 1972, when the Hydro-Electric Commission threatened to sue the author of a report by the Australian Conservation Foundation which was critical of the company's plans to flood the Lake Pedder National Park.[79] The most notable of the recent cases are as follows:

In 1991 the North East Forest Alliance (NEFA) received a letter from logging contractors threatening legal action for damages due to NEFA's anti-logging campaign in the Chaelundi State Forest.[80] Chaelundi is an area of high conservation forest. Two years later, the Wilderness Society (TWS) received a letter from solicitors representing North Broken Hill Peko's wholly owned subsidiary Associated Pulp and Paper Mills (APPM). It stated that APPM would be taking legal action to recover any supposed damages arising out of TWS's campaign against export woodchip licences from native forests.[81] Despite the threat, no action was taken.

Later that year, three environmentalists were sued by a developer for alleging to conspire and damage the commercial interests of his company. The environmentalists had objected to plans for development for the upper Hacking River, an area predominantly within the Royal National Park.[82] Moreover, when a student, Alexandra De Blas, planned to publish her thesis into the environmental effects of the Mount Lyell copper mine on the west coast of Tasmania, the company involved reacted vehemently. They warned that either she or her department, The Centre for Energy Studies, or anyone else who was involved in the publication could be sued. They also asked De Blas to undertake a written assurance to 'acknowledge receipt of this letter by immediate return and to include in that acknowledgement a written undertaking that neither your thesis nor any part of it will be published'. At first the University refused to assist in the publication, a position it later reversed and the thesis was published by the Centre for Independent Journalism. There was widespread publicity of the attempt at suppression by the company.[83]

The same year, when the Clean Seas Coalition from the north coast of New South Wales criticised the local council over sewage outflow into the sea, they were sued by the Council for damages, after they refused to apologise for their remarks. However, in a victory for the environmentalists, the NSW Court of Appeal ruled that local councils did not have the right, power or authority to sue for defamation. The court ruling was seen as a victory for democracy.[84] In 1993, the Byron Environment Centre were threatened with legal action over so-called defamatory remarks they made about the ecological impact of Club Med's resort at Byron Bay. No further action was taken after the initial threat.[85]

The following year, 1994, protesters who were concerned about the environmental impact of a $6.4 million bridge in Adelaide, were threatened with legal action by the developer. People picketing the site were warned that they could be sued for more than $47 million. 'We are not trying to frighten anyone,' said the developers' lawyer.[86] But legal action is meant to frighten people into inactivity. As Burton told an audience of concerned environmentalists, 'legal and violent harassment is aimed at increasing the personal costs to such a level that individuals withdraw from advocacy'.[87]

Conservation and environmental groups in New Zealand have also been intimidated. The Royal Forest and Bird Protection Society has been threatened numerous times, by Timberlands (a state-owned enterprise operating in native forest logging), the Fishing Industry Board and mining companies. Every time the proposed SLAPP has not been implemented.[88] In 1993, the gold mining company Coeur Gold sought a planning consent to be allowed to discharge water into a river on the Coramandel Peninsula. Two environmental organisations lodged objections over the threat of increased pollution. One of the groups, Peninsula Watchdog proceeded, whilst the other, Ohinemuri Earthwatch, withdrew after possible threat of damages by the mining company. When the planning decision went in favour of the company, Coeur Gold New Zealand announced in July 1994 that they were seeking costs against Peninsula Watchdog.[89]

DIRTY TRICKS

Burton's research has documented 130 incidents of harassment aimed at environmentalists and people who are perceived to be siding with them, such as journalists and public employees. And like his counterparts around the world, he is still counting.[90] As in North America, it is the rural activists who are in the front-line of intimidation and harassment. There is 'virtually no harassment at all in major cities, a few instances, nothing consistent', says Burton, 'Whereas once you get out into the rural areas it is quite overwhelming. I have actually been staggered at the scale of it.'[91]

In rural areas, Burton believes that harassment, which was once purely unorganised, now has an organised element to it. 'It's really hard to know, or disentangle, what you think might be an organised component to it, compared to what is just random harassment,' he says, adding:

> The sense I have, and I can't put it down to anything beyond it being a bit of a gut instinct and just seeing some of the patterns emerging, but there is now an organised component to it. The source of that organisation is not obvious.

This leads Burton to ask, 'The symptoms across a number of states seem remarkably similar, which begs the question, if the same sorts of harassment are occurring, with the same sorts of timing for different instances, then why is this happening?'[92]

Other parallels can be drawn with the USA. With the increasing use of direct action in the forestry debate in the early 1980s, so came the language backlash, attempting to stigmatise the environmentalists. Bob Burton believes, for the most part, this attempt failed and people became tired of the mudslinging. This said, there is one label that anti-greens have used time and time again against the environmental movement. It is the one tactic they still use today – it is the accusation of terrorism. As in the USA, as if by calling the environmental movement 'terrorists', this somehow justifies a violent response against it. Furthermore, the 'terrorist' smear also creates the opportunity for state-sanctioned harassment such as the use of the police and security forces against environmentalists.

Moreover, what the authorities and industry are aspiring to do is to negate the effects of direct action and to equate it with terrorism. This is exactly what Ron Arnold attempted to do on his tour to New Zealand in 1986. Environmentalists become the terrorists, whilst the loggers become the victims of these acts of outrage. Burton says:

> The eco-terrorism stuff is where I think they are at their most sophisticated, with dirty tricks and media management. If you work back from the incidences and the way in which sequences are constructed, the only plausible explanation is that it is only someone who has got a very good understanding of public relations, and someone who has a very good understanding of what is being done in North America.

One of the fastest growing sectors of the PR industry in Australia has been in countering environmental activism. The two leaders in the country are also the global pioneers: Burson-Marsteller and Hill and Knowlton.[93] Once again, it is in the forestry debate where the level of grassroots activism has been at its most intense that the fury of the backlash has especially been felt, and where the dirty tricks have been most dirty.

'Environmental Terrorism', was the press release issued by the Tasmanian Forestry Commission on 4 February 1992, claiming that they had found a sign in the East Picton Forest Area that said the forest had been spiked. East Picton had long been a source of controversy between environmentalists and loggers, having being revoked from the National Park for logging operations, but that did not mean that someone was spiking trees. The media had an anti-green circus. 'Loggers Fear Spate of Green Terrorism', ran the *Sydney Morning Herald*. The accusations were dismissed by the Australian Conservation Foundation as 'misinformation' aimed at achieving draconian new laws to inhibit people's right to peaceful protest.[94]

The police and the logging industry reiterated this theme just over a year later when they held a press conference to discuss 'Eco-terrorism'. Mr Steve Guest, the Public Affairs Manager for the Victorian Association of Forestry Industries, appeared with representatives of the Victoria police counter-terrorist branch.

They warned of the international 'eco-terrorist' group, Earth First! and that tree-spiking was likely to happen on the island. Once again, despite the scare tactics, no 'eco-terrorism' had occurred in Tasmania and there was a commitment from the environmental groups to protest non-violently.[95]

Just under a month later, and crucially two days before the federal election, a hoax bomb was placed on a railway line in north-western Tasmania. Left without a detonator, but coincidentally next to a banner that read 'Earth First, Save The Tarkine', referring to Australia's largest rainforest Wilderness Area being campaigned for by environmental groups. 'Railway Bomb, Environment Group Link,' read the headline in one newspaper the following day as the media focused on Earth First! and the environmental movement. The Federal Forest Minister got in on the act too: 'It is most regrettable that some more extreme elements of the Conservation movement may be willing to use the threat of violence to pursue their cause,' said Ray Groom.[96] This led the leader of the Greens, Bob Brown, outraged by the coincidental timing of the device to retaliate, calling the episode 'A blatant and shoddy set-up. It reeks of a set-up by sympathizers of the loggers.'[97] In the election, the Tasmanian Green candidate who was running for the Green's first Senate seat lost out by 1.5 per cent. Elected instead was a strong supporter of the logging industry. Months later, long after the election, the police cleared environmentalists of any involvement with the fake bomb.[98]

A succession of incidents that Burton has pieced together occurred in Tasmania in 1994 which culminated in late March. On the very day that logging gangs moved into the extremely controversial Jackeys March Forests in Tasmania, a hoax tree-spiking letter was widely distributed throughout the area. The letter, claiming to be from the Evan Rolley Fan Club (after the Chief Commissioner of Forestry Commission) and printed on a high quality laser printer, was written by an extremely competent media specialist with sections mimicking the style of environmental press releases.[99] The decision to log the forest was bound to cause a non-violent protest by activists, who would need public support on their side if they had any hope of stopping the logging. It seemed that this was the first of a series of activities which happened whose sole purpose was to discredit any non-violent campaign.

Burton says:

> That was very curious timing. When you follow the rest of the sequence through, it's brilliant diversionary stuff, timed impeccably to divert attention from something else that was going to be happening on that day. The sequence kept being altered to be plausible in the eyes of journalists.

Burton tracked:

> maybe seven or eight different incidents, the first couple were letters, then a nail would be found in a rather prominent position. They had to keep the nails story alive and they had to keep migrating into the mills and they had to start hitting saws and stuff like that. The interesting thing I found was when I started making phone calls to people in the timber industry, and government agencies it all suddenly stopped. Like instantly. It was quite extraordinary, there has never been another spiking in Tasmania since.[100]

Hysteria over 'eco-terrorism' has swept the country, led by the standard bearer of the eco-terrorism band-wagon, the Forest Protection Society.[101] By the end of 1994, the Minister for Land and Water Conservation in NSW had even set up exclusion zones around logging sites, banning unauthorised access 'as an anti-sabotage precaution'.[102] Burton argues that these hoax eco-terrorist incidents have been used to justify anti-environmental legislation. For example, following 'sabotage' claims in Tasmania, there were moves to make environmentalists liable for damages occurring in projects -something that the British government is attempting with anti-roads protesters in Britain (see chapter 12).[103]

There have been dirty tricks against the environmental movement in other areas of Australia and New Zealand. In 1991 when the Coode Island chemical terminal in suburban Melbourne was ripped apart by a huge explosion, a fire raged for two days causing $20 million of damage.[104] Six weeks later, just days before the commencement of the public hearings into the fire and a week before a major TV investigative report into safety problems in the chemical industry, Victoria police convened a press conference to announce that they had conclusive forensic evidence that pipes had been cut with oxy-acetylene equipment. They speculated that a small group of people may have been trying to light a fire as a protest or make a statement to the government about safety concerns at Coode Island. No local environmentalists were questioned. The local police was deluged with calls from 'welders who stated emphatically that oxy-acetylene equipment could not be used to cut stainless steel pipes'.[105]

Following the claims, local environmentalists received a number of threats. However, eight months later, the police had to admit it had been an accident, and that the investigating government department had believed this all along.[106] Years later, Brian West, the Chief Executive of PR Firm Hill and Knowlton, Australia, revealed that they had been advising the company on the Coode Island explosion. Using Coode Island to illustrate general themes about crisis management, he told an industry audience that crisis management is designed to protect or rebuild goodwill, pre-empt publicity, build equity and manoeuvre the company to be viewed as the victim rather than the culprit. West said:

> It makes a big difference as to how you are perceived and reported in the short term and how you are ultimately viewed in the long term as to which box you are in . . . make sure that you are the victim and that you are clearly seen to be in the victim box not in the culprit box.[107]

The same year as the Coode Island blast, Greenpeace were accused of being a bunch of 'banana terrorists' in New Zealand, when the environmental organisation highlighted the potential contamination of imported bananas by the pesticide aldicarb.[108] Although the 'terrorist' accusation had been printed in the *National Business Reporter*, one of the banana companies caught up in the controversy, Chiquita, was also using Hill and Knowlton to manage the potentially embarrassing situation for the company.[109] Hill and Knowlton had used dirty tricks against environmentalists the year before in the USA by circulating a false press release which proclaimed that environmentalists were prepared to resort to violence.[110]

In November 1995, the Goongerah Environment Centre in Victoria announced that they would be starting a blockade of the state's East Gippland forest. Over the weekend, machinery at the proposed site was damaged, leading to some forest workers ramming a car with two environmentalists inside, who were subsequently assaulted. A leading local environmentalist, Jill Redwood, also received a threatening call that her house was going to be burnt down. Redwood had written to a major newspaper highlighting how attacks on forest equipment often coincided with 'important political events such as elections or major decisions on forests . . . accusations that conservationists are responsible for sabotage creates situations where innocent people are assaulted and property vandalised'. Further threatening calls were received by environmentalists.[111]

Despite all the talk of eco-terrorism and violence by environmentalists, the only people charged and convicted of violence in the forest debate in the last decade have been from the forest industry.[112] The victims have been the environmentalists.

ANTI-GREEN VIOLENCE

There have been sporadic episodes of violence aimed at environmentalists over the last twenty-five years, dating back to the 1972 Lake Pedder Campaign, when two environmentalists disappeared under suspicious circumstances.[113] Nine years later, as part of the ongoing campaign to save the Franklin River, environmentalists conducted a bike ride around Australia.[114] After having been invited to a meeting by Hydro-Electric Commission workers, some dozen activists were beaten up by locals in the town of Tullah. No action was taken by the police.[115] As the campaign to protect Franklin intensified, so did the local anti-environmental sentiment, with anti-green bumper stickers appearing on many cars. In 1983, Bob Brown, the leader of the anti-dam activists was voted in to the state parliament. Within a week he had been assaulted by four youths, who were actually charged with the attack.[116] The Wilderness Society, the environmental group leading the campaign, also had its offices vandalised. 'Damn you greenies', was also sprayed on the walls of their shop.[117]

Other acts of aggression were also perpetrated against activists fighting to save the Franklin river. One activist had the wheel nuts on their car removed which resulted in a wheel coming off whilst the vehicle was in motion. Others had their cars vandalised and tyres slashed and windows smashed.[118] Some activists arrested by the police on demonstrations developed exposure and hypothermia having been kept without adequate shelter and food for over ten hours. On one occasion, by the time they were actually put in prison, many activists had been denied food for over a day.[119] But in the end it was victory for the environmentalists. On 1 July 1983 the High Court of Australia handed down its decision upholding the federal government's legislation which prevented the construction of the Franklin River dam.

Once again it is the forestry debate which has been at the forefront of the violent campaign against environmentalists. Much of this violence has been met with indifference by the police, whose inactivity could be seen as sanctioning the use of violence. For example, after a bulldozer had ploughed into a group of anti-logging demonstrators in the Lemonthyme Forest knocking one unconscious and another in the face, the police refused to assist in the situation despite repeated calls for help.[120]

At the same time, when many environmentalists were beaten or assaulted by workers at demonstrations at Farmhouse Creek in the 1980s, the police once again refused to intervene or prosecute anyone for assault.[121] The Tasmanian Premier, Robin Gray, called the accusations by environmentalists a 'concocted

Plate 9.3 Bob Brown is dragged from a protest at Farmhouse Creek.
Source: The Mercury, Hobart, Tasmania

. . . reckless irresponsible stunt'.[122] After the official inactivity of the authorities a private prosecution was lodged by a female protestor. Although the case was thrown out by the presiding magistrate, the Supreme Court found the Managing Director of the logging company and three of his men guilty, although he refused to fine them. Announcing his verdict, Mr Justice Cox commented that the men may have been 'emboldened' in their actions because they had carried them out in front of police without any interference from them.[123] 'This ruling shows that despite all these crimes of assault committed,' said Dr Brown, leader of the Greens, 'the Government stood by and did not prosecute – thereby failing to ensure justice for many people.' Brown had himself been shot at whilst protesting at Farmhouse Creek, for which two people had actually been arrested although not charged with assault.[124]

Fellow MP Dr Bates said the attempted shooting illustrated exactly what happened when a government set a precedent of allowing people to take the law into their own hands. He said the 'mentally aggressive approach by the Premier, Mr Robin Gray, in his relations with the conservation movement' had directly translated into physical violence perpetrated by people who had felt that they had the authority and support of the government.[125] At best, government officials have shown complete contempt for the anti-environmental violence. During a period of renewed tension between loggers and greens, Members of Parliament even took part in 'toss a greenie' competitions at a pro-logging rally where they tossed a dummy in a pair of green overalls stuffed with straw with a pitchfork.[126] The winner of the 'toss a greenie' competition is now Australia's current Premier. Even worse, the government's attitude has had serious ramifications for the environmental movement.

Bob Burton believes that the government is as much to blame as anyone for the violence against environmentalists, having amassed evidence that proves that the government has incited violence against environmentalists, and refused to condemn it once it has happened. Moreover, it has sought to blame environmentalists for violence when there was no supporting evidence or attempted to blame them when they themselves have been the actual victims. The government has attempted to deny violence has occurred, and it has downplayed violence when it has actually happened. More sinisterly, it has even sought to harass environmentalists from their employment.[127]

Police indifference to the violence is not isolated to the incidents at Farmhouse Creek, either. In February 1992, there was a series of violent incidents against protesters at the Wilderness Society's East Picton Road peaceful blockade. Activists who were protesting against logging in the area were threatened, had rocks thrown at them, were chased and had their vehicles vandalised. By the end of the month the violence had escalated into shooting attacks and two firebombs which completely gutted two vehicles. Once again the police said that no action would be taken.[128] Moreover, the press blamed the victims for the attacks: *The Mercury* editorialised, 'it is not surprising that a worker whose job is threatened should want to strike back at those causing him or her distress'.[129]

Apart from refusing to investigate violence and harassment, once the police actually refused to take statements from environmentalists who had been shot at. Moreover, some openly express support for the logging industry. 'I'm on their side' is what the Senior Sergeant of Tasmania Police, wrote in the local newspaper, openly endorsing the forest industry in their conflict with environmentalists. Others have gone further. On one occasion in the South East forests, a group of protesters were attacked by loggers. A policeman actually joined the fight and punched a female environmentalist in the chest cracking her sternum.[130]

This line that the victims are to blame has its roots deep in the forestry industry, who have attempted to politicise the media and the press and pit them against environmentalists. For example, a Director of NAFI, David Bills, wrote in 1993, 'if violence does emerge, before passing judgement, we should take time to understand the perspective of somebody being driven to financial ruin'.[131] 'This seems very clearly to be excusing violence,' says George Monbiot from Green College, Oxford, in England. Monbiot and other British environmentalists were expressing concern at Bills' appointment as the new Director General of the British Forestry Commission, the nation's largest landowner, in December 1995.[132]

To some within the industry violence against environmentalists is both forgivable and justified. Mr Jack Kile, the head of the Esperance Progress Association, a pro-logging organisation, talking about environmentalists to the national press commented that 'offer the locals $10 a head and we'd soon clean them up'.[133] 'If we have to fight, if we have to physically confront these people who have opposed us for so long, then so be it,' said Mr Col Dorber, the Executive Director of the NSW Forest Products Association, on ABC TV. 'I also say to people in the industry, if you are going to do that, use your common sense and make sure it's not being filmed when you do it.' Dorber was effectively inciting covert violence against environmentalists.[134] Not surprisingly, he was forced to apologise after his remarks caused a public outcry. Some sections of the press, though, were less repentant; 'It may not be palatable to say so publicly but violence can sometimes be good,' wrote Miranda Devine in the *Telegraph Mirror*, 'it's a pity that Dorber felt he had to apologise . . . there comes a point in any disagreement when diplomacy ceases to be of any use. That is when violence has its place.'[135]

With public attitudes like these it is hard to see how the increasing use of violence against environmentalists will end, and so it continues. 'Tell that Chris Sheed to pull his head in or he will get it blown off', was one such warning that outspoken environmentalist Chris Sheed received in 1992. Someone threatened to burn down his house two years later in January 1994. 'You're dead, you slut' was the message the person at the Wilderness Society office received when they picked up the phone three months later. Activists were assaulted and harassed in 1992, 1993, 1994 and 1995 in different incidents around the country.[136]

Perhaps the most extreme harassment was against opponents of a mine on the West Coast of South Island in New Zealand. The mining company, a subsidiary of North Broken Hill, proposed to mine mineral sands along a coastal strip and into a local reserve. One activist and her family have been subject to systematic harassment. She had been told by a local contractor that a 'price' had been put on her head. Her daughter was subjected to sexual abuse at school. The police, she felt, were indifferent and dismissive every time she reported incidents to them. On 15 November 1993 at 4 p.m., the activist and her husband bought a house on adjourning land to their current plot of land. By midnight the house had been burnt to the ground. The local media refused to run the story, and there have been no prosecutions for the arson attack. The children are still subject to abuse at school.[137]

Other harassment continues unreported and unnoticed apart from those people at the blunt end of the intimidatory violence. For example, leading Tasmanian environmentalists living in rural areas have had their letterboxes destroyed on a regular basis in the last few years. Some have been demolished up to thirty times. One Victorian activist had a log truck blast its horn every morning at 4 a.m. for months as it drove past her front gate. One morning two severed goatheads appeared in her letterbox. The local policeman was notified, but subsequently admitted he could take the matter no further. In 1993 a car driven by a protestor pulled over to let a truck pass on a plateau road. The truck slowed and carefully hooked the car with its bull bar and pushed it over the edge with the driver inside.[138]

Another prominent woman environmentalist's dog disappeared in suspicious circumstances, her fence was adorned with obscene graffiti and property was stolen from her house. Her car was vandalised and human faeces was left on her front door mat, whilst she received numerous threatening phone calls in the middle of the night. One of the calls was a person who wrongly informed the woman that her son had just been killed in an accident.[139]

THE WAY AHEAD

Alarmed by what he sees as a vicious circle of violence and harassment that could spiral out of hand, Bob Burton has started to inform other environmental activists of what he sees as a significant problem. A problem not to be shied away from or feared, but one to be exposed and learnt from. He believes that there are five key sections of society that need to act: the police, the industry, politicians, the media and lastly, environmentalists.

Industry and government agencies, he believes, must be 'willing to and publicly be seen to discipline its supporters against harassment and incitement to violence'. The police too must publicly state that incitement and harassment of environmentalists are unacceptable. Moreover, they must retain their independence and stop being involved in joint public relations strategies with industry front groups. Both the media and politicians can also assist by refusing to blame the victims of harassment and violence for its actual occurrence. Politicians should go further, argues Bob Burton, and 'accept that as opinion leaders their words and actions can be critical in determining whether their supporters go beyond healthy debate to viewing harassment and violence as acceptable'.[140]

Environmentalists, as they are beginning to do in the USA, must take steps to mitigate violence and harassment. They should highlight that violence is occurring and support the victims of such acts.[141] In response to the increasing harassment, moves are underway to establish a support service for environmentalists in Australia and New Zealand to prevent and mitigate the impact of legal and violent harassment and dirty tricks campaigns. 'It is readily apparent that many of the ways of preventing harassment and dirty tricks have already been done but not widely recognised,' adds Burton. 'I'm very confident that within a short space of time much of the harassment can be prevented and, where it does occur, mitigate its effects. Most of the dirty tricks campaigns can be prevented with a little investigation.'[142]

10

SOUTH ASIA AND THE PACIFIC

Where dissent can mean detention or death

Land is our life. Land is our physical life-food and sustenance. Land is our social life, it is marriage, it is status, it is security, it is politics; in fact, it is our only world. When you take our land, you cut away the very heart of our existence.

Residents of Bougainville[1]

Many of the countries in Asia and the Pacific differ in one crucial respect to the ones examined so far. Dissent in some of these countries is barely tolerated and the backlash against an organisation for speaking out on an issue – whether political, economic, ecological or religious – could be severe. In some countries the simple expression of free speech is not tolerated, let alone political discussion or dissent. It can be classified as a subversive activity, punishable by harassment, imprisonment, torture and death.

MALAYSIA

OPERATION LALLANG

Although Malaysia differs from some of the countries in the region by allowing some degree of environmental activism, there are times when the government has cracked down on dissent in the country *per se*. 'Operation Lallang' took place between October and November 1987, when at least 106 people were detained under the Internal Security Act (ISA), to silence criticism and pre-empt political dissent. Originally introduced in Malaysia in 1960 to flush out 'communist terrorists', the ISA has been invoked by all four Prime Ministers since independence to detain people because of their political activity. The supposed crime of the hundred detainees, who included prominent politicians, trade unionists, community workers, educationalists, church officials and environmentalists, was that they threatened national security. However, no specific charges were brought against them and no evidence produced that they had ever advocated or used violence.[2]

Under the Act, any person can be detained for sixty days without charge. All were held in solitary confinement, many denied access to family and relatives. All were refused legal representation. Forced to sleep without a mattress, the lights were permanently left on. Heavy-handed interrogation techniques, interspersed with more standard formal questioning was used against the detainees. Confessions were extracted from people who were often interrogated for periods of over ten hours and who were put under severe mental and physical duress. People were beaten, stripped naked, slapped, punched or pulled by the hair. One man was stripped and threatened by a police officer with a lit rolled-up newspaper that he was about to burn his genitals. All the detainees reported being interrogated for hours in supercooled rooms, making them shiver with cold, many were placed directly in front of the cooling system or sprayed with freezing water.[3] Amnesty International condemned the 'systematic use of such interrogation methods as a violation of international legal standards'.[4]

Among those held were prominent environmentalists within the country, including Tan Ka Kheng, the Vice President of the Environmental Protection Society of Malaysia (EPSM); Meenakshi Raman, a legal advisor to both Sahabat Alam Malaysia (SAM which is Friends of the Earth Malaysia) and the Consumers Association of Penang; Harrison Ngau, a leader of SAM Sarawak and Arokia Das, a member of SAM. Hew Yoon Tat from the Perak Anti-Radioactive Committee (PARC), who publicly came to the defence of Tan Ka Kheng, was himself arrested two days later with four members of his committee.[5] Ngau received a two-year restricted residence order and Arokia Das, a two-year sentence in a detention camp.[6] It was due to her involvement in the ARE radiation case that Meenakshi Raman was arrested, and put in jail for forty-five days.[7] Although the four members of PARC were also released they were placed on a two-year restricted residence order.[8]

The EPSM is a non-profit NGO working for ecological enhancement with the poor and disadvantaged in Malaysia. Tan Ka Kheng had been supervising a number of environmental studies and headed the Papan Support Group, who were helping people campaigning against radiation from a monazite factory owned by the company, Asian Rare Earth (ARE). He was also opposing the

Bakun Dam and lecturing in the Department of Civil and Environmental Engineering.[9] The ARE plant and the Bakun Dam have been two of Malaysia's contentious ecological issues for over a decade.

BAKUN DAM

The Bakun Dam is the largest public works project ever undertaken by the Malaysian government, at a cost of RM15 billion (US$6.25 billion). Set to submerge an area of forest the size of Singapore and displace some 7,000 tribal people in Sarawak on the island of Borneo, the plan is highly controversial.[10] For questioning the need for the dam, Tan Ka Kheng was charged with being engaged in 'activities propagating communist influence and ideology among specific sections of the population in Malaysia'. His activities had 'undermined national security'. For his supposed crime, he was served with a two-year detention order.[11]

Due to continuing controversy over its size and impact, the original Bakun Dam proposal was shelved, only to resurface in 1993, with different design specifications. The new proposals have not been without conflict, with the government severely criticised for proceeding with the dam several months prior to the publication of environmental impact assessments (EIAs). When the first EIAs were released in July 1995, they showed that the dam would not only lead to the clearing of more than 69,000 hectares (170,430 acres) of tropical forest but would also have a negative impact on the flora and fauna of 1.5 million additional hectares (3.7 million acres) of land. Environmentalists also castigated the credibility of the EIA and how much impact it would have on the government's promise to stop the project if the EIA's results were negative.[12] Minister Datuk Seri Dr Mahathir Mohamad described the dam as 'very good', adding that its benefits far outweighed its environmental or social costs.[13] However, in August, representatives of the 7,000 tribespeople who would adversely be affected by the dam demonstrated against the project, because they had not even been consulted about it. In December 1995, the state government gave the dam 'environmental approval'. However, in June 1996, the High Court announced that the dam was in violation of the Environmental Quality Act, because it had a flawed EIA. Despite this, work continued unabated.[14]

ARE YOU OUTSIDE THE LAW?

Meenakshi Raman, another detainee of Operation Lallang, was a lawyer assisting in the case against Asian Rare Earth Sdn. Bhd (ARE) filed by eight residents of Bukit Merah and supported by 10,000 local residents. The eight had filed suit in February 1985 in the Ipoh High Court to stop ARE producing, storing and keeping radioactive waste near the village. At Bukit Merah, the company processed monazite, a by-product of the tin industry, to produce the rare earth metal yttrium. The process is highly toxic, producing both radon, a radioactive gas and two radioactive wastes, thorium hydroxide and barium radium sulphate. In fact, due to the risk of radiation and pollution, the importing and processing of raw monazite had been banned in Japan. Despite this, Mitsubishi, the Japanese multinational, is the joint primary shareholder in ARE. The effects of the radiation on the local population have been devastating. Health surveys had found levels of childhood leukaemia over forty times greater than the surrounding area. Miscarriages and infant mortality were also 'abnormally high'. Eight international experts declared the storage sites unsafe.[15]

In October 1985, the High Court judge granted an injunction to the residents of Bukit Merah to stop ARE producing and storing radioactive material, until the storage facilities were deemed safe. In February 1987, the national Atomic Energy Licensing Board disregarded the court decision and granted the company a licence. Finally the case came to court in 1990, and lasted thirty-two months until the High Court announced that the plant should be closed down. However, the Supreme Court subsequently ruled that closure of the plant would cause undue hardship for the 183 workers and ARE's international shareholders and the decision was reversed against the community. 'It was a political decision,' says Philip, an environmental activist who has worked in Malaysia for a number of years.[16] In January 1994, SAM announced that Asian Rare Earth had finally decided to permanently close their operations at Bukit Merah.[17]

The ARE case highlights the difficulties environmentalists face with the legal system. Although the lower courts are felt to be non-partisan, the Supreme Court appointees are political. This works against environmentalists in two ways. 'NGOs recognise that it is difficult to win damages or get a favourable outcome in court because of undue political influence in the judicial process,' says Philip.[18] The law is also used against environmentalists and the Consumer Association of Penang (CAP) had been sued by the government for material which they published in the *Utusan Konsumer*. Moreover, a recent move by the Supreme Court to fine four journalists some RM $10 million (US$3.8 million) in damages in a libel case brought by Vincent Tan, CEO of the Berjaya Group, due to an article concerning his business practices, sent shivers down the spines of people concerned with the expression of free speech.[19]

A CLIMATE OF REPRESSION

The repressive legislation and oppressive techniques against environmentalists are nothing new to activists and indigenous people who have been attempting to preserve some of Malaysia's fast disappearing forest. They are used to dealing with a corrupt, repressive regime that has a vested political and economic interest in making sure the forests are felled with as little national disruption and international condemnation as possible. People speaking out on forestry issues face a ferocious backlash for doing so.

'Any type of alternative viewpoint, any kind of position which is critical, and I do not mean critical in a major sense, but which asks questions about the wisdom of government policy is not tolerated,' says Philip. 'The Government can just shut you down and there are strict censorship laws in Malaysia.' Philip continues:

> Every year, if you want to publish a magazine you have to have a licence to do so and an NGO has to get a licence to operate, and they can arbitrarily just refuse that licence. The same goes for having meetings. If you want to hold a public meeting of more than ten people, you have to have a police permit to do so. It is a very legal tool that they can control people. It is difficult to operate as an NGO to say the least.[20]

People are fearful and have to be careful. Philip is not the activist's real name.

The authorities are also permanently monitoring the group's activities. Philip adds:

> They watch and they listen and if you touch a raw nerve, and the particular raw nerves at the moment are deforestation, logging, anything about tropical timber and how it shouldn't be bought and sold or traded, as well as indigenous rights, you could be in serious trouble. If you publish too much they will then pull your magazine or even close you down.

An NGO's source of funds can be scrutinised and sequestered by the authorities.[21]

Plate 10.1 A typical logging camp in Sarawak at Tubau.
Source: E. P. L./Rod Harbinson

Logging is a sensitive issue with the government, because many leading politicians own logging concessions in Sarawak, the state which has the largest remaining tracts of rain forest and is home to thousands of indigenous peoples. These communities are often ignored in the industrialisation development model imposed by the federal and state governments, which benefits private business at the expense of both local communities and the environment.

Philip contends:

> In Malaysia and in that region generally, the issues of indigenous rights, the rights of peoples, the rights of communities are a total anathema to the government. They are very scared of it and they don't understand it. They really do think you are challenging the legitimacy of the state government or the national government by campaigning on indigenous rights.[22]

The government has a genuine paranoid belief that environmentalists working on forestry are just exercising another form of imperialistic control by the First World on their neighbours in the South.

For many countries in South Asia there is a massive preoccupation with development and industrialisation at almost any cost. The government rationale is that development can only be achieved by economic growth led primarily by privatisation, foreign investment and freeing barriers to trade. For this to happen, there has to be foreign investor confidence, hence the preoccupation with political stability. 'The only way you can ensure political stability is to ensure there is no dissent,' says Philip. 'To be labelled as anti-development is almost to commit high treason, in the eyes of certain politicians.'[23]

THE FORGOTTEN PEOPLE

The indigenous people are stigmatised by the authorities. They are characterised as backward and 'eating monkeys and suffering from all kinds of diseases, and in desperate need of development themselves'.[24] They must suffer for the benefit of all. 'Outsiders want the Penan to remain nomadic and I will not allow this because I want to give a fair distribution of development to all communities in the state,' says the Chief Minister of Sarawak, Datuk Patinggi Abdul Taib Mahmud, who directly controls 10 per cent of Sarawak's logging concessions and is set to make a personal fortune from timber. His relatives control around one-third of all logging concessions on the island.[25]

One of these group of indigenous people with whom the government have long been at loggerheads are the Penan, who have resisted the rampant deforestation by the logging companies and have been fighting to protect their land. The Penan are semi-nomadic and subsist by hunting, fishing and gathering in the rain forest, and although some Penan would admit that they could benefit from better standards of living, rapid deforestation of the forest, with no consultation with them, threatens their very survival. Predominantly illiterate and without any kind of political representation, the Penan have used the only option open to them to stop the logging. They have been manning non-violent blockades across the logging roads. It was for supposedly organising these blockades that Harrison Ngau was arrested.[26]

The government and timber industry have attempted to blame the deforestation, not on the illegal logging activities, but on the Penan themselves for practising the centuries-old tradition of shifting cultivation. 'Europeans should blame the Penans instead of the Government for destroying the forests,' said Prime Minister Datuk Seri Dr Mahathir Mohamad, at the time of the mass arrests in 1987. Shifting cultivation is widely regarded by forest experts as posing

Plate 10.2 Penan blocking logging road in rain forest.
Source: E.P.L./Jeff Libman

no threat to the forest.[27] The plight of the Penan has made international news for years, especially through the dedication of one man: Bruno Manser. A Swiss national and shepherd by trade, Manser lived with the Penan for a number of years in the 1980s. Basically becoming an adopted Penan, living with the tribe in the forest and assisting them in their anti-logging campaign, Manser became 'public enemy number one' in both the eyes of the government and the logging industry. So threatening was Manser to the industry and authorities that the government called him a 'white tarzan' and 'subversive Zionist and communist', whilst the timber companies offered a $30,000 reward for his capture.[28]

SMEAR CAMPAIGNS

Using smear campaigns and disinformation techniques such as those against Manser is a tactic that the authorities have adopted against anyone working on ecological issues or the protection of indigenous peoples in Malaysia, especially if they are foreign. Over the last ten years there has been a concerted effort to marginalise and scapegoat environmentalists, ecologists and anthropologists who have been working on the Malaysian forestry debate. In language that is remarkably similar to that used elsewhere on the globe, Harrison Ngau had been labelled a 'communist stooge', whereas in general anti-logging activists have been called 'the number one traitors'.[29] Environmentalists have been accused by the Sarawak Deputy Chief Minister of exploiting the Penan for selfish aims.[30]

'Environmental NGOs have for a long time been classified as agents to bring down the government,' says Philip, because they 'challenge the government in their development programme'. International environmentalists are similarly dismissed as 'eco-imperialists' or racists. During the UNCED Conference in Rio in 1992, one of Malaysia's delegation, Ting Wen Lian, labelled environmentalists as 'Nazis'.[31]

The stakes in the forestry debate were further raised when European environmentalists, appalled by the indiscriminate logging practices and high deforestation rate, called on a boycott of Malaysian tropical timber products in the late 1980s. Friends of the Earth's *Good Wood Guide* blacklisted Malaysian timber. To head off the boycott, a delegation of officials went to Europe, accusing environmentalists of being 'patronising' and of 'belligerent idealism'.[32] Dr Lim Keng Yaik, the Minister of Primary Industries, and leader of the delegation accused soft-wood producers of being behind the anti-hardwood campaign.[33] He also defended Datuk James Wong, the Sarawak Environment Minister, who is also a timber baron, who owns over 300,000 hectares of forest concessions. 'Logging is good for the forest,' says Wong. 'We get too much rain in Sarawak.'[34]

But the boycott was beginning to have an impact. 'I know the reason they went for a full-on PR campaign was that it was so successful,' says Philip. 'With environmental groups saying to the public and lobbying their politicians, "stop buying tropical woods from Sarawak, it had such a huge impact". Sales of tropical timber plummeted, and they were really, really scared.' Moreover, the authorities were immensely angry as they saw the boycott as a further example of hypocritical eco-imperialism.[35]

PUBLIC RELATIONS VERSUS PUBLIC REPRESSION

The authorities acted by investing in a counter-campaign. In 1992, $4 million was donated by the Malaysian Timber Industry Development Council to send a European taskforce to 'repel falsehood and lies spread by evil-intended environmentalists' who have supposedly been 'brainwashing' people, with the sole purpose of damaging Malaysia's reputation abroad.[36] The following year they were reported to be setting up an office in the United Kingdom.[37] Both the Malaysian Timber Industry Development Council and BC Forest Alliance are clients of the global anti-green PR company, Burson-Marsteller. Another PR company used by the Malaysians is Lowe Bell Communications.[38]

So while the PR companies legitimise Malaysia's forest industry to the world, the oppression of critics continues. Anti-logging activists also have to contend with intimidation, violence and widespread repression. Anderson Mutang Urud, Executive Director of the Sarawak Indigenous People's Alliance was arrested and detained for his peaceful work on deforestation. Although later released on bail, he was charged with directing an illegal society, an offence that carries up to five years in jail. His real crime though was to have brought a visiting Canadian politician to witness logging blockades by indigenous people in Sarawak. 'Prior to his release on bail in March 1992, he was subject to gruelling interrogation and allegedly threatened with torture. He is now in exile in Canada,' says Philip. By 1992 some 500 tribal people had been arrested.[39]

The Sarawak government has tried to prevent anti-logging activity by restricting outsider access to logging areas, and by controlling the rights of indigenous people to travel outside of Sarawak to highlight their plight. 'Most of my colleagues who have represented indigenous communities in court against land invasion are just simply banned from going to Sarawak. Banned for life,' says Philip, who describes the case of Thomas Jalong, a native who works for SAM in Sarawak. Jalong had his passport confiscated to prevent him attending an ITTO (International Tropical Timber Organisation) meeting in Japan, where he was going to describe the native communities' fight for their land against logging.[40] In July 1993 poet Cecil Rajendra had his passport seized before leaving on a poetry tour to Britain. The authorities said he was restricted because he was a prominent anti-logging activist, who 'could damage the country's image overseas'. Rajendra maintains his only crime was that some of his poems touched on ecological issues.[41]

In 1991, three journalists from Germany, England and America were detained and interrogated by the Sarawak police, and eventually deported. The previous day they had been covering an anti-logging demonstration.[42] The same year another British journalist was deported for attempting to cover the timber issue. Even if journalists can actually get into the country, they are now banned from visiting the sites of anti-logging activities.[43] In 1992, the Malaysian authorities were reported to be setting up a special police unit which would 'monitor western and local environmentalists'.[44]

There is also concern amongst environmental groups that Malaysia is jumping on the greenwashing 'eco-labelling' bandwagon, attempting to portray to the outside world its green image. The Tourist Development Council tells visitors of the 'wealth of natural unspoilt beauty,' 'the world's oldest unspoilt jungles' and that the 'many indigenous people of Sabah and Sarawak' are all reasons to visit the country.[45] The Malaysian Timber Industry Development Council (MTIDC) have also stepped up their international PR effort, with adverts in respected journals. One in the *International Herald Tribune*, headed 'Malaysia Evergreen Managing for Perpetuity', talked of 'Forest is Forever and Malaysia Forever Green'.[46]

Burson-Marsteller, PR advisors to the MTIDC are also reiterating the greenwash for their clients. 'The industry is committed to, and making progress towards sustainable forestry. We think there is a positive story to tell,' says B-M's Simon Bryceson, who used to work for Friends of the Earth.[47] In June 1995, another timber delegation was dispatched: this time to Australia, New Zealand and the South Pacific countries. 'We have to keep telling the world that our forests are well managed and we are good forest managers,' said Primary Industries Minister Datuk Seri Dr Lim Keng Yaik, the leader of the mission.[48] The MTIDC have also started organising press tours of the forests, where journalists are taken on luxury, all-expenses-paid 'familiarisation' tours of selected forests, which the industry want people to see. This is a standard PR ploy used by industry worldwide, and one where balance is replaced by bias.

The truth depicts a far less sustainable future. Anderson Mutang Urud, from the Sarawak Indigenous People's Alliance, released from detention earlier in the year, addressed the General Assembly of the United Nations in December 1992.[49] He said:

> Sarawak is less than 2 per cent the size of Brazil, yet it is currently producing almost two-thirds of the world's supply of tropical timber. Even if the current rate of logging were immediately reduced by one half, all primary forests in Sarawak would be destroyed by the year 2000.

In 1994, a Malaysian government report concluded that five states had over-logged whilst not providing new ground for timber growth.[50] According to the

Rainforest Information Centre in Lismore, the volume of logging 'far outstrips even the most conservative estimates for a "sustainable" harvest'. Warned in 1990 by the International Tropical Timber Organisation (ITTO) to cut annual logging production to 9 million cubic metres per year, the rate has stayed at between 16 and 19 million for the last five years.[51]

INDONESIA

Indonesia joined Malaysia in the fight against the anti-timber boycott in the late 1980s, and although the country has a chronic illegal logging problem, its forest industry has stepped up the PR war as well. Bob Hasan, chairman of the Association of Logging Companies of Indonesia announced that he was donating some US$2 million to fight against the environmentalists' campaign.[52] One advertisement in 1989 in *The New York Times* used the slogan 'Indonesia – Tropical Forests Forever'.[53]

The facts, however, show a less sustainable future. In 1990, a World Bank study reported that Indonesia's harvest from tropical forests was about 50 per cent higher than sustainable levels. Analysis of satellite pictures by environmentalists have shown that Indonesia's total forest area has plummeted from 120 million hectares in 1985 to 89 million hectares in 1992.[54] Industry is also involved in an ingenious scam. Clearfelling for pulp in industrial forest estates is theoretically only allowed if less than 20m^3 of timber exists per hectare. Companies selectively log natural rain forest for timber, then clear-fell the area once the quantity per hectare has fallen below the permitted minimum.[55] Meanwhile the industry's critics are just closed down. In 1989 the environmental group, Skephi was ordered to stop publishing its bi-monthly journal, *Forest News*, which focused on the social and political implications of environmental issues.[56]

In 1994, Bob Hasan once again declared war on NGOs working on the socio-economic and environmental impacts of deforestation. In May, a major glossy international advertising campaign was launched by the Indonesian Timber Community, an organisation Hasan chairs. The advertisements, depicting lush tropical forests, were shown in Indonesia, The Netherlands, Germany, Japan, France, Britain and on the CNN news channel. They claimed that clear-cutting was not permitted in Indonesia, that 280 million acres of forest had been permanently protected and that 9 billion trees had been replanted. There was only one thing wrong with the advertisement: the facts and images used were wrong. In August, the Independent Television Commission, having been approached by various environmental groups, banned the advert for depicting inaccurate images and distorting the true facts of forest management in Indonesia.[57]

Hasan's company has been accused by many local and international NGOs of causing environmental devastation in Kalimantan, on the island of Borneo. His company is also charged with breaching Indonesian logging regulations in the area and of violating the human rights of the local Bentian people. Local ancestral graves were dug up by Hasan's company operating under armed guard and their traditional lands were destroyed as were their sustainable agroforestry schemes. Their complaints were met with intimidation and threats by the authorities, with some leaders being interrogated and subject to severe harassment.[58]

Environmental NGOs working in Indonesia are subject to severe restrictions that limit their freedom of speech and ability to campaign without harassment by the military. They can also be closed down at a whim. A new presidential decree in 1994 made it even harder for the groups to operate. Along with other NGOs an environmental organisation now faces suspension if it 'undermines national unity and integrity (or) undermines the authority of the government and/or discredits the government'. They could also be dissolved if they engage 'in activities which threaten public security and order and/or receive foreign assistance without the prior approval of the central government and/or provide assistance to foreign parties of a damaging nature (to) state or national interests'.[59] The decree was widely seen as a blunt way of curtailing NGO activity and effectiveness, and of undermining basic rights to freedom of speech.

Other people in Indonesia are fighting ecological campaigns as well as ones of political self-determination, or self-rule. They are waging a joint campaign against ecological destruction of their homeland and for a right to participate in decisions and policy making that affect their lives and their land. Although their political and environmental struggles are inseparable, the stakes involved are much greater because the government views such dissent as a threat to national cohesion and security, whereas the local communities see themselves as having to fight an army of occupation for their own survival. Although different from other people suffering from the green backlash, it is necessary to highlight the plight of people campaigning for ecological justice and political self-determination.

WEST PAPUA

The people of West Papua, as they call themselves, or of Irian Jaya, as Indonesians have labelled them, are fighting such a campaign. Their international plight was raised in January 1996 when they kidnapped twenty-four Westerners who were rescued later in the year by the Indonesian authorities. They not only oppose the Indonesian authorities whom they view as illegally occupying their land, but also dispute activities that threaten their environment. Their protests have historically been met with force. In the early 1980s police shot dead a person in the village of Sawa-Erma who was attempting to organise support to halt logging on their ancestral lands. The policeman concerned was never prosecuted.[60] Four years later, local indigenous people were jailed for seeking compensation from local authorities for oil exploration and plantation activities that were being carried out on their lands. They were jailed for at least three years.[61]

However, the current cause for concern is the Freeport Copper mine jointly owned by the New Orleans-based Freeport McMoRan Copper and Gold Corporation, the Indonesian government and the British mining monolith RTZ. The mine, the world's single largest mining operation, has copper reserves estimated at some US$23 billion and reserves of gold of US$15 billion, and contributes 47 per cent of Irian Jaya's GDP. Over US$1.7 billion of British money is being invested in the mine.[62]

To exacerbate an already tense situation, in 1994 the company's concession area was expanded from 10,000 hectares to 2.6 million hectares. 'The potential is only limited by the imagination', says the Chairman of Freeport, James Moffett.[63] Kelly Kwalik, the leader of the local resistance movement, maintains that this new concession contract will further degrade the local environment and culture. He sees no future for the indigenous people.[64] The mine has had a devastating effect on the local people, many of whom will be forcefully repatriated, having lost all their ancestral land. Tribal elders report an increase in social and cultural fragmentation due to the mine and heavy pollution in the Ajkwe River.[65]

In fact, the Ajkwe River is said to be so polluted that people have been warned not to drink it, or even eat crops growing near it.[66] In 1993, 50,000 tons of tailings from the mine were washed into the Ajkwe River, further exacerbating the pollution, which has been linked to increasing incidence of diarrhoea and miscarriages.[67] The 'massive deposition of tailings in the Ajkwe River and Minajeri River severely degraded the rainforests surrounding (them)', and was a contributory factor to the US government agency, the Overseas Private Investment Corporation (Opic), withdrawing $100 million in political risk insurance from the mine in 1995.[68] The agency also believes that the project 'has created and continues to pose unreasonable or major environmental, health, or safety hazards with respect to the rivers that are being impacted by the tailings, the surrounding terrestrial ecosystem and the local inhabitants'.[69]

The Indonesian Environmental Organisation, Walhi, also alleges that the Freeport mine operates to double standards, as it would not be allowed to discharge potentially toxic waste directly into rivers in Europe or the USA. 'This huge mine is massively damaging the rich biodiversity of the area and harming the health and sustenance of local indigenous communities,' concludes Walhi. The Freeport company has attempted, unsuccessfully, to stop US government funding for Walhi.[70]

Protests against the mine have been met with brutal force. Over thirty-seven villagers have been killed in the last two years. Eye witnesses report horrific acts of violence against indigenous people, including routine intimidation and torture. One person recounts being beaten with rifle butts, having his body sliced with razor blades, and being given electric shock treatment. Another detainee was so badly beaten that he threw up blood, coughed up blood for over a month, and lost numerous teeth during the torture.[71] On one occasion a peaceful protest was met with security people just firing into the crowds at random with no warning and no reason. Despite the fact that the killings have been verified by the BBC and the Human Rights Office of the Australian Council for Overseas Aid, Freeport mine officials have denied the killings, adding that the company's presence 'had benefited the indigenous people'.[72] RTZ says of the demonstrations that 'we understand all the villagers were released and accounted for subsequently'.[73]

In May 1995, the local Catholic Bishop, Mgr H. F. M. Munninghoff reported that troops were actually torturing the locals inside Freeport's own containers, and in the company's security positions. The following month, TAPOL, the London-based Indonesian Human Rights Organisation declared 'it is clear that there is a major operation under way to crush all resistance to the company and to physically "cleanse" the territory of local people who are regarded as a threat to the activities of the company'.[74]

BOUGAINVILLE AND THE SOLOMON ISLANDS

Another mining operation in the Pacific that is caught up in a struggle for political independence is the mine on the island of Bougainville. In 1963, the Australian subsidiary of RTZ was granted permission to explore for minerals on the island of Bougainville, at the time under Australian control. Although Bougainville is part of the Solomon Islands, Australia passed it to Papua New Guinea (PNG) in 1975, when PNG became independent. 'In essence,' says Martin Miriori, Secretary to the Bougainville Interim Government, 'Australia gave Bougainville and her people as an independence gift to Papua New Guinea.'[75]

The local people were not consulted on the decision to mine the island and the Australian government gave land that belonged to the islanders to the mining company. The Bougainvillians, having seen their land and rivers polluted and their rights destroyed, have begun a campaign for ecological justice and political self-determination. The environmental and social devastation, coupled with the inequitable benefits of the mine have also added the impetus for secession from a country with which they have no ancestral connection.

Over 400 hectares were chemically sprayed with chemicals such as arsenic pentoxide solution and the herbicide 'Bush killer', to make way for the mine. No environmental impact assessment was ever undertaken. The mine is one of the largest holes on the planet, measuring an area six kilometres long, by four kilometres wide, and half a kilometre deep. Operations have caused chronic air and water pollution. Over a billion tonnes of chemical wastes and tailings have been dumped into rivers, which in turn have polluted over 4,000 hectares of once fertile river valleys. In fact 3,000 hectares of land have been totally destroyed. Wildlife and fish have been devastated. The area suffers from chronic heavy metal pollution, including mercury, cadmium, lead, zinc and arsenic.[76]

Despite this, the mine was said to be the 'Jewel in RTZ's crown'. It had brought riches to the RTZ and the PNG government. Between 1972 and 1989 export earnings were US$6 billion, and the mine had accounted for 45 per cent of PNG's export earnings.[77] Protests though have been met with violence. When in 1965 the Bougainvillians pulled down the first survey stakes in protest, 200 of them were jailed, with many being beaten in custody. Subsequent peaceful protests were met with force, including tear gas. Women and children were clubbed senseless by riot police sent in to remove the protesters who were defending their traditional land.[78]

When in 1988, the local villagers believed that a report whitewashed the environmental and health problems of the mine, young militants started acts of sabotage to close down the mine. The response of the authorities was to order a shoot to kill policy against protesters. The following year the PNG army itself was called in to quash the dissent. Over 6,000 homes were destroyed and 24,000 people (out of a total population of only 200,000), were displaced from their homes. Villagers were ordered into 'care centres', which are little more than concentration camps. Some 40,000 people still live in such centres.[79]

The struggle to shut down the mine has been superseded by the struggle for independence, although the two remain intricately linked. In 1990, the PNG army and police, defeated in their campaign, left the island, and mounted a total blockade of Bougainville. On May 1990 the people of Bougainville declared independence, but their so-called freedom has come at a heavy price. Over 8,000 people are known to have died as a result of the blockade. The Red Cross estimated in 1992 that this included some 2,000 children who had died due to lack of medicines. Furthermore, by 1990, malaria had increased by 180 per cent and the cases of stillborn births by 80 per cent, attributed to the blockade.[80]

Despite the army's withdrawal, they have periodically returned to the island to loot, bomb, burn, rape, pillage, degrade, torture, and kill the local Bougainvillians. Some 500 people have been reported killed by these attacks.[81] Although a peace agreement was signed in 1994, and the blockade was meant to be lifted, in reality it still exists. Although the mine remains closed, its reserves are estimated at nearly 500 million tonnes of copper and gold, and Bougainville remains a testimony to a people whose struggle could only be solved by the political will so often found lacking in today's world.

Bougainville is historically and geographically associated with the Solomon Islands. As timber supplies dry up elsewhere in the world, transnational logging companies, especially from Korea and Malaysia, are exploiting the forests of the Solomon Islands. These concessions are often given without the consent of local islanders, and the companies have been accused of bribery and offering prostitution in order to gain access to ancestral lands.[82]

As islanders see their livelihood destroyed, conflict has followed. In 1994, paramilitary soldiers were sent to one of the Solomon Islands to protect employees of a logging company after islanders armed with knives and axes threatened them. In April 1995, Greenpeace called on the Solomon Islands' government to withdraw military troops sent to Pavuvu Island to counter the protests by indigenous islanders against logging on their lands. The following month the Solomon Islands' Prime Minister, Solomon Mamaloni, accused foreign NGOs of stirring up trouble and warned them not to meddle in the development of the Islands.[83] In June, the Solomon Islands' Foreign Minister accused activist groups of spreading disinformation that foreign companies were ravaging his country's environment. 'We regard logging and the forestry industry as part and parcel of the total development plans of Solomon Islands,' said Minister Danny Philip. 'We cannot isolate logging and treat it differently.'[84]

On 30 October 1995, Martin Apa, a well-known Russell Islander opposed to the logging on Pavuvu Island, was killed. The postmortem found his neck was broken and a sharp object had pierced it. 'Here is a community that has been trying to move forward to provide a future for their forests and children, only to be thwarted by an aggressive logging operation,' said Grant Rosoman, Greenpeace New Zealand Forests Campaigner. 'Martin Apa was a key supporter of ecoforestry and we fear that he may have been targeted for the stand he took. Pavuvu landowners are being intimidated into agreeing to further unsustainable logging.'[85] The government seemed reluctant to investigate the killing. In a separate development, three of the seven Solomon Island government ministers appeared in court on charges of receiving bribes from the Malaysian logging company embroiled in the controversy.[86]

THE PHILIPPINES

The Philippines is another country in the region where environmentalists pay a heavy price for their activism, fighting against a system where political power is vested in those who control land or own logging concessions. Communities that have tried to protect their land and resources by opposing the logging or mining industries have suffered from militarisation, imprisonment and human rights abuses. Furthermore, logging in the country is out of control. A Philippines Senate Committee has estimated that the country loses an estimated US$5,000,000 per day from illegal logging.[87]

There is evidence that environmental advocates, including priests, journalists, and indigenous groups have been attacked by armed agents of businessmen and politicians involved in the illegal logging trade.[88] Labelled as 'communists', or charged with 'subversion', they have been subjected to both legal and physical intimidation and harassment and some have been killed.[89] The involvement of the military in many of the offences against activists makes the situation even more brutal.

Members of the clergy have paid the highest price for being critical of the logging industry. In 1988, Father Mario Estorba was shot, two weeks after he had filed a complaint arguing for better working and remuneration conditions with the local logging company.[90] The same year a priest who had expressed strong anti-logging and anti-mining views was murdered in South Cotabato. 'We were accused of being communist because of the stand we were taking on ecological issues,' said Father Sean McDonagh, a colleague of the dead priest. Other clergy at the mission had also received death threats and one had survived an attempt on his life.[91]

In 1991, Father Nery Lito Satur, who was an outspoken critic of illegal logging was ambushed walking home after saying mass. His three attackers, who shot him five times and crushed his skull with a rifle butt, were identified as being involved in the illegal logging activity: an army intelligence officer and five members of a local militia. Two other outspoken anti-logging priests have also received numerous death threats.[92]

The military were also involved in another killing the same year when Henry Domoldol, who headed a community association fighting to retain the local forests under tribal control was shot dead outside his home. Witnesses identified the gunmen as members of the army and the paramilitary organisation the Civilian Armed Forces Geographical Units. Tribal leaders believed that the killing was an attempt to force them away from their land so that logging groups could move in.[93] Other people have been killed too for opposing the illegal corrupt logging practices. In 1988 Antonio Dimpas, a town councillor in Palawan, was shot dead after he stopped a truckload of logs belonging to a local policeman.[94]

In 1988 the largest Philippine environmental organisation, the Haribon Organisation, started a campaign to save the forests of Palawan, the remote island which houses the last areas of virgin forest in the Philippines. Because of their campaigning, nine members were arrested and interrogated by the police in 1991. A further five members were charged with subversion. Their crime, according to Global Response, was that the 'group had questioned the involvement of a Marine battalion in the case of 3 million pesos worth of endangered timber found near the Marine department'. The illegally cut timber was subsequently believed to be shipped to Malaysia. The arrests were 'designed to silence them in their campaign for environmental conservation,' said the Haribon group, 'particularly in the monitoring/reporting of logging activities in Palawan involving the military'.[95]

The country's journalists also bear the brunt of much of the legal and physical harassment. The problem is, according to journalist, Marites Vitug, that writing about environmental issues 'is not about pollution, it's about power'. Vitug received a series of death threats after an article she wrote depicting the deforestation in Palawan. She was also sued by the person depicted in her story for 25 million pesos ($1,000,000) or equivalent to about sixteen years' wages. Her co-author on the story received a death threat.[96]

Journalists say these two responses are now typical. Both death threats and libel suits, meant to silence or 'chill' the press, are becoming common. Seven journalists were threatened for investigating illegal logging by the military in 1991 and 1992 alone. 'The chilling effect of a libel suit is real,' says Vitug, 'It serves as a constant reminder not to tread on dangerous ground.'[97]

Other journalists in the region have died or been harassed for covering the illegal logging trade in Cambodia, which is closely linked with the military. Chan Dara was fatally shot in the back, whilst riding his motorcycle in December 1994. A few days earlier, Dara had told his newspaper, the *Koh Santepheap* that he had received warnings from the military police to stop investigating military involvement in the illegal timber trade in Kompongh Cham province. Another newspaper, the *Preap Norm Sar* has also published articles documenting the illegal involvement of provincial military officials in both the timber and rubber industries. Reporters there are also concerned for their safety.[98]

In 1994, in a move that was seen as totally illegal, Cambodian contracts overseeing lucrative logging deals with Thailand were transferred from the Finance Ministry to the Defence Ministry. 'It makes a mockery of the whole environment question,' a senior official close to the government was reported as saying.[99] So bad is the illegal logging that in July 1994, a senior government official reported that Cambodia's forests could be eradicated within five years. Furthermore, two months before Chan Dara had been killed, the King of Cambodia had written to the government warning that 'Cambodia risks becoming a desert country in the 21st century.'[100]

TRANSNATIONAL SLAPPING IN THE PHILIPPINES

In what is seen as a worrying trend, multinationals are using libel or SLAPPs (Strategic Lawsuits Against Public Participation) as an intimidation tactic in countries in which they operate around the world. In April 1993, the German multinational chemical company Hoechst SLAPPed an $813,000 lawsuit on Dr Romy Quijano, a toxicologist from the General Philippine Hospital, who was also a pesticides activist and panel member of the Pesticide Technical Advisory Committee of the Philippines Department of Agriculture. The *Philippines News and Features Service* agency were also SLAPPed by the company. Dr Quijano's crime against the company was to have given a paper at a conference on the 'Effects of Pesticides on Women', organised by Sibol ng Agham at Teknolohiya (SIBAT) and the Pesticide Action Network Asia and the Pacific (PAN-AP). The allegedly libellous conduct by the news agency amounted to covering a statement from Dr Quijano's speech, which was subsequently printed in a number of Filipino newspapers.[101]

The controversial remark made by Dr Quijano during the conference, was where he reportedly stated that endosulfan, a Hoechst pesticide, more commonly known by its trade name 'Thiodan', *may* cause cancer. Dr Quijano was accused by the company of 'wilfully, maliciously and falsely stating that Thiodan causes cancer'.[102] PAN-AP calls Hoechst's actions of SLAPPing legal suits on critics of its products, 'unacceptable'.[103] It has also led to a widespread view throughout many South Asian NGOs that they can no longer discuss the extremely important issue of pesticides and their impact on public health and the environment, without fear of severe legal repercussions.[104] But this is exactly what SLAPPs are designed to do.

There was evidence to back action on curtailing pesticides. Between 1980 and 1987, some 4,031 pesticide poisoning cases had been reported in government hospitals.[105] Another study by Dr Quijano himself, coupled with the National Poisons Control and Information Service (NPCIS) and the University of the Philippines showed a further 1,302 cases of poisoning between January 1992 and March 1993 in the National Capital region alone.[106] Endosulfan, in part because of its widespread use, had been identified by the NPCIS as the leading cause of poisoning in the Philippines.[107] After the pesticide was banned by the Philippine Fertiliser and Pesticide Agency (FDA) in 1992, the company wrote to the President arguing that 'mishandling of the regulatory agencies would have serious undesirable consequences in terms of attracting foreign investors to come to the country'.[108]

Lawyers for Hoechst have also threatened legal action against the *Philippine News and Features Service* for reporting allegations of harassment against another anti-pesticide activist, Ermina Abongan, who had been at the same conference as Dr Quijano. During the proceedings, Ms Abongan had described the symptoms she had experienced, including skin lesions and shedding fingernails, several years after she had used Brestan, another Hoechst product which had been banned by the government in 1990.[109] Ms Abongan says that she was later questioned extensively and videotaped by a number of men who claimed to be from the company, although they refused to divulge their names. The *Philippine News and Features Service* had reported how she had felt these tactics were intimidatory. Lawyers acting on Hoechst's behalf deny any wrong-doing and are demanding a retraction from the paper.[110]

In 1994 Hoechst's case against Dr Romy Quijano was thrown out of court. Citing that there was no libellous intent, the judge said the article was 'clearly a matter of high public concern and interest, and therefore comes under the protective mantle of press freedom'. Hoechst had not even been mentioned directly in the article.[111] The tactics by the company have led the PAN-AP to question the company's commitment to the Chemical Industry's Responsible Care Programme, and its adherence to the FAO's International Code of Conduct on the Distribution and Use of Pesticides. PAN-AP have also called for the immediate withdrawal of endosulfan.[112]

Corporate tactics such as the ones outlined above, where company profit is pursued at the expense of free speech might become all too common across the region.

DAM THE PEOPLE

Another issue that has dominated ecological campaigning over the last two decades in Asia is the building of dams and the ecological, cultural and social impact such projects inflict. Many of these dams have been financed by World Bank loans or financed by the Bank's development aid. The Bank has a history of indifference to the people or land impacted by its policies, although recently it has shown some reluctance to fund certain mega-dam projects. But its policies have left an appalling legacy. For example, in the 1980s in the Philippines the Bank-funded Chico Dams project threatened to displace 80,000 tribespeople from their ancestral lands. Their protests were met with violent force, with the Marcos regime repeatedly bombing their villages. Large-scale counter-insurgency operations were also carried out in the area.[113]

At the same time, the World Bank were also funding the Kedung Ombo Dam. Despite the fact that 8,000 affected people had refused to move, the government closed the dam gates flooding their land. Road blocks were set up around the area to stop the press reporting the unfolding human tragedy.[114] Meanwhile in Indonesia, four people, including women and children, were killed and three others injured when security forces opened fire on 500 peaceful protesters at a dam on the island of Madura, East Java, in September 1993. A mission from the Indonesian Legal Aid Foundation found that the security forces had opened fire without warning and without provocation on the peaceful protest. It was also reported that seventeen people suspected of organising the demonstration were detained by the authorities.[115] Amnesty International reported that this was just the latest in a series of repressive acts by the authorities, including detention, ill-treatment, torture and imprisonment, used against activists in Indonesia protesting against real estate or development projects. As elsewhere, these people are then deemed subversive.[116] Amnesty has also highlighted how police beat demonstrators protesting against the environmental and social impact of the Pak Moon Dam in Thailand in March 1993. Three people were reported to be seriously injured.[117]

NARMADA

In recent years, the most controversial ecological and development issue for the region has been the building of the Sardar Sarovar hydro-electric dam on the Narmada River in Central India. The dam, conceived in the dinosaur development era that deemed mega-projects the panacea for the world's energy problems, has been nothing short of a human and ecological catastrophe, and a national and international hot-spot of controversy. Narmada is an example of the neglect of local communities' needs and views in the development equation and the state repression against dissent when communities want to have their voices heard.

The Sardar Sarovar Dam (SSD) project is part of the largest and the most expensive multi-purpose hydro-electric scheme ever attempted in India and the largest irrigation scheme on earth. The original plans were for 30 mega dams, including Sardar Sarovar, the largest of them all, 135 medium dams and a further 3,000 small dams, plus canals and irrigation projects.[118] Its cost – US$11.4 billion in total: US$3 billion for the Sardar Sarovar Dam alone.

The SSD project, if successful, and that is a large *if*, should provide drinking water to 40 million people, irrigation for 4.4 million acres and generate 1,450 megawatts of electricity.[119] But environmentalists have called the project an ecological and human catastrophe. It will create a huge artificial lake swamping 248 villages, displacing more than 200,000 tribespeople from their ancestral homes whilst destroying fragile eco-systems, fertile farmland and rich forests. In total, the project destroys 37,000 hectares (91,400 acres) of land, of which 13,700 hectares (34,000 acres) are predominantly good quality forest. Furthermore, people believe that the cruelty of relocating whole communities away from their traditional ancestral lands itself constitutes a powerful argument against the dam. Moreover, it is compounded by the absence of decent alternative land for the resettled people.[120] Over 135 metres high and 1,210 metres long, the dam is due for completion either next year or in 1998. Alternative schemes do exist which would still harness the same amount of water for irrigation, and power available for peak consumption but considerably reduce the ecological and human impact.[121]

Protests against the dam have been led by the Narmada Bachao Andolan NBA (Save the Narmada Movement), a broad coalition of non-governmental organisations and individuals who are opposed to the dam and who have used non-violent protest and civil disobedience in their campaign. When the NBA began in 1985 their primary focus was ensuring proper resettlement for the people. However, as they became more informed about the ecological and human consequences of the project, the protest has widened out to encompass the whole issue of sustainable development of such projects funded by the World Bank, and the NBA subsequently declared its total opposition to the SSD.[122]

The role of the World Bank in the scheme has received international condemnation. The Bank had initially donated US$450 million to the project in 1985, claiming that the dam would bring widespread benefits for all. Two more bank payments worth a further US$440 million were due as well. However, the NBA published a report showing that the irrigation schemes would only enrich the rich, that the electricity benefits and water drinking benefits were largely illusionary, and that the scheme would be an ecological and human catastrophe. The project became so controversial that the Bank commissioned an independent review of the project, the first time it had ever done so in its history. The review concluded the Bank had scorned its own directives on resettlement of people and had not even bothered to undertake any correct environmental assessments. In damning language, it concluded that it was 'clear that engineering and economic imperatives have driven the Projects to the exclusion of human and environmental concerns'. One of the Bank's own consultants described parts of the project as 'death traps'.[123]

Still the Bank funded the flawed project, a decision that so outraged the review's authors, that, according to Susan George and Fabrizio Sabelli, in their criticism of the World Bank, they accused the Bank's management of 'lies, fraud and wilful misrepresentation of their report to its own executive directors'.[124] Finally the Bank did drop the funding of the project in 1993, at the request of the Indian government, who were increasingly suffering from bad public relations. Lewis Preston, the new Bank President was still unrepentant. 'The Narmada project is not one to be ashamed of,' he said.[125]

Activists attempting to stop the dam went on *satayagraha*, which is non-violent direct action. Many felt so angry that the only course open to them was to drown in the rising waters behind the dam, rather than be resettled with little compensation for the loss of their homes, their land, their community and life as they knew it. In fact the Bank had admitted that forced relocation at Narmada can be 'expected to cause multidimensional stress', including 'psychological stress' and 'feelings of impotence associated with the inability to protect one's home and community from disruption'.[126]

But relocation was not all the protesters had to deal with. During their campaign, they have been harassed, beaten, arbitrarily arrested, tortured and even shot. In one of the worst cases of state violence, in an apparent unprovoked attack, a 15-year-old activist was shot dead by police whilst on an anti-dam protest in 1993. A further three people on the demonstration were injured.[127] On other demonstrations protesters have been beaten by police with canes, assaulted by the police and tear-gassed.[128]

The police have also used the law to intimidate and frustrate activists, by repeatedly arresting them, filing criminal cases as a form of legal and financial harassment. After three arrests, magistrates can impose severe bail conditions which can effectively end someone's involvement in the protests. Hundreds of activists and people who have simply refused to leave their homes have been detained or assaulted. The authorities, it seems, are particularly keen on dragging women protesters from their homes by their hair. Once in prison, many have been ill-treated or subjected to de-humanising conditions. Some have been kept in handcuffs in detention, although this is illegal. Others have been denied medical attention or legal assistance.[129]

Other activists complain of being forced to sign documents in custody, even though they have not read them. If they have refused to sign they have been beaten. A key village activist was sexually assaulted and raped by two policemen and then kept without food for a week in custody. She was also threatened in case she told anyone about the rape. This was seen by activists as a sordid attempt by the authorities to intimidate women activists and break their morale. The district police chief responded to the complaints by the woman as 'a cooked-up story' and 'mischief'. Other women have also been sexually molested by police whilst being arrested.[130]

During 1994, tensions heightened in the region. The NBA office was attacked, reportedly by members of Congress and the Bharatiya Janata Party. Medha Patkar, one of the leaders of the NBA, and other activists were manhandled and verbally abused during the raid, which resulted in 50 per cent of the office contents being destroyed.[131] Demonstrations were banned surrounding the dam and police were given powers to shoot on site at 'unruly' demonstrations.[132] Any assembly of more than five people was banned. So too was 'dancing', 'music', 'gesticulating', 'slogan shouting' and 'dramatising'.[133]

So upset with the government's lack of compensation for resettled people was Medha Patkar that, along with three colleagues, she began a hunger strike in November 1994. On their twentieth day they were arrested by the authorities, where they were fed intravenously against their will.[134] The following month, whilst the activists remained on intravenous drips in hospital but remained on hunger strike, the Supreme Court finally ordered the release of a report into the dam project, withheld by the government since 1993. It revealed that there were serious problems with resettlement and that the conditions under which environmental approval for the project had been granted had been broken. It called on the government to review the height of the dam, recommending that it should be reduced from 455 feet to 436 feet.[135] Due to the release of the report, and a change in policy by the Madhya Pradesh state government in a reduction in the dam height, the protesters called off their fast on the twenty-sixth day of their hunger strike.[136]

In May 1995, a confidential World Bank memo on the dam, written by the Bank's Operations Evaluation Department (OED) was leaked. It admitted serious shortcomings in the environment, resettlement and rehabilitation components, and project appraisal, as well as supervision performance. It described the future and sustainability of the project as 'uncertain'.[137] 'The World Bank Project Completion Report and OED memo are a welcome evaluation,' responded Himmanchu Thakker of the NBA, 'now the OED should look deeper into critical issues of project viability, hydrology, and economic impacts. We believe a serious study of these "uncertainties" will reveal the total unsoundness of the project.'[138] It was a vindication of the NBA's non-violent struggle against the dam. The fight for Narmada continues and, despite police repression and against the odds, the NBA, under Medha Patkar's leadership, have raised the issues of sustainable, practicable solutions to India's need for energy and water but ones which are fair to the ecology and needs of its people. One where the people should also have a say. The World Bank and the authorities should have listened first time around.

When asked who is ultimately responsible for the violence against the protesters, Vandana Shiva, Director of the Research Foundation for Science, Technology and Natural Resource Policy, in India answers:

> The whole free trade group of institutions, the transnational corporations and the development institutions like the World Bank, who are also acting on behalf of corporations, because for every dam they build they are basically generating contracts for the turbine manufacturers and the construction companies etc., etc. Once there is that kind of greed behind any exercise, whether it is development or free trade, greed stamps out people's chance of survival, and people protest. There will be violence and there will be backlash, and people have to constantly create new forms of resistance to make that backlash look what it is: one of brutal force.[139]

ANTI-GATT

Farmers from India have rallied against the social and ecological impact of free trade and GATT. On 2 October 1992, the anniversary of Gandhi's birth, over 500,000 farmers launched the 'seed satyragraha', aimed at resisting handing over food and seed production to transnational companies. Using direct action, the farmers served 'Quit India' notices on the offices of seed giants Cargill and other foreign transnationals. The government attempted to suppress the growing movement by arresting tens of thousands of farmers, and organising counter-demonstrations by large farmers, led by an ex-World Bank official. A year after the movement started, on 2 October 1993, a huge rally was held in Bangladore with farmers' representatives from across the globe pledging to fight against increasing transnational corporate domination of agriculture and trade.[140]

In January 1995, police fired on peaceful protesters who were campaigning against Du Pont's plan to construct Asia's largest nylon factory in Goa. One protester was shot dead by the police. The protesters, angry at the death, subsequently went on the rampage, burning down Du Pont's project office. Activists alleged that the $200 million plant would pollute rivers and deplete drinking water supplies and desecrate sacred Hindu land. Claude Alvares, from the Goa Foundation summarised their frustration with Du Pont. 'Not only is Du Pont completely dishonest and cynical regarding its concern for India's environment, it is cheap as well,' said Alvares. 'Du Pont is willing to invest tens of millions to make a hefty profit, but it doesn't want to invest a paltry $100,000 to figure out how to avoid poisoning the local Goans.'[141]

In June, due to the continuing protests against the plant in the Goa region, Du Pont announced that it was moving the site of its plant to the neighbouring state of Tamil Nadu.[142] According to Vandana Shiva, the violent retaliation against the protesters:

> explains to us that free trade and corporate investment comes with that kind of violent backing, and it's a good message to those who think about free trade as only a positive exercise. Because it is so clear that the free trade order is a very violent order.[143]

11

'A SHELL-SHOCKED LAND'

Ogoni is the land
The people, Ogoni
The agony of trees dying in ancestral farmlands
Streams polluted weeping filth into murky rivers
It is the poisoned air coursing the luckless lungs of dying children
Ogoni is the dream breaking the looping chain
Around the drooping neck of a Shell-shocked land.

Ken Saro-Wiwa's poem, Ogoni! Ogoni![1]

On 10 November 1995, Ken Saro-Wiwa, along with eight other Ogoni, were hung for a crime they did not commit. Their murder by the ruling Nigerian junta caused an international outrage. Just over three years before, Saro-Wiwa, writer, ex-Nigerian TV producer, President of the Movement for the Survival of Ogoni People (MOSOP) and environmentalist, had warned that, 'It's just going to get worse, unless the international community intervenes.'[2] His warnings were not heeded as the Ogoni community were failed in their hour of need that was to darken in the months to come.

Saro-Wiwa's only 'crime' was to campaign against the ecological destruction wrought on Ogoni by Shell and to ask for a greater share of the oil wealth that has been drilled from under Ogoniland. Criticising Shell hurtled the Ogoni on a collision course with the Nigerian military and the company. It has led to over 1,800 Ogoni being killed, 80,000 left homeless, the much publicised killing of Saro-Wiwa and other Ogoni, and continuing repression and violence.

THE IMPORTANCE OF OIL TO NIGERIA

Oil is the life-blood keeping the military junta alive. Oil revenues account for 90 per cent of Nigeria's export earnings and 80 per cent of the Federal Government's total revenue.[3] Shell Petroleum Development Company of Nigeria Limited (SPDC), a subsidiary of the Royal Dutch/Shell Group operates a joint venture with the Nigerian National Petroleum Corporation (NNPC), with Elf and Agip holding minority stakes.[4] SPDC is by far the largest exploration and production venture in Nigeria, producing some 910,000 barrels of oil per day, out of Nigeria's total of some 2,000,000 barrels.[5] Nigeria is also significant to Shell's profits, accounting for 13 per cent of total worldwide group production. Shell's position in the country is both powerful and unique. According to an Ogoni activist who wishes to remain anonymous for fear of retribution, 'With such an illegitimate political system, each bunch of unelected military rulers that comes into power, simply dances to the tune of this company . . . Shell is in the position to dictate, because Nigeria is economically and politically weak.'[6]

Shell have been producing oil in the Niger Delta since 1958. Since then, communities in the region have accused the company of destroying both their lands and their livelihood. The Ogoni have been the leading voice of dissent, with Saro-Wiwa their most vocal spokesperson and critic of Shell. Others are also critical of the company's operations: 'Shell Petroleum Company wants a free hand to maximise profits even at the expense of the protection of the basic rights of people and considerations of environmental sustainability,' says Claude Ake, an internationally renowned academic, the Director of the Centre for Advanced Social Studies, a former UN advisor and member of the Shell-commissioned Niger Delta Environmental Survey Team, until he resigned in the aftermath of Saro-Wiwa's death.[7]

COMPENSATION

The complaints of the communities against Shell basically concern compensation and pollution. It is widely reported that when Shell acquired land in the Delta it only paid people for their annual crop and not for the land itself. Not only did the farmers lose that year's income, they also lost their future revenue. Furthermore, three-quarters of Ogoni cannot read or write and cannot understand the compensation forms. Others maintain that they had to accept money 'under duress' from the military or that promised monies were never paid.[8] Following the 1991 Judicial Commission of Inquiry after the Umuechem massacre (see later), the Nigerian government raised the derivation fund to oil-producing states from 1.5 per cent to 3 per cent, despite recommendations of 15 per cent.[9] In reality, very little reaches those communities. Most is creamed off into the pockets of corrupt officials.[10]

The lack of compensation is exacerbated by the competition for land in the Delta. Ogoni is an area of some 404 square miles, which has been described as one of the most fragile and important biological eco-systems in the world, home to both rain forest and mangrove swamps. Due to intense industrial activity, namely oil and gas exploration and production, the Delta can be considered the most endangered in the world. Although over 6,000,000 people live in the Delta, in Ogoni, a traditional subsistence farming and fishing community, demand for land is particularly high with a population density of 1,250 people per kilometre compared to the Nigerian average of 300.[11]

ECOLOGICAL IMPACT

Furthermore, construction of roads and canals for the oil industry 'has precipitated some of the most extensive environmental degradation in the region' by encouraging migration of people into the Delta, including farmers and loggers. This influx exacerbates the ecological and social problems of the already crowded Delta, according to a study by World Bank specialist David Moffat and Professor Olof Lindén from Stockholm University.[12]

Oil operations have had a significant effect on the ecology and people of the Niger Delta, through water, air and noise pollution from pipelines, wells, pits, waste-water, flow-stations and flares. A delegation from the Delta presented a report at the Earth Summit in 1992, which stated:

> Apart from air pollution from the oil industry's emissions and flares day and night, producing poisonous gases that are silently and systematically wiping out vulnerable airborne biota and otherwise endangering the life of plants, game and man himself, we have widespread water pollution and soil/land pollution that respectively result in the dearth of most aquatic egg and juvenile stages of the life of fin-fish and shell-fish and sensible animals (like oysters) on the one hand, whilst, on the other hand agricultural lands contaminated with oil spills become dangerous for farming, even where they continue to produce any significant yields.[13]

Professor Claude Ake says:

> When you go to any of the rivers on the entire Port Harcourt coast, anywhere you go you see a film of oil in the water. This is a permanent feature. The pollution of underground water is also obvious . . . Flaring of gas is constant. There are flares all over the place. It is night and day pollution. It is unthinkable that such a thing would happen in Britain.[14]

Until Shell claimed it stopped its operations in Ogoni in January of 1993, the company had constantly flared gas close to villages for over thirty years, a practice that would be unacceptable in Europe or America.[15] Flares emit constant air and noise pollution and a constant flame of light. 'Some children have never known a dark night even though they have no electricity,' wrote British environmentalist, Nick Ashton-Jones after a visit to the Delta in 1993.[16]

Flaring causes other serious problems. Gas flaring in the Delta emits more greenhouse gases than from fuel burning for Britain's homes.[17] Some 35 million tons of CO_2 and 12 million tons of methane are produced by gas flaring. Up to 76 per cent of gas is flared in Nigeria compared to 0.6 per cent in the USA and 4.3 per cent in the UK.[18] Given that subsidence is occurring in the Delta due to oil operations and that global warming will cause sea level to rise, oil and gas operations are going to leave a deadly legacy for the inhabitants of the Delta. Moffat and Lindén estimate that up to 80 per cent of the Delta's population will have to move because of sea-level rise, with property damage estimated at $9 billion.[19]

However, Moffat and Lindén had omitted to study the effect of subsidence on the Delta. A memorandum produced by the River Chiefs of the Niger Delta and presented at UNCED had studied the problems of sea-level rise coupled with subsidence. The report said:

> If we superimpose the predicted accelerated sea-level rise on the gradually subsiding Niger Delta, the net effect is that, within the next two decades, about a 40 km wide strip of the Niger Delta and its people would be submerged and rendered extinct.[20]

The Ogoni activist blames the special relationship between the government and Shell for allowing flaring to take place:

> The government and Shell have put in place a law that makes it more comfortable for Shell to flare gas, because the amount they pay for flaring gas is peanuts compared to the amount that they should pay in the law court. They have never been fined because the government is behind them.[21]

Plate 11.1 Pipelines on farmland at Yorla, Ogoni, Nigeria.
Source: E.P.L./Tim Lambon

The company's high-pressure pipelines criss-cross the region, passing over farmland and within feet of people's homes. 'I don't know what you do with these pipes – they're old, they're rusted, they're totally neglected,' says Sister McCarron, the contact person for the Africa-Europe Faith and Justice Network between 1990 and 1994. 'I think that they were just unaware that there were people there at all,' she adds.[22] Sister McCarron's network works on social justice issues in Nigeria that have their origins in European policies and businesses. McCarron was assisting farmers affected by oil exploration. Claude Ake agrees:

> The K-Dere flowstation [in Ogoni] was built in the middle of a community, a settlement, with high pressure pipes in front of the houses. Shell now turns around and tells people that the community have built houses there. Tell me how they would allow people to build a house with a doorstep two feet from pipelines. This kind of irresponsible propaganda gives the impression of not taking the people seriously.[23]

Oil spills from pipelines and other oil operations in the region have become a routine occurrence. Official estimates record some 2,300 cubic metres of oil spilt annually in some 300 pollution incidents, although the true figure could be ten times higher.[24] From 1982 to 1992, 1.6 million US gallons of oil were spilt from Shell's operations in twenty-seven separate incidents. Forty per cent of all spills from the company's worldwide operations were from Nigeria.[25] Indeed, Shell's own figures report that since 1989 there have been 190 spills per year, involving on average 319,200 US gallons of oil.[26] Unlined waste pits are said to litter the region. Pits found in the middle of villages have little or no security.[27] At Bonny terminal, where Shell separates water from crude, the concentration of oil in river sediments has been described as 'lethal', at 12,000 ppm. A Shell environmental survey found an average hydrocarbon content of 53.9 ppm in a creek near the Bonny terminal, (which compares to concentrations of oil sampled in water close to the *Braer* wreck off Shetland in 1993). Such figures indicate 'poor or no treatment of effluents', according to Moffat and Lindén.[28]

'Shell started operating to double standards from day one,' says the Ogoni activist, 'all the things they have been doing in the area, they are exactly the things they cannot do in other countries in which they operate.'[29] 'The accusation of double standards against Shell is real,' adds Oronto Douglas, a Nigerian lawyer with the Civil Liberties Organisation.[30] Greenpeace has also repeatedly accused Shell of operating to double standards. It is Shell's apparent double standards that led Saro-Wiwa to accuse Shell of racism.[31] The European Parliament has labelled the effects of oil extraction in Ogoni an 'environmental nightmare'.[32] Shell, for its part, has conceded that its environmental standards are lower in Nigeria than in America or Europe.[33]

Finally in May 1996, a senior ex-Shell employee, who was head of Environmental Studies in Nigeria for two years, publicly stated that:

> Shell were not meeting their standards, they were not meeting international standards. Any Shell site I saw was polluted, any terminal that I saw was polluted. It was clear to me that Shell were devastating the area.[34]

IMPOVERISHMENT

In thirty-eight years an estimated $30 billion of oil has been extracted from the Niger Delta, but still Ogoni lacks adequate basic facilities such as pipe-borne water, electricity, sanitation, health clinics and schooling.[35] Next to the ex-patriot oil workers living a life of luxury, the communities remain impoverished, their frustration and anger very real. 'Years of exploiting oil have left a debris of sorrow, of pollution, environmental degradation all around,' recorded the *Daily Sunray* in 1994, complaining that none of six villages they visited had 'functional pipe-borne water, a health centre or electricity'.[36] In response to criticism, Shell has publicly stated that it has given large amounts of money to the communities. 'Privately officials conceded that these donations never reached their intended destination,' highlighted the UK *Guardian* in November 1994.[37] David Moffat and Olof Lindén describe as 'minimal' the impact of industry initiatives to improve the quality of life of communities.[38]

It is not just Ogoni which is impoverished. 'I have explored for oil in Venezuela, I have explored for oil in Kuwait, I have never seen an oil-rich town as completely impoverished as Oloibiri,' remarked a BP engineer in 1990 about the state of the town where oil was first found in the region over thirty years before.[39] Having been in the town of Warri in 1969, Sister McCarron returned in 1993, to be 'shocked' at the site of town, and how little the infrastructure had improved in the intervening years. 'I figured out that the school system in Warri had completely broken down,' says McCarron, who agreed with Saro-Wiwa when he wrote that the oil industry has dehumanised the communities in the Delta.[40]

COMMUNITY PROTESTS MET BY FORCE

Due to pollution and impoverishment, many communities have demonstrated against Shell, and by 1990, Shell officials admitted that there were some sixty-three protests against the company by different communities.[41] But disenchantment had been growing for a decade. The Iko people had written to Shell in 1980 demanding 'compensation and restitution of our rights to clean air, water and a viable environment where we can source for our means of livelihood'. The pressing demands were for roads, water, a health centre and school. Two years later, the community organised a peaceful rally demanding that Shell be a 'good neighbour to us'. Shell responded by calling the police who arrested some of the demonstrators.[42]

In 1987, the community once again held a peaceful demonstration. This time the notorious Mobile Police Force (MPF) were called, being transported to the demonstration in company speedboats. Two people were killed, forty houses were destroyed and 350 people made homeless by the MPF's attack. Although a panel of inquiry was set up to investigate the 'unwarranted behaviour of policemen drafted to maintain peace in Iko', the results have never been made public.[43]

Three years later, the Etche people peacefully protested against Shell at Umuechem, 'because they had seen Shell continually exploit their land without adequate compensation in any form of paying for the land, crops or providing basic social amenities' according to one villager.[44] Despite knowing what havoc the MPF could cause, Shell specifically requested their attendance. Headed 'Threat of Disruption of our Operations at Umuechem by Members of the Umuechem Community', the Shell letter said 'In anticipation of the above threat, we request that you urgently provide us with security protection (preferably Mobile Police Force) at this location'.[45]

The MPF subsequently massacred eighty people and destroyed 495 homes. 'They just unleashed terror on the peaceful demonstrators and started killing and maiming whoever came their way,' recalled one terrified witness.[46] The Inquiry blamed the police for the massacre, but the community frustration was evident in the official report. The Umuechem community said in their evidence to the official inquiry:

> These [Shell Petroleum Development Company] drilling operations have had serious adverse effects on the Umuechem people who are predominantly farmers, in that their lands had been acquired and their crops damaged with little or no compensation, and are thus left without farmlands or means of livelihood. Their farmlands are covered by oil spillage/blow-out and rendered unsuitable for farming.[47]

Shell has distanced itself from the killings, stating that 'the problems which gave rise to the demonstrations and the consequent police action were not really of Shell's making at all'.[48]

Umuechem was a chilling precursor to the military might that would be unleashed against the communities, but still they voiced their concerns. The year after Umuechem, villagers were protesting at Yenegoa in the Gbaran field. In March 1992, the Omudiogo community were complaining of a lack of community assistance from the company. An independent report into the conflict between Shell and the community concluded that 'Shell Petroleum Development Company has been indifferent, insensitive to the requests and plight of the people of Omudiogo.' It continued that 'the conflict could have undoubtedly been averted if Shell Petroleum was responsible and conscious of her social obligation to the host community'.[49]

Four months later, despite the lesson of Iko and Umuechem, the MPF were once again sent to silence Shell's critics. This time killing a 21-year-old man, shooting thirty people and beating 150 individuals at the town on Bonny who were complaining that Shell had not provided the local community with basic facilities – water, roads, electricity – despite being in the area for over twenty years.[50] Undeterred by the violence, the Uzure community, whose thirty-nine oil wells produce some 56,000 barrels of crude a day, demonstrated later that month. 'Uzure has no light, water and hospitals' complained one protester.[51]

The Movement for the Survival of the Izon (Ijaw) Ethnic Nationality in the Niger Delta (MOSIEND), produced their own charter in October 1992, demanding control over their natural resources and the restoration of ecological damage caused by oil companies. Earlier in the year, they blocked a Shell flow-station protesting about oil spillage by the company.[52] The Ogbia community in Oloibiri also produced a 'Charter of Demands' in November, complaining that 'due to intentional, negligent and/or inadvertent oil blowouts and spillages, our entire land, creeks and rivers and atmosphere have been polluted over these long years'. They demanded compensation for environmental damage, for the flaring of gas to stop and the burying of pipelines. The Izon also protested against Shell in 1992, as did the Igbide. Both communities were demanding basic amenities.[53]

In December 1993, tension rose between Shell and the Nembes and the Kalabaris, resulting in an attack on a Shell flow-station. Two months later, four men were arrested for carrying an air conditioner in the vicinity of Shell's facilities. At the police station, the four men were beaten with knotted rope whips called *Kobokos*, as well as guns. When people started demonstrating for their release outside the prison, two men were shot and wounded by the police.[54]

In February 1994, a peaceful demonstration in Port Harcourt by the people of Rumuobiokani, against Shell's facility in the town, resulted in armed soldiers, mobile police, navy and air force personnel all arriving on the scene. When the notorious Major Okuntimo (see below), the head of the River State Internal Security Force arrived, he ordered his men to 'Shoot at anyone you see'. Tear gas and bullets were fired at the demonstrators in an indiscriminate fashion resulting in five people being injured. One person, shot from a distance of one metre, was trying to prevent his father being beaten. Others were battered and arrested.[55]

Community protests have not just been targeting Shell, either. In October 1993, 5,000 people demonstrated against the local Elf refinery in Obagi, which resulted in a peace agreement being signed by the community and the company. However, when police came searching for missing Elf property four months later, a fight broke out in which a policeman was killed and an Obagi severely injured. The resulting crackdown by the mobile police was swift and severe, with houses being burnt, looted and destroyed. People were beaten and shot indiscriminately. The villagers, terrified by the police, hid in the bush for six months. A community leader, J. G. Chinwah was repeatedly harassed by the authorities, and, although innocent, has twice been detained for the death of the policeman. The authorities have attempted to have him sacked from his job at the Rivers State University of Science and Technology and evicted from his university home.[56]

When 3,000 peaceful protesters from the town of Brass demonstrated the following month outside the local Agip terminal, they were fired upon with tear gas by the mobile police and navy. Demonstrators were attacked with *Kobokos* and clubs, resulting in numerous injuries. All roads to the village were then blockaded for nine months as punishment. The following year a demonstration against Chevron at Opeukebo in May resulted in the protesters' sixteen boats being rammed by the police.[57]

MOSOP: A MOVEMENT FOR ECOLOGICAL AND SOCIAL JUSTICE

It is the Ogoni who have become the vanguard movement for compensation and ecological self-determination. Three months after Saro-Wiwa's visit to London, on 4 January 1993, the Ogoni mobilised in the largest peaceful demonstration against oil companies ever. To usher in the UN Year of Indigenous Peoples, they danced and sang, believing that through non-violent protest and grassroots empowerment they could influence the oil companies and military. We have 'woken up to find our lands devastated by agents of death called oil companies. Our atmosphere has been totally polluted, our lands degraded, our waters contaminated, our trees poisoned, so much so that our flora and fauna have virtually disappeared,' said an Ogoni leader to the crowd:

> We are asking for the restoration of our environment, we are asking for the basic necessities of life – water, electricity, roads, education; we are asking above all for the right to self-determination so that we can be responsible for our resources and our environment.[58]

It was the mass mobilisation of the grassroots that made MOSOP a movement, in Saro-Wiwa's words, for 'social justice and environmental protection',[59] that

Plate 11.2 Ken Saro-Wiwa addresses the crowd on 4 January 1993.
Source: Greenpeace/Lambon

was so threatening for the authorities and to Shell. 'No other person in Nigeria can get 100,000 people on the streets,' said a member of the Nigerian Civil Liberties Organisation about Saro-Wiwa.[60] 'MOSOP is genuinely grassroots,' adds a Lagos-based diplomat. 'It cuts across élite lines, it's extremely effective. It is a fundamental threat to the way politics is done in Nigeria.'[61] Also worrying was the Ogoni Bill of Rights which the Ogoni presented to the Nigerian government and the international community in 1990. For the first time the Ogoni had a document articulating their concerns and demands for environmental and social justice and control over their affairs and resources.[62]

Saro-Wiwa had been the driving force behind the Bill. 'I think the marvellous thing about Saro-Wiwa,' said Sister McCarron, a year before his murder, 'is that he's been able to put a language on all of this, and a language for all the Delta communities as well.'[63] What the authorities feared most was that the 'Ogonis' well-organised protests could be emulated by others among Nigeria's 250 ethnic groups,' concluded Amnesty International.[64]

Ogoni protests are not new, however. They have been campaigning against Shell's ecological impact since 1970, when the company was a joint venture between Shell and BP (whose assets were subsequently nationalised). Letters written by Ogoni leaders in 1970 complain of Shell-BP 'seriously threatening the well-being, and even the very lives' of the Ogoni. They also protested about how much the company paid in compensation then.[65] A letter from the Dere Student's Union to Shell-BP shows how little has changed over the last twenty-five years. They complained of constant gas flaring and noise from Shell's operations and pipelines being laid over their farmland.[66]

That same year there was a major blow-out in Bori from one of Shell-BP's well-heads. Not stopped for three weeks, it caused widespread pollution. 'Our rivers, rivulets and creeks are all covered with crude oil,' wrote the Dere Youths Association:

> We no longer breathe the natural oxygen, rather we inhale lethal and ghastly gases. Our water can no longer be drunk unless one wants to test the effect of crude oil on the body. We no longer use vegetables, they are all polluted.[67]

Twenty-two years later, the areas remained a wasteland, according to Saro-Wiwa.[68] The animosity felt towards Shell was reflected in an Ogoni song that was written in 1970, and which became a popular protest song:[69]

> The flames of Shell are flames of hell,
> We bask below their light,
> Nought for us serve the blight,
> Of cursed neglect and cursed Shell.

A BRUTAL BACKLASH

Since 4 January 1993, the backlash wrought on the Ogoni has been devastating. Ogoni has been driven to the abyss of annihilation, crushed by a military regime who have had one simple aim: to silence Ogoni, and to stop other communities from voicing their legitimate concerns.

Excessive force on the Ogoni has continued as the collision course between community and company has intensified. Due to continuing unrest in the region, the company withdrew all personnel from Ogoniland in January 1993. According to Brian Anderson, the Managing Director of SPDC, Shell 'have not attempted to resume operations since that time'. However, in April 1993, the US company Willbross, under contract to Shell and under the protection of the Nigerian military, cleared land in order to continue work on the Rumuekpe-Bomu pipeline through Ogoniland. While attempting to collect what remained of her crops, Karalolo Korgbara was shot, later losing her arm as a result. 'My farm was destroyed,' she said, ' . . . There is nothing I can do now, I can't farm. They've paid me no money, they have done nothing.'[70] Thousands of protesters gathered at the site to protest at the shooting and the absence of compensation for the land. On the fourth day, a demonstrator, Agbarator Otu was shot in the back and killed and at least twenty villagers were wounded. Amnesty International issued an 'Urgent Action' request concerned about possible extrajudicial executions by the military.[71]

Days later in May, the government passed the Treason and Treasonable Offences Decree 1993. A simple act of advocating minority rights could now be interpreted as treason, punishable by death.[72] It immediately became known as the 'Ken Saro-Wiwa Decree', with people suggesting it was a direct attempt to silence him and the Ogoni.[73] The authorities also repeatedly arrested Saro-Wiwa and other Ogoni leaders during the first few months of 1993. He had his passport confiscated on at least two occasions, one time being prevented from attending the United Nations World Conference on Human Rights. Ten days later, in June 1993, Saro-Wiwa was arrested again, and with two MOSOP leaders was charged with six counts relating to unlawful assembly, seditious intention and seditious publication.[74] Having been moved to his third prison, Saro-Wiwa suffered a third heart attack, remaining unconscious for two hours, having been subjected to, in his words, 'psychological torture'. He was subsequently refused treatment for his heart condition.[75] Greenpeace condemned Saro-Wiwa's detention as 'a way of silencing a critic, not only of the Nigerian authorities, but also of Shell's disregard for the environment'.[76]

Starting in the summer of 1993, there was a series of brutal ethnic clashes between the Ogoni and neighbouring tribes, namely the Andoni, Okrika and Ndoki. All independent evidence of the clashes shows that they were instigated and partly implemented by the military, under the guise of 'inter-communal conflict'. In mid-July, some 136 Ogoni were murdered by supposed 'Andonis', returning from a trading trip to Cameroon. Children were put in sacks and thrown into the sea to drown. Two people were released and told to warn the Ogoni that they had been killed by 'Andoni.'[77] Although this matter was reported to the Rivers State Police Command, they refused to act on it.[78] At least 124 people were killed at the town of Kaa in early August, with the town effectively destroyed. Three weeks prior to the attack, all Ogoni police had inexplicably been removed from the area. This factor, together with the sophisticated machinery, such as grenades, mortars and automatic weapons, used in the attack, and 'the military's failure to restore order in Kaa', infer that the attack was not a neighbourly conflict but was instigated and undertaken by the military.[79]

'These clashes are really organised attacks by the military government using otherwise peaceful neighbours as proxies, as fronts,' said an Ogoni who survived the attack at Kaa.[80] 'There is really no reason why it should be an ethnic clash,' concluded Claude Ake, who was appointed by the government to examine the conflict. He continues:

> As far as we could determine, there was nothing in dispute in the sense of territory, fishing rights, access rights, discriminatory treatment, which are the normal causes of these communal clashes. One could not help getting the impression that there were broader forces which might have been interested in perhaps putting the Ogonis under pressure, probably to derail their agenda.[81]

Human Rights Watch backed up Ake's sentiments, concluding that 'evidence now available shows that the government played an active role in fomenting such ethnic antagonism, and, indeed, that some attacks attributed to rural minority communities were in fact carried out by army troops in plain-clothes'.[82] Their investigators interviewed two members of the Nigerian military who had participated in raids on Ogoni. One soldier, allegedly sent to the region to maintain the peace, described what happened next. 'They suddenly changed the orders,' he recalled. 'They said we were going across to attack the communities who had been making all the trouble.' The soldiers subsequently cordoned off the village of Kpea, firing indiscriminately at civilians, burning and looting people's homes.[83] Footage smuggled out of Nigeria shows the aftermath of the attack: smouldering buildings, corpses, and torsoes hacked to pieces with machetes. Wailing relatives mourn their dead.[84]

Residents from Kaa told Human Rights Watch that they were attacked by people in military uniform.[85] The use of heavy weaponry in these attacks, such as explosives, mortars and machine guns, unknown in a traditional fishing community, is further confirmation that these clashes were not inter-communal in nature.[86] Just before Christmas in 1993, the Ogoni were once again attacked, supposedly by the Okrika. Although some sixty-three people were killed by an assortment of weaponry which included explosives and machine guns, the police did nothing to stop the carnage.[87] An Ogoni was turned away from a local police station, even though the sound of explosions could be heard in the background. A commission of inquiry concluded that the attack was 'coordinated with military expertise and precision'.[88] There are also allegations that the security forces actively encouraged another tribe, the Ndoki to attack the Ogoni.[89] 'These so-called disturbances, were orchestrated by the military,' says an Ogoni. 'They have used the Andoni, they have used the Okrika, and they have used the people in Ndoki area and we have never had any so-called ethnic clashes . . . it only started in 1993 when our struggle was at its peak.'[90] In all, some 30,000 people were displaced in the clashes and between 1,500 to 2,000 killed.[91]

During 1993 there had been another development in Ogoni: an internal political struggle for the leadership of MOSOP, between Ken Saro-Wiwa and more traditional leaders. Saro-Wiwa represented the popular leader of the younger generation, who wanted a more radical, more democratic movement to represent Ogoni, whereas the traditional leaders adopted a more conservative and conciliatory approach, based on a top-down hierarchical leadership. There was open dissent between the two. The authorities had actively been working with the more traditional element of the Ogoni, to undermine MOSOP and had even given money to these Ogoni for the 'purpose of stopping' Saro-Wiwa.[92]

The authorities attempted to label Saro-Wiwa a 'terrorist', by, amongst other techniques, circulating a booklet, entitled *Crisis in Ogoniland: How Saro-Wiwa Turned Mosop into a Gestapo*,[93] with a picture of a swastika and Saro-Wiwa on the front cover. 'Members of the international community have been misinformed about the true nature and scope of the crisis in Ogoniland,' says the anonymous author, who adds that Saro-Wiwa 'established the NYCOP (National Youth Council of Ogoni People) Vigilante as a terrorist organisation well-trained and equipped in the art of terrorism'. According to the book:

> Clearly, unlike what the international do-gooders and confusionists, like Amnesty International, would like the world to believe, there has been no extrajudicial execution and detention by the security forces. The fact of the matter is that the security forces are the saving grace in Ogoniland.

The military also maintain that 'the crisis in Ogoniland revolves around only one person – Ken Saro-Wiwa' and the 'Federal Government has established due processes to ensure that he and his cohorts have a fair trial'.[94]

The Ogoni activist says:

> What the government and Shell are doing is to demonise MOSOP to damage its credibility, but they have not been successful in Ogoniland among the grassroots. But the grassroots see MOSOP as the only single organisation that can bring to them their emancipation.[95]

While the government hope to subvert MOSOP's aims by calling it a terrorist organisation, they are also attempting to coerce the more traditional elders: a simple divide and rule tactic. It is easy to see why the authorities are worried. The younger generation of Ogoni are more educated and literate than their elders, and have a greater understanding of their rights to compensation, and are less corruptible and more radical.

WASTING NOW NECESSARY

When General Abacha took control of Nigeria in the autumn of 1993, the situation worsened for MOSOP. Abacha appointed Lieutenant Colonel Komo as the Head of Rivers State, and Major Okuntimo, to head the newly formed Internal Security Task Force, whose simple remit was to effectively destroy dissent amongst Ogoni. In April 1994, an internal memo, entitled 'Restoration of Law and Order in Ogoniland', gave details for an extensive military presence in Ogoni, drawing resources from the army, airforce, navy, and police, including both the Mobile Police Force and conventional units. In a move seen to facilitate the re-opening of oil installations, one of the missions of this operation was to ensure that those 'carrying out business ventures . . . within Ogoniland are not molested'.[96]

A month later a memo was sent from Okuntimo to Komo. It took a year to be leaked to the international community, and showed the ruthless methods by which the military were determined to re-open oil production in Ogoni. If the authorities had been inciting 'ethnic clashes' against the Ogoni, they now had a problem, they had run out of potential attackers.[97] Okuntimo remarked that, 'Shell operations still impossible unless ruthless military operations are undertaken for smooth economic activities to commence.' To counter this, Okuntimo recommended: 'Wasting operations during MOSOP and other gatherings making constant military presence justifiable; . . . wasting operations coupled with psychological tactics. . . . restrictions of unauthorised visitors especially those from Europe to the Ogoni.'[98]

Just such a 'justifiable' occurrence transpired on 21 May 1994, only nine days after Okuntimo's memo had been written, when four traditional Ogoni elders were brutally murdered. This event, horrific as it was, was the green light for the military to enter Ogoni as they had planned the previous month and outlined in Operation Law and Order in Ogoniland. Under the guise of looking for the killers, they would systematically terrorise, torture, rape and kill across the region. The fact that it was a so-called 'intra-communal' dispute between Ogoni that triggered the military occupation of the region, is extremely disturbing if more of Okuntimo's memo is examined: 'Division between the elitist Ogoni leadership exists . . . Intra-communal/Kingdom formulae alternative as discussed to apply . . . Wasting targets cutting across communities and leadership cadres especially vocal individuals of various groups,' the memo read.[99] A strategy that would 'justify' a military presence, with which to undertake 'wasting' operations against Ogoni had been hatched.

Other suspicious happenings occurred that day, too, which have led MOSOP to believe that the whole event was a complete set-up. Eye witness accounts talk of Ogoni 'filled with soldiers' in the morning before the killings, as if they were waiting for something to happen. These security forces did nothing when alerted of the disturbance to prevent the killings, although they were asked to quash the growing dissent. 'The information we received,' says the Ogoni activist, is that, 'the event in which these people who died were involved, was under the eyes of the police.'[100]

Saro-Wiwa, who was campaigning for an election as an Ogoni representative to a Constitutional Conference, had given his schedule to the military authorities days beforehand. The Nigerian Conference, which Saro-Wiwa supported, was a forum for redress and a chance for the Ogoni to work towards a solution to the conflict. Due to address four rallies on May 21, he was prevented from speaking at the first two and under military escort was taken on an unplanned trip to the vicinity where the killings would take place. He did not stay long in the area before being driven off by the military and was miles away by the time the murders actually occurred.[101]

Ogoni youths are believed to have thought that Saro-Wiwa had been arrested, but it cannot be ruled out that outside forces facilitated a mood of dissent into actions which led to the death of the four elders. There are too many other coincidences to suggest that *agents provocateurs* were not used, although conclusive proof will probably never be discernible. It was only after the murders had actually occurred that 'we Ogonis ever knew that there was a grand design to cause disturbances in Ogoni in order to create an excuse for the government to send in more troops', said one witness.[102]

Considering the apathy of the security forces to investigate the deaths of the 1,000 or so Ogoni who had died over the last year, the extreme measures that they have undertaken since May 21 is further evidence that the whole event was planned. The Internal Security Force, 'ostensibly searching for those directly responsible for the killings', were in fact, according to Amnesty International 'deliberately terrorising the whole community, assaulting and beating indiscriminately'.[103]

The day after the murders, Ken Saro-Wiwa and Ledum Mitee, the President and Vice-President of MOSOP, were arrested and detained without charge.[104] Saro-Wiwa had been quick to condemn the killings, once again reiterating his commitment to non-violence, 'I have stood against violence, all those responsible for this should be brought to book,' he said.[105] In fact Saro-Wiwa and MOSOP had long taken steps to expose and prevent any vigilante activity in Ogoni.[106] Despite this, Lieutenant Colonel Komo held a conference blaming the killings on MOSOP and calling Saro-Wiwa a 'dictator who has no accommodation and no room for any dissenting view'.[107]

Amnesty International believes that Saro-Wiwa's arrest was 'part of the continuing suppression by the Nigerian authorities of the Ogoni people's campaign against the oil companies'.[108] Amnesty declared Saro-Wiwa a 'prisoner of conscience – held because of his non-violent political activities' and that 'solely because of his campaign against environmental damage and inadequate compensation by oil companies operating in Ogoniland and because of his influence both within the Ogoni community and internationally', Saro-Wiwa has been subjected to repeated arrest and harassment. The Ogoni activist says:

> If Ken had not been campaigning against Shell, he wouldn't have been arrested in the first place. All this trouble that started in the area, wouldn't have taken place. If the Ogoni people had not risen up to complain about their predicament, none of the things would have happened.[109]

SANITISE OGONI

With the leaders of MOSOP arrested, the security forces, under the orders of Major Okuntimo, set out, in his own words, to 'sanitize Ogoni'. 'For "sanitize" read "ethnically cleanse",' says Wole Soyinka.[110] Moreover, just as his memo had suggested, Okuntimo boasted that he was now engaged in 'psychological warfare' in Ogoni.[111] 'You require a psychological approach to rewind them out of mobilisation,' says Okuntimo,[112] outlining his methods used against Ogoni:

> I operated in the night so nobody knew where I was coming from. I just take a despatchment of soldiers . . . about 20, and give them grenades, explosives, very hard ones. The machine gun with 500 rounds will open up. And then we throw the grenades into the bush and it goes boom. We have already put a roadblock on the main road. We don't want anybody running so the option we made was that we should get . . . all these people into the bush with nothing except the pants and the wrap they are using that night.[113]

Under his personal orders, the military systematically raided villages at night, randomly shooting, looting, torturing and raping. Soldiers would storm into a house, firing and beating the occupants. They would also demand 'settlement fees' from terrified inhabitants. As people ran from their villages, some would be shot, including the old, the young and the frail. Money, property, livestock and food would all be stolen by the troops, who would then set fire to their homes.

Within a month some thirty villages had been attacked, fifty people killed and 180 injured, according to Amnesty International. By late June, the human rights organisation was again concerned: 'These attacks appear to be part of the continuing attempts by the Nigerian authorities to suppress the Ogoni people's campaign against the oil companies,' said Amnesty.[114] Some of the most harrowing testimonies are from victims who were gang raped by soldiers. 'They tore my dress,' recalled a girl of 14. 'One soldier held each of my legs. Then each of the four soldiers took turns. I was lying in a pool of blood when they left, unconscious.'[115] Women and children were gang-raped whilst being beaten, pregnant women were raped, too.[116] Victim's fingers were cut off and people were threatened with having their eyes ripped out or with being killed, in order to terrify them into silence.[117]

Soldiers told Human Rights Watch that Okuntimo had himself committed rape, and would point out Ogoni women to be brought to him.[118] Much of the torture committed against prisoners was also under the personal command of Okuntimo. The reports of atrocities committed against the Ogoni have been dismissed by both Okuntimo and Komo as 'propaganda'.[119]

Those arrested fared no better; held without charge, they would be beaten, tortured, humiliated. The dehumanising conditions were said to be 'life-threatening' by Amnesty International.[120] People were kept in small, airless cells, with no food or amenities, such as toilet facilities, and water to drink. If their relatives did not bring food, people simply starved. Some cells, measuring two by three metres, were so cramped all people could do was stand. Nearly everyone in detention was flogged by *Kobokos*, some on a daily basis, others so many times that 'you couldn't see their skin'. Many people had to be hospitalised because of their injuries. By the end of the year some 600 people had been detained.[121] An unknown number of people died in prison.

It was not just the Ogoni who were beaten up or imprisoned though, so too were lawyers, observers and journalists. When two Nigerian lawyers, Oronto Douglas and Uche Onyeagucha went with British environmentalist, Nick Ashton-Jones, to see Ledum Mitee at Bori camp in June 1994 they were arrested and detained by Okuntimo for four days. They were beaten and flogged with an electrical cable. Uche Onyeagucha was also kicked in the face by a soldier.[122] Journalists from the *Wall Street Journal* and other Nigerian newspapers were detained for writing stories on Ogoni. Geraldine Brooks, a reporter from the *Wall Street Journal* was deported for 'sneaking around'.[123] 'I curse the day Shell found oil on our land,' Brooks had been told by an Ogoni villager. 'It has killed the gardens and the fish, and now it is killing the people.'[124] When she refused to hand her interview notes over, Brooks was told, 'Let me remind you that you have had 38 years of very good life. You have a husband. Don't risk it all now.'[125]

Okuntimo ordered that Saro-Wiwa 'be taken to an unknown place and be chained legs and hands and not to be given food . . . I cannot allow anybody to come and spoil my job'.[126] He was routinely tortured in prison and put in leg-irons and denied access to family, friends, a lawyer and medication. When in November 1994, Saro-Wiwa and MOSOP were awarded the 'Right Livelihood Award', known as the alternative Nobel Peace Prize, Lt-Col. Komo ordered more repressive measures against Saro-Wiwa.[127] The Award was given for Saro-Wiwa's 'exemplary and selfless courage and in striving non-violently for the civil, economic and environmental rights of his people'.[128] Saro-Wiwa won two more awards – the Goldman Award – the world's largest grassroots environmental prize, for 'leading a peaceful movement for the environmental rights of the Ogoni people whose land has been ravaged by multi-national oil companies' and the Hellman/Hammett Award of the Free Expression Project of Human Rights Watch.[129]

WHO'S ON TRIAL?

In November 1994 it was announced that a three-member tribunal was being convened by the military to try the fifteen Ogoni who were detained. Instead of standing before a civil court, the defendants' case would be heard before what amounted to nothing more than a kangaroo court, established by an illegal military regime. Denied access to civil courts, they had no right of appeal and faced the death penalty if convicted, in blatant breach of fundamental rights which should be guaranteed by both Nigerian and international law.[130]

After a wait of eight months, though, finally Saro-Wiwa and the other Ogoni were charged with either murder or 'counselling and procuring' the killings.[131] So concerned about the trial were the Bar Human Rights Committee of England and Wales and the Law Society of England and Wales, that they asked Michael Birnbaum QC to attend as an observer. His report is a damning indictment of the injustice of the court. Birnbaum criticised the presence of a military officer, the fact it had been convened by a military government, and the fact that defendants could not appeal against any judgement, which included a mandatory death sentence for murder.[132]

His fears were increased when the tribunal made 'two decisions which to my mind strongly suggest that it is biased in favour of the prosecution and the federal government'. Although, by law, the tribunal was meant to be satisfied that the defendant had committed an offence before the trial started, there was no case against some defendants. Second, the tribunal decided to hold simultaneous trials of different Ogoni, seriously impairing the defendants' chance of a fair trial.[133] None of the defendants were allowed to see their counsel before the beginning of the trial, on which Lt Col. Okuntimo had undue influence.[134]

There was a high military presence around the court, as well as inside, for which Birnbaum 'could see no reason'. Supporters of the defendants were harassed, beaten and denied entry into the court, including Saro-Wiwa's mother and wife and even two of the key defence lawyers – Femi Falana and Gani Fawehinmi, as well as Oronto Douglas again. Falana was actually arrested and held incommunicado for eight days. Fawehinmi had his passport confiscated and required a court order to allow him to travel abroad. He has subsequently been re-arrested.[135] All this led Birnbaum to conclude, 'It is my view that the breaches of fundamental rights are so serious as to arouse grave concern that any trial before this tribunal will be fundamentally flawed and unfair.'[136]

'I am like Ogoni, battered, bruised, brutalised, bloodied and almost buried,' Saro-Wiwa wrote during the trial.[137] 'There is no doubt that the authorities want me to die,' he continued, 'I have caused them too much trouble in my struggle for the Ogoni people.'[138] Although the verdict was not handed down until the end of October, Saro-Wiwa's defence pulled out of the trial in early summer. The final straw came when evidence which suggested that the military and Shell were bribing the chief prosecution witnesses, was ruled as 'too good to be true'.[139] Michael Birnbaum reported that the two chief prosecution witnesses against Ken Saro-Wiwa, Charles Danwi and Nayone Akpa, and the only two to implicate Saro-Wiwa directly, had signed affidavits alleging they were bribed to give evidence against him.[140]

Extracts from Charles Danwi's affidavit, signed 16 February 1995, are as follows:

> He was told that he would be given a house, a contract from Shell and Ompadec and some money . . . He was given 30,000 Naira . . . At a later meeting security agents, government officials and the Kobani, Orage, and Badey families, representatives of Shell and Ompadec were all present.[141]

The other chief prosecution witness against Saro-Wiwa, Nayone Akpa also alleged that he was offered '30,000 Naira, employment with the Gokana Local Government, weekly allowances and contracts with Ompadec and Shell' if he signed a document that implicated Saro-Wiwa.[142] At the end of the trial Shell vehemently refuted any allegations of bribery in the Tribunal as well as having anything to do with the deaths of the four Ogoni leaders.[143]

On 31 October, Ken Saro-Wiwa, and another eight Ogoni were sentenced to death. Despite the international outcry at the sentences, they were hanged on 10 November 1995. John Major described the killings as 'judicial murder'. 'I and my colleagues are not the only ones on trial,' Saro-Wiwa wrote as his closing testimony at the Tribunal, 'Shell is here on trial and it is as well that it is represented by counsel said to be holding a watching brief. The Company has, indeed, ducked this particular trial, but its day will surely come.'[144]

PR RESPONSE

Since January 1993, when MOSOP first mobilised, Shell has launched a counter-PR offensive, and its preoccupation has been to keep its image clean rather than address the real concerns of the communities. 'Shell went into siege mentality and it became engrossed in damage limitation,' says Claude Ake.[145]

'Shell is unwilling to face the environmental and social problems that its presence has created in Rivers state, beyond a cosmetic attempt to tidy up its Nigerian and international environmental image,' adds Nick Ashton-Jones.'[146] Adding to Shell's troubles, in November 1992, a Channel Four documentary entitled *The Heat of the Moment* had been shown on British television, which was highly critical of Shell. Leaked draft minutes of a meeting on Community Relations and the Environment held in The Hague and London in February 1993 highlight that:

> The problem is not restricted to Nigeria; it has been thrust into UK sitting rooms through the TV programme *The Heat of the Moment* and the information has spread, most recently in the Netherlands and Australia . . . International networking . . . is at work and gives rise to the possibility that internationally organised protest could develop.[147]

Company officials discussed the need for environmental improvements especially in relation to spills, flares, air and water quality. They also proposed that:

> SPDC and SIPC PA departments keep each other more closely informed to ensure that movements of key players, what they say and to whom is more effectively monitored to avoid unpleasant surprises and adversely affect the reputation of the Group as a whole.[148]

In response to this Claude Ake alleges that:

> Shell wants Ken to be closely monitored, because Ken is at best a nuisance, at worst a great danger, because he has tabled before the nation and to some extent before the international community the matter of the irresponsible behaviour of the oil companies, including Shell. This is in their disregard for environmental protection, and for their reluctance to apply the normal standards of environmental sustainability in oil prospecting. I think that they fear that if everybody begins to raise the kinds of questions that Ken is raising and to agitate for change of behaviour, it is going to make Shell's operations much more costly and to reduce their profit margin substantially.[149]

Shell has consistently attempted to distance itself from the conflict, believing 'that most of MOSOP's demand are outside the business scope of oil operating companies and within the government's sphere of responsibility'. Furthermore, contend Shell, 'their campaign is overtly political and Shell is being unfairly used to raise the international profile of that campaign through disruption of oil operations, and environmental accusations.'[150] For over twenty-five years the communities throughout the Delta have complained of pollution, exploitation and appropriation of their resources, and have severely suffered because of Shell's operations. For Shell to state that the Ogoni are 'using' them is simply misleading, and such corporate arrogance just adds to the bitterness felt by the people of the Niger Delta.

There is also a political edge to the Ogoni campaign, because the Ogoni are asking for a degree of political self-determination. 'We have been fighting for the environment for a long time,' said Saro-Wiwa. 'Nobody listened because the environment was not a serious issue with anybody except those of us who were suffering. But when we made a political case, ah, then that began to draw some attention.'[151] 'You can't really separate the environment and the political, in this case anyway,' argues Sister McCarron, 'they're so intertwined because it's a political responsibility to protect the environment.'[152]

Shell also downplays the reports of environmental devastation and double standards. 'Allegations of environmental devastation in Ogoni, as elsewhere in our operating area, are simply not true,' says the company. 'We do have environmental problems but these do not add up to anything like devastation. Where environmental problems in the Niger Delta do stem from oil operations, we are committed to dealing with them.'[153] Claude Ake and environmentalist, Shelley Braithwaite, strongly disagree. 'They are causing a lot of devastation, it is not only Shell but all the other companies,' argues Ake.[154] Braithwaite says:

> One only has to visit the site of the 1970 pipeline explosion near Ebubu to realise the paucity of the conviction behind this statement. Here is a two to three acre site, twenty-five years on, that remains encrusted with crude oil several metres thick. In June 1993, a spill at Bomu-Tai reportedly continued for forty days before Shell stemmed the flow.[155]

The company also refutes that it operates to double standards and uses Principle 11 of the Rio Declaration to justify disparate standards of operation in Nigeria, which states that 'standards applied by some countries may be inappropriate and of unwarranted economic and social cost to other countries, in particular developing countries'.[156] This argument does not wash with Professor Ake. 'Who decides inappropriate?' asks Ake:

> The point is it is decided by the corporations, and that is part of the problem, that they are not subject to the rule of law. They do not decide this by negotiation, they decide it by their own interest. This is lawlessness. It is not an acceptable way of doing things.[157]

Shell has made some shrewd adjustments on pollution figures in the Delta. In 1992, the company line was that 'as for oil spills and leakages, some 60 per cent are caused by sabotage.'[158] By 1995, the company had changed its tune, admitting that 'some 75 per cent of oil spills throughout our operations are from corrosion', although they maintained that 'in the Ogoni area, investigations show that 69 per cent of all spills between 1985 and the start of 1993 had been caused by the communities'.[159] 'Most of the pollution on Ogoni land is not caused by sabotage,' says Claude Ake.

I think that this is the kind of irresponsible propaganda that the oil companies are putting out in order to discredit those who are trying to do something about the environment . . . if Shell put out this, it can only be a smokescreen.[160]

Ake points out that it is the companies themselves that award the compensation and claims can take five years or more, so it is extremely unlikely that people will actively pollute their own land.[161] 'The evidence of vandalism and the evidence of sabotage are very, very weak,' says Sister McCarron, 'because the people will say, "who better knows what oil will do to us than ourselves, so are we likely to punish ourselves?".'[162] According to an independent ten-year spill record of Shell's Nigerian operations, only four of the twenty-seven spills were from sabotage.[163]

In late 1994 the TSB announced that it was selling its Shell shares from its Environmental Investors Fund, because of Shell's 'environmental and social policies in Nigeria'.[164] This was the first hard evidence that the company's shareholders were losing faith in Shell. Something proactive had to be done to quash the growing dissent, and plans were hatched for a two-year $2 million 'independent' environmental survey to be carried out in the Delta, which would be announced just after Saro-Wiwa's trial began in January 1995. Rather than concentrating on the ecological and social impact of oil operations, the Nigerian Delta Environment Survey (NDES) would 'catalogue the physical and biological diversity' of the whole Delta.[165]

Although Shell immediately denied that the NDES was pressure driven,[166] leaked minutes of a meeting Shell had with a potential contractor for the survey showed that the purpose of an independent survey 'would solve the dual purpose of absolving themselves of all responsibility and addressing the local and international accusations that not enough is being done to mitigate environmental problems created by the oil and gas industry'.[167] The potential contractor also felt that Shell 'had already made up their minds' and that 'I detected some cynicism in their tone whenever the local people were mentioned.'[168]

Furthermore, people questioned the independence of a survey whose steering committee Chairman was the Head of Dunlop Nigeria, a major user of Shell's products. Other committee members included the top public relations person from Shell Nigeria. The steering committee would not include anyone from MOSOP, although it had been promised that 'this committee has been specifically chosen to include representatives of all major stakeholders of the region, including the most important stakeholder of all, the communities of the area'.[169] Consultation with MOSOP is crucial to resolve the conflict in the Delta. MOSOP responded by hoping 'that this is not another attempt to hoodwink the Nigerian public into a belief that Shell is environmentally conscious' and called on the 'company to enter in negotiation with MOSOP with a view to conducting environmental and social impact assessment studies/audits in Ogoni'.[170]

Other commentators felt that Shell was looking to blame the population for the problems of the Delta, rather than its own operations. All just a sophistical public relations counter-strategy that annoys Oronto Douglas from the Civil Liberties Organisation. 'They should not take it as a public relations thing. How many press conferences will they organise, but to explain what?' he asks:

To explain that my community was not destroyed, to explain that there is no cholera in my community, to explain that they have not been looting the resources in my community, no electricity, no roads, no water, nothing in my village, is that what they want to explain? Why is it that they are not applying the same standards in the North Sea, in England in other parts of the world, in Nigeria?[171]

The survey began to seriously backfire when Professor Claude Ake resigned in the aftermath of Saro-Wiwa's murder. Asked by Saro-Wiwa to serve on the survey, Ake had agreed 'with a leap of faith', and had been representing the interests of the communities. He resigned because:

It is clear now that NDES is too late and does not represent a change of heart. To begin with, it does not enjoy the enthusiastic support of the oil industry at large. Clearly there is nothing in the recent performance of the oil companies notably Shell, NOAC, Elf and Mobil to suggest that NDES is associated with increasing sensitivity to the plight of the oil-producing communities.[172]

Ake is also convinced that Shell:

did not do enough to encourage conciliation because I have been at them for well over a year, because it did not take much intelligence to see all these things coming. I felt there was a need for a comprehensive dialogue, and urged them repeatedly, and I told them that I could deliver the communities to talk to them, including MOSOP.[173]

Shell did not heed Ake's advice, rather, it seems that it was discussing PR strategies with the military. A leaked memo shows that a meeting took place between four senior Shell officials, Malcolm Williams, A. J. Brack, D. Van dan Brook and A. Detheridge and the Nigerian High Commissioner at Shell Centre in London on 16 March 1995. Also in attendance were representatives from the Nigerian army and the police.

Public relations strategies and counter-publicity measures were discussed. At one point the High Commissioner, having 'expressed dismay at the apparent network of misinformation orchestrated by Gordon and Anita Roddick of The Body Shop . . . wondered if a counter-measure, like producing posters, putting adverts in papers as well as sponsored TV programme could not be explored'. Malcolm Williams, Head of Regional Liaison replied that he 'was wary of any approach that will play into the hand of the propagandists' and informed the minister that 'his company was embarking on its own film production'. The High Commissioner expressed his 'delight' at this and 'promised to assist them overcome any bureaucratic problems they may encounter'.[174] At no time during the meeting did the Shell officials express any concern for Ken Saro-Wiwa, who had been hospitalised the week before. Shell has admitted that a meeting took place, but it refuses to comment on what was discussed.

The Nigerian government, during the atrocities committed in Ogoni, took out adverts in *The New York Times* and *Washington Post*.[175] The PR and lobbying firm they employ, Van Kloberg & Associates, has a reputation of improving 'the image of unpopular countries', such as El Salvador, Haiti, Iraq, and Myanmar (Burma).[176] In the immediate aftermath of Saro-Wiwa's murder, the military stepped up their disinformation campaign against the Ogoni, spending a reported £5,000,000 on public relations strategies in London, and hiring a further seven PR companies in America. They placed adverts in both the American and British press, justifying the executions of the Ogoni. As the Ogoni protests continued into 1996, with six people being shot dead on Ogoni day, the military accused MOSOP of being 'mercenaries' paid by foreign governments to destabilise Nigeria.[177]

THE MILITARISATION OF COMMERCE

Instead of conciliation with the communities, Shell continues to pursue what Claude Ake calls the 'militarisation of commerce' and the 'privatisation of the state'. Oil extracted in the Delta does so under military protection. Professor Ake said in December 1995,

> The situation right now is that all the flow stations, that is the operational bases of the oil industry, operate under armed presence. This is a process of the militarisation of commerce and the privatisation of the state, and I have actually used these phrases in discussion with Shell executives.[178]

It is Shell's reliance on the military for protection and to quash peaceful dissent that angers the Ogoni, who believe Shell is colluding with the government. 'It is demonstrated in their actions and in their relationship with the government,' argues the Ogoni activist:

> That is to say, once there is any situation which SPDC sees as uncomfortable, they don't waste any time in inviting the army, not even the police. Each time the army are brought in they just come in to shoot and kill, they don't use peaceful methods.[179]

This argument is backed up by Human Rights Watch:

> Because the abuses set in motion by Shell's reliance on military protection in Ogoniland continue, Shell cannot absolve itself of responsibility for the acts of the military . . . The Nigerian military's defence of Shell's installations has become so intertwined with its repression of minorities in the oil-producing areas that Shell cannot reasonably sever the two.[180]

Sister McCarron is equally damning about Shell's use of the military: 'As a business you've failed completely if you now have to call in armed troops to protect your interests,' she says, 'You've lost, it's a complete acceptance of failure, or ignorance if you like, of the fact that people really matter at all.'[181]

Events at Iko in 1987, Umuechem in 1990, Bonny in 1992 and Bomu-Tai in 1993 are classic examples of the intertwined nature of oil exploration and state violence that has become endemic in Nigeria. Furthermore, at Iko and Bomu-Tai there is evidence that Shell provided logistical support for the military by providing transport for the soldiers. At Bomu-Tai, a few days after a spillage, soldiers appeared in the local village and opened fire on local youths, killing one of them with a shot in the back of the head. They had been transported by Shell, in the company's words, to 'dialogue' with the community. In retaliation for this, local people set fire to two trucks belonging to Shell. However, Shell in a meeting with environmentalist Nick Ashton-Jones suggested that the attack on the village by the army was reasonable retribution for the loss of the water truck, even though the trucks were only damaged after a local person had been killed. Shell officials showed no regret at the loss of life.[182]

So what is the relationship between the oil companies and the military? In the May 1994 memo, under the heading 'Funding' Okuntimo suggested to put 'pressure on oil companies for prompt regular inputs as discussed'.[183] Evidence that the military could have been paid by Shell emerged when Oronto Douglas and Nick Ashton-Jones were detained by Okuntimo. Douglas writes of the event that:

> The conversation went on and the Major said, 'Shell Company has not been fair to him in these operations.' He said he has been risking his life and that of his soldiers to protect Shell oil installations. He said his soldiers are not been (*sic*) paid as they were used to.[184]

Asked if he believed that this was an admission by Major Okuntimo that the military received money from Shell, Douglas replied, 'Shell does pay the military.'[185]

Nick Ashton-Jones has confirmed that the conversation took place, and recalls what Okuntimo had said: 'That he was doing it all for Shell . . . But he was not happy because the last time he had asked Shell to pay his men their out-station allowances he had been refused which was not the usual procedure.'[186] Saro-Wiwa agreed with Douglas and Ashton-Jones: 'This confirms a fact that is well known, that Shell . . . has been paying protection money to the Nigerian security agencies.'[187] Owens Wiwa, Saro-Wiwa's brother, who was also arrested by Okuntimo, was also told by the Major that his wages were being paid by Shell.[188] In an interview for *The Sunday Times*, Okuntimo himself admitted, though later denied, being paid by Shell. 'Shell contributed to the logistics through financial support,' said Okuntimo.[189]

Shell also met regularly with Okuntimo. 'A highly placed government source in Rivers State told Human Rights Watch that SPDC representatives meet regularly with the director of the Rivers State Security Service and Lieutenant-Colonel Paul Okuntimo.'[190] The Unrepresented Nations and Peoples Organisation maintains that 'indications are that the company was in close communication with the military authorities at important times in the sequence of events affecting the Ogoni'.[191] In response to these allegations

Shell has stated that, 'It is not true, as has been claimed, that Shell has supported any alleged operations of a Task Force under Lt Col. P. Okuntimo in Rivers State.'[192] Lt Col. Komo, the Rivers State Military Administrator has admitted that Shell gave his government financial assistance to 'enhance security operations in the state'.[193]

Furthermore, according to Human Rights Watch, 'Shell executives acknowledged that they hire members of the Nigerian police to provide internal security.'[194] Claude Ake, confirming what he calls the 'privatisation of the state', has corroborated Shell's payment of the police. It is indicative of this process, believes Ake, that Shell has admitted importing weapons into Nigeria to arm the police. Eric Nickson from Shell Public Affairs attempted to justify Shell's actions to the *Observer* saying that this was common practice in Nigeria and that 'the Nigeria police do not have sufficient funds to equip themselves'. This was rejected by a former Chief of Defence Staff, Lieutenant-General Alani Akinrinade, who countered that the Nigerian police were 'well equipped and do not need anyone to import arms'. The General also pointed out to the *Observer* that the Mobile Police Force, used against communities at Umuechem and Iko were 'armed to the teeth', and that 'there was no excuse for anyone to have a private army in Nigeria'.[195]

Two weeks later the *Observer* revealed that, at the height of the atrocities against the Ogoni, Shell had been negotiating to buy more than half a million dollars worth of 'upgraded weapons', which included semi-automatic rifles, Beretta pistols, pump-action shotguns and 130 sub-machine guns. The *Observer* also reported how Shell executives put pressure on Nigeria's Police Chief to approve the arms purchase, warning that 'the importance of our organisation on the nation's economy cannot be over-emphasised'.[196]

Claude Ake argues:

> You cannot privatise the state and at the same time say that your operation is immune from politics and oppression. Shell operates under the umbrella of a repressive apparatus. This increases the alienation and makes it impossible for environmental issues and human rights issues to be discussed at all. Force pre-empts all those things so that the issues are not addressed, much less resolved, so the alienation is increasing not diminishing.[197]

Shell has told its shareholders that accusations of it being in 'collusion with the Nigerian military authorities' is 'deliberate misinformation'.[198]

TIME FOR CHANGE

The situation in Nigeria would be different if Shell had listened and negotiated with the communities of the Delta. 'Matters could have been dealt with before they even got to the point of the confrontation that subsequently ensued,' maintains Claude Ake, who also alleges that, 'Shell, in my opinion, in terms of my own moves, did not do enough [to secure the release of Saro-Wiwa].'[199]

Shell has finally admitted that the whole Ogoni crisis could have been averted.[200] Brian Anderson, Head of Shell Nigeria now concedes to a 'black hole of corruption', which is 'acting like a gravity that is pulling us down all the time'. Up to twenty Shell staff face dismissal with one accused of organising a 1,000,000 Naira compensation claim against the company for a bogus oil spill.[201] 'Shell,' says Ashton-Jones, 'has grown into a corrupted human organisation which is driven by its own staff, by some local people in powerful positions and by the military establishment, and does not know how to reform itself.'[202] 'For Nigeria to have peace,' adds Oronto Douglas, 'companies like Shell will play very great role in it, particularly to achieve democracy, but presently they are enjoying the loot, the corruption that the military have brought off, in as much that they are not prepared to do anything.'

For eighteen months, Saro-Wiwa's supporters asked Shell to intervene in order to secure his release. Shell publicly maintained it could not get involved. When Saro-Wiwa was sentenced to death, the company maintained it could not intercede. Finally, after the sentences had been confirmed, Shell issued a statement in Lagos calling for clemency. 'We believe that to interfere in the processes, either political or legal, here in Nigeria would be wrong,' said Brian Anderson, Managing Director of the Shell Petroleum Development Company of Nigeria. 'A large multinational company such as Shell cannot and must not interfere with the affairs of any sovereign state.'[203]

However, during the summer of 1994, Saro-Wiwa's brother, Owens Wiwa met Brian Anderson from Shell three times, to see if Shell could use its influence to get Ken out of prison. According to Wiwa, Anderson told him that it was 'difficult, but not impossible, but that the international campaign is hurting Shell and the Nigerian government and if we can stop the campaign then he might be able to do something'.[204]

Shell's decision to press ahead with a planned $4 billion natural gas plant in the immediate aftermath of the Ogoni hangings was seen as further indication of Shell's insensitivity to the peoples of the Niger Delta, but it was greeted with relief by the Abacha regime. 'Shell understands the facts of this matter and has reacted appropriately unlike the emotional outburst by others,' said an Abacha official.[205] Shell justified its decision in adverts placed around the world. 'Whatever you think of the Nigerian situation today, we know you wouldn't want us to hurt the Nigerian people. Or jeopardise their future.' Shell had no intention of changing its operations, apart from spending more on public relations, a stance that angered Professor Ake.

> I suspect that 10 per cent of what has been spent in worldwide advertising and public relations by Shell since this thing happened [Saro-Wiwa's death] could already have made a decisive difference in Nigeria, if it wanted to do the proper thing.[206]

In May 1996, the week before the company's AGM, Shell announced 'a plan of action in Ogoniland,' in which Brian Anderson publicly stated that, 'it is vital that we have the support of the communities in which we work. That is why we are adamant that we will not work behind a security shield'.[207]

Independent observers believe the proper course of action would be to concede to mistakes of environmental pollution, and to confess to the militarisation of its operations in Nigeria. Shell should start talking to MOSOP and the other communities, as the tragedy that has become Ogoni could be repeated throughout the Delta. Claude Ake says:

> I have let Shell know that the real danger for them is not even an oil boycott, but ordinary Europeans and Americans catching on to this process of the militarisation of commerce, and the fact that the oil that they are getting at the petrol stations is collected through an armed presence. This is not something that any amount of propaganda is going to be able to explain.[208]

In a plea for Saro-Wiwa's life to be saved, Nigerian Booker prizewinner, Ben Okri, had written, 'There are some things on Earth that are stronger than death, and one of these is the eternal human quest for justice.'[209] Social and ecological justice for the people of the Niger Delta still remains to be achieved.

12

THE ROAD TO NOWHERE

Societies, cultures, community structures thrive on scapegoats. The group of people who are saying it's time for change, who are saying look at our alternative, are going to be scapegoated, and increasingly I think that this is going to be the green lobby.

Jason Torrance, co-founder of Earth First! in England[1]

In the spring of 1995, ironically the European Year of Conservation, the green backlash hit Britain. 'Greenlash' or 'ecobacklash' as some journalists called it and the views of the 'contrarians', were suddenly the new *Zeitgeist*. 'It's now very fashionable to be very questioning about green issues,' said Michael Grubb of the Royal Institute for International Affairs.[2]

The reason the press danced to the anti-green tune was the publication of three books, all in a month, and all attacking the environmental movement. The most controversial, and arguably the most heavyweight in nature, was *Life on a Modern Planet: A Manifesto for Progress*, as it was written by former environmental correspondent, the ex-*Independent* journalist, Richard D. North. Soon to follow were *Small is Stupid: Blowing the Whistle on the Greens* by Wilfred Beckerman, Emeritus Fellow of Balliol College, Oxford, and *Down to Earth: A Contrarian View of Environmental Problems* by Matt Ridley, a columnist for the *Sunday Telegraph* and published by the Institute of Economic Affairs, a leading right-wing think-tank.

'Britain is on the verge of a green backlash,' reported Sue Pennington on the BBC Radio's *Special Assignment* programme. 'A new breed of thinkers accuses environmentalists of scaremongering and apocalyptic shroudwaving,' Pennington continued, ' . . . the voices of the new contrarians, shamed into silence in the last years by right-on environmentalists, are now fighting back'. Two of the authors appeared on the programme: Dr Beckerman accused environmentalists of making 'false apocalyptic predictions'. Richard North laid into Greenpeace, denouncing it as:

pandering to a taste in the public for excitement and drama, Greenpeace presents a particularly dramatic, simple world view . . . They often exaggerate the scare of the chemicals they are worried about, some of this green thinking has been a luxury that we can't really afford.[3]

THE DEFENCE OF CHLORINE

Six months' research for North's book was funded by ICI, Britain's largest chemical company.[4] ICI's tactics should not be seen in isolation. The global chemical industry has geared up its counter-attack against Greenpeace and other environmental organisations that have been highlighting the ecological problems associated with their products. Greenpeace has borne the brunt of this opposition due to its campaign to phase out chlorine. Leading chlorine manufacturers such as Dow, Solvay, Bayer and ICI were said to be 'mounting an unprecedented public relations drive to trumpet chlorine's merits,' according to Haig Simonian in *The Financial Times* in February 1995. 'The manufacturers argue that the point is to find a balance between ecology and economy,' continued Simonian.[5] This balance will be based on greenwashing and a business-as-usual scenario.

The European Chemical industry has founded industry lobby organisations and corporate front groups to defend chlorine. In an amalgamation of the ECDC (the European Chlorine Derivatives Council), Euro Chlor and ECSA (European Chlorinated Solvent Association), the Euro Chlor Federation was formed in 1991, to provide, amongst other functions, 'a focus for communication on the benefits of chlorine'.[6] A major group within CEFIC, or the European Chemical Industry Association, Euro Chlor's members include Solvay, Rhone-Poulenc, ICI, Akzo, EMC, Atochem, and the French industry group SHD.[7] Based in Belgium, Euro Chlor is conveniently placed to argue chlorine's case in the European Union and Commission.

Some companies have set up their own front organisation, Chlorophiles, comprising employees of the industry in Belgium and Holland. The Chlorophiles profess to be 'totally independent of our employers', but 'want to show the other side of the chlorine story to the general public'. Their first demonstration was, not surprisingly, at Greenpeace's Belgium headquarters in January 1994.[8] Later that year, 100 members of Chlorophiles went on a bike ride, sponsored by LVM, Tessenderlo Chemie and Solvay.[9] Despite the stunts, the chemical companies have also been flexing their muscles: Solvay has threatened to sue, or has sued, Greenpeace in Belgium, Austria, the Netherlands and Austria for PVC public education campaigns and protests.[10]

LIFE ON A DIFFERENT PLANET

In his book, *Life on a Modern Planet*, Richard North ploughed a favoured furrow of environmental critics that predictions made years ago by certain people have not come to fruition. He then uses this to criticise the current environmental movement, referring to environmentalists as 'despairists' and 'doomsters'.[11] As is now the fashion, North goes on to question global warming and other current ecological problems. Describing himself as a 'post-Luddite', he believes that old technologies such as nuclear power and new ones such as genetic engineering will eventually solve the world's current problems. Free trade is also supposedly a panacea for the planet's poor.

The future offered by North is simplistic in its attitude and dangerous in its assumptions. Whilst praising the benefits of biotechnology, North glosses over the hazards of genetic engineering and the complex issues of intellectual property rights. Third World countries are increasingly concerned about the piracy of their genetic resources by transnationals.[12] North supposes we will be able to feed ten billion people adequately when we do not even feed five billion now. While talking about future food patterns, he misses out two areas of increasing conflict, where there are already significant problems and which will only be exacerbated by increased population: adequate clean drinking water and declining fish stocks. Even the World Bank believes that the wars fought in the twenty-first century will be over water.[13]

Wilfred Beckerman's book, *Small is Stupid*, reiterated much of Richard North's argument. He attacked the 'semi-hysterical eco-doomsters' and 'self-righteous obscurantism of the environmental extremists' who peddled 'pseudo-scientific scare stories'. Beckerman argued that environmentalists were wrong about population, finite resources, biodiversity, and global warming.[14] He was equally dismissive of the case of sustainable development, calling it 'morally repugnant or logically redundant' and called for the need for a 'balanced debate'.[15] Geoff Mulgan, writing in *The Independent*, called the book 'itself both small and stupid; sloppily written, intellectually weak' and 'philosophically naive'.[16]

Matt Ridley was described by John Blundell, the General Director of the

Institute of Economic Affairs (IEA) as 'in the vanguard of one of the greatest intellectual battles of our time'. He also accused Greenpeace of spreading 'propaganda' and labelled environmentalists 'Gestapo'.[17] It is a dangerous game calling people 'Gestapo', and not really the language of serious intellectual debate. Ridley also launches an attack on the science of global warming and 'ozone exaggeration', because, as he sees it, many 'green' arguments are just old socialist ones dressed in new clothes.[18]

There is no doubt that the new greenlash triumvirate have found a little niche to exploit. There is money to be found at the end of the counter-rainbow. There is also political mileage to be made.

THE BRENT SPAR BACKLASH

Both North and Ridley were quick to resurface during the spring of 1995 when Greenpeace ran its campaign to stop the British subsidiary of the Royal Dutch/ Shell company from dumping its redundant rig, the Brent Spar, into the Atlantic.

Labelling Greenpeace 'hypocrites', Ridley, writing in the *Sunday Telegraph* dismissed Greenpeace's 'silly games in wetsuits and helicopters' which were 'not to save the environment, but to reverse an alarming fall in their membership from nearly five million people in 1990 to fewer than three million this year'.[19] This line was repeated in an editorial in the *Daily Telegraph*, entitled 'Green for Danger'.[20] This is rather an idiosyncratic interpretation of Greenpeace's *raison d'être*. If every action Greenpeace undertakes concerns membership, every word that is written in a newspaper concerns readership: not a very rational argument.

Greenpeace's success against a powerful company such as Shell sent shock waves reverberating through the big polluters of industry and the government, and this is where much of the backlash came from, as well as the print presses of the political Right. If Shell could not withstand pressure from the environmental community, who could? The response was predictable. Despite Shell's climb-down, Greenpeace had been 'wrong', Shell had been 'right'. 'Exaggeration, emotion and hysteria' had prevailed 'over science and reason'.

'The backlash fashion is being allowed to frame the environmental issue in a particular way as one of "fact versus emotion" which we believe is a fundamental misreading of what is at stake,' argue Brian Wynne and Claire Waterton, from

Plate 12.1 Shell's Brent Spar rig is sprayed with water during Greenpeace's second occupation of the rig.
Source: Greenpeace/Jurgens

the Centre for Study of Environmental Change, at Lancaster University. They continue:

> In this case, it is not that a gullible public has been misled by 'unscientific' emotional posturing, as eco-backlash writers like North and others claim. This increasingly influential stance only reproduces the self-defeating and misbegotten framing that only instrumental knowledge and literalistic claims matter.

The two academics argue there is:

> profound and justified public anxiety about the narrow technical criteria which continue to restrict and dominate the ways in which political parties, government bureaucracies, scientific establishment, and avowedly far-sighted industries think about environmental policy and its wider implications for society's direction.[21]

The British government was obviously furious with Greenpeace. According to the Energy Minister, Tim Eggar, Greenpeace's campaign was one where 'blackmail succeeded against well-proven scientific evidence'. However, Eggar went further, accusing Greenpeace of 'environmental terrorism' which was a 'threat to democracy'.[22] In one sentence Greenpeace, a non-violent organisation, were being tarred with the same brush as the militia who blew up the federal building in Oklahoma earlier in the year. Eggar was only reiterating a theme that had rebounded in the right-wing press for a while. Environmentalists were, according to the *Daily Telegraph*, threatening democracy itself.[23] *The Financial Times's Energy Economist* accused Greenpeace of 'eco-terrorism' which is 'just as irrational and dangerous as any other forms of terrorism'.[24] The *Daily Telegraph* even accused Greenpeace of enciting violence. 'Violence, too, is always liable to result from this type of protest,' wrote its leader writers.[25] The pages of the *Daily Telegraph* quoted their Industrial Editor warning of the 'environmental jihad', the Adam Smith Institute warning of 'extremists', and their environmental correspondent labelling Greenpeace 'fundamentalists'.[26]

The British press, having fed off the excitement of the Spar campaign, attacked Greenpeace for having being 'manipulated'. Editors from the BBC and ITN told a gathering of collected journalists that they had been duped by Greenpeace. 'Greenpeace Used Us, TV Editors Say' ran the headline in *The Guardian*.[27] It was an admission of failure for editors to acknowledge that they could not do their job properly. As one person from Greenpeace put it, the TV companies, having got drunk on the exhilaration of the Brent Spar campaign, were now suffering from the hangover.[28]

THE 'APOLOGY'

The backlash that Greenpeace had suffered so far was tame compared to the near universal condemnation it received when it admitted that a figure had been wrong. Greenpeace also apologised to Shell for the mistake, which made the error seem worse than it really was. The press had a field day, running the line that Greenpeace's whole campaign had been an error. 'Red-Faced Greens Admit: We Got it Wrong', ran the *Daily Mail*, as the *Daily Express* warned of the 'Dark Side of Greenpeace do-Gooders', and an editorial in *The Times* told Greenpeace to 'Grow Up'. Only *The Independent* ran a balanced editorial entitled 'Better to Blunder than to Lie'.[29]

Tim Eggar, the Energy Minister, accused Greenpeace of 'scaremongering',[30] Shell was far more shrewd in its response. In a brilliant PR counter-attack, handed to Shell on a plate by Greenpeace's 'apology', Shell's Christopher Fay seized on Greenpeace's 'misinformation', maintaining that Shell 'had tried to take a balanced view, as against just taking one aspect'. Fay said:

> The real issue is, . . . do you want to take a balanced view to the environment, or do you want to succumb to a separate and single issue, which says that perhaps the sea is clean but we don't actually mind if people die, or the land is dirty or whatever, but this single issue is fundamental. I am sorry, but I think society, if we are going to have sustainability, has really got to take a balanced view on the environment.[31]

In Fay's words were the implications that it was the oil companies who were 'offering balance' and a vision of 'sustainability' whereas Greenpeace, a single-issue campaigning organisation was 'fundamental' and really did not mind if people died in pursuit of its goals. It was a case of paradigm shift to perfection.

Both Eggar's and Fay's statements have wider implications for the environmental movement. Eggar reiterated that the only reasonable medium with which to hold environmental debate was science. By this token, any moral or social arguments will no longer be examined in the equation of rationality. However, what Fay was implying is that single-issue pressure groups could no longer be trusted to achieve 'balance', whereas 'balance' could be got from industry. Richard North had also attacked the single-minded nature of Greenpeace's campaign. 'The wider concern,' wrote Wynne and Waterton, 'voiced by Richard North and others is whether Greenpeace's actions herald a new era of single-issue "anarchy".'[32]

Attacking single-issue politics is becoming common. In September 1994, Michael Portillo, the standard bearer of the UK political Right, said that single-issue pressure groups were a 'threat to the very fabric of society'.[33] The environmental movement have been attacked in the USA along the same lines. 'What they throw at me is that I am a single-issue person,' says Lois Gibbs, from the Citizens' Clearinghouse for Hazardous Waste, one of the most successful grassroots environmental organisations in the USA. She continues:

> Yeah, I'm a single issue person. I look at that issue of people being poisoned and it makes me mad . . . All they understand is profit and loss. When the cost is high enough corporations will decide to recycle wastes, reclaim materials, substitute non-toxics in their products and eventually change their processes of production.[34]

The reality is that it is only single-issue groups that are now prepared to challenge the power of the corporate and political élite and to challenge these processes of production. Increasingly these groups will be attacked, but to mount effective campaigns they will have to organise in broad-based coalitions. Some commentators argue that there can be no greater a single issue than the politics of pure profit, and the reason why there was a backlash against Greenpeace for stopping the Brent Spar, was because it had threatened the *status quo* and was getting too powerful.

THE RIGHT-WING 'CONTRARIANS'

The Institute of Economic Affairs (IEA), the publishers of Ridley's work, is a leading neo-liberal right-wing think-tank in Britain, along with the Centre for Policy Studies and the Adam Smith Institute (ASI). It has been linked to the Earth Day Alternatives Coalition since 1990. The same year saw the downfall of their political mentor, Mrs Thatcher and the IEA already warning that the policies of the environmental movement and the Green Party would lead to an 'ecological hell' and a plummeting standard of living.[35] Wilfred Beckerman also wrote a report for the IEA, advocating pollution credits rather than green taxes.[36] Beckerman has maintained a close relationship with the IEA since then and is currently an Honorary Fellow of the Advisory Committee of the IEA Environment Unit. Two other notorious green sceptics on the Advisory Council include Dr John Emsley from Imperial College and Dr Jo Kwong, from the Atlas Foundation in the USA.[37]

Other right-wing think-tanks have also denigrated the environmental movement. The Institute for European Defence and Strategic Studies strayed outside their normal field to attack the 'New Authoritarians' in a report in 1991. Author Andrew McHallam warned about the threat of 'eco-terrorism' of greens who had a fundamentally 'anti-capitalist and anti-societal outlook'.[38] Jonathan Porrit, a leading environmental commentator, called the report 'dizzyingly unstructured, ill-informed and facile'.[39] Two years later, Russell Lewis, the former general director of the IEA, had teamed up with the ASI to publish the *Environmental Alphabet*. According to Lewis, 'environmental myths' were on 'a par with pixies, sea monsters and weather gods'.[40] So-called environmental fictions were being perpetuated by self-interested scientists, zealous environmentalists and misguided governments. He advocated that the free market would solve environmental problems and called for the re-introduction of the banned pesticide DDT. Disposable nappies were more 'green' than their cloth counterparts, added Lewis, who also questioned the value of recycling.[41]

In 1994, the IEA marked the launch of their new 'Environmental Unit' with the publication *Global Warming: Apocalypse or Hot Air?* by Roger Bate and Julian Morris. The foreword to the book was by Wilfred Beckerman, who dismissed environmentalists as 'eco-doomsters'. The bias was evident in their argument. In the chapter entitled an 'outline of the science of global warming', over half of the references of cited studies were of known climate sceptics. The recommended reading list was a wish list of right-wing counter-science.[42] 'Hard to take seriously', was how Sir Crispin Tickell, the Prime Minister's chief environmental adviser described it. Sir John Houghton, Co-Chairman of the Scientific Working Group of the Intergovernmental Panel on Climate Change, called it 'uninformed'.[43]

The first seeds of a political scientific backlash became evident, when an article appeared in the right-wing *Spectator* magazine questioning the dynamics of the ozone hole. 'There are,' wrote the leader writers at *The Independent*, 'growing signs of an ideological backlash.'[44] With the introduction of three political anti-green books in March 1995, the scientific backlash added momentum, but there are other issues that these new contrarians want to counter as well.

Inheriting the new right-wing economic thinking of cost-benefit analysis on all environmental issues, (one of the arms of the 'unholy trinity' so argued in the USA), industry and the Right are out to negate the concept of 'environmental risk'. Environmentalists, they will argue, always exaggerate the risks of contamination and pollution. If this is the case, industry is being unfairly penalised and regulated for a disproportionately perceived risk on society. If industry can convince the press and politicians and eventually the public that the environmental risk is overstated, they can argue that there should be less regulation. At a conference on Environmental Risk, held by the IEA in October 1995, the networking between groups and individuals became apparent. Apart from industry speakers, from, amongst others, Proctor and Gamble, BHP, Scottish Nuclear, Chevron UK, there were the well-known US anti-greens or sceptics, Fred Smith from the Competitive Enterprise Institute, Dr Jo Ann Kwong from the Atlas Foundation and Patrick Michaels, from the University of Virginia. Wilfred Beckerman was there too.[45]

THE POLITICAL BACKLASH

The right-wing of the Conservative Party have copied some of the language and ideas emanating from the think-tanks. Teresa Gorman has called environmentalists in the House of Commons 'eco-terrorists'.[46] John Redwood, a close colleague of Gorman's, lumped environmentalists together with 'European neo-Nazis'.[47] But it was Redwood's actions that infuriated environmentalists more than his words. During his tenure at the Welsh office, Redwood secretly drew up plans to privatise the Snowdon National Park and fifty other prime nature reserves, a scheme that would completely dismember nature conservation in the country.[48] The Labour Party dismissed the plans as 'dangerous and loony' and Redwood's Cabinet colleagues called him 'barmy'. Others warned that they would emasculate the Welsh conservation agency and lead to Britain defaulting on its obligations to international treaties. Environmentalists labelled the changes as 'politically motivated'.[49]

Despite this, privatisation of the nation's natural heritage had been mooted for the first time at Cabinet level. It must only be a matter of time before right-wing politicians and think-tanks, mindful of ever necessary budget cuts, will privatise the national parks and gut the statutory conservation bodies into sparse skeletons of ineptitude. This was the essence of the Republicans' 'Contract With America', which key right-wing politicians such as John Redwood, Peter Lilley and Michael Portillo, would love to emulate.[50] In August 1995, John Redwood announced the introduction of a think-tank to further his political ideology. The following month, whilst touring the USA he won backing to fund a US branch of his think-tank, the Conservative 2000 Foundation. He also met Newt Gingrich, representatives from the Heritage Foundation and the Moonie paper, the *Washington Times*.[51] There now exists an Anglo-American think-tank that will pursue a radical right-wing agenda that will include anti-environmental policies.

If the Conservatives were to lose the General Election in 1997, it is believed that they would realign the party to the Right, advocating how rampant spending cuts are necessary which would include closing government departments.[52] Gutting government-funded conservation bodies and the privatisation of the nation's national parks will be back on the political agenda, following in the tradition of their hero, Margaret Thatcher.

The deep psyche of Thatcher's premiership still pervades many Conservative policies and forms the cornerstones of the current backlash in Britain. First, the neo-liberal view that the dominance and the freedom of the motor car could not be challenged, despite dire predictions of unsustainable growth. Second, the neo-conservative belief that people opposed to road-building are subversive in nature and should be punished accordingly.

THREATENING NATIONAL SECURITY

Seen from a neo-conservative perspective, nuclear weapons are a matter of important national self-esteem[53] and anti-nuclear protesters are considered a threat to national security. Anti-nuclear campaigners and other proponents of social justice are 'no more than a Trojan horse for egalitarian totalitarianism'.[54] Environmental campaigners who sit in front of bulldozers on road construction sites are also seen as subversive in nature.[55] The wording 'threat to national security' becomes an important phrase that determines whether certain sections of protesters fall under the remit of the security services. This includes, according to Gary Murray, writing in *Enemies of the State*, 'just about anyone who is considered to be a threat not so much to the State as to a small diehard faction's view of the *status quo*'.[56]

It is two groups of campaigners, the anti-nuclear activists in the 1980s and the anti-roads protesters in the 1990s, who have suffered the most from the state's attempts to silence them. Both sets of protesters are still subject to state surveillance, harassment, infiltration, dirty tricks and violence. The security services and a government agency have used private detective agencies to spy on protesters.[57] Moreover, the state has attempted to demonise both sets of protesters, either as 'communists' in the case of anti-nuclear protesters or 'terrorists' and 'fascists' in the case of anti-road organisations. Incorrectly labelling people as communists, terrorists and fascists justifies a different response to that of a mere protester. They can be deemed a threat to national security, whereas protesters are not. It can also vindicate violence, harassment and surveillance of them by the state, as has happened with the anti-nuclear movement.[58]

Anti-nuclear activists were spied on and infiltrated during the Sizewell Inquiry, in the early 1980s. One of the prominent anti-Sizewell activists, Hilda Murrell, was found murdered days before she was due to give evidence to the Inquiry. Murrell's objections to Sizewell, though, could be one of two plausible reasons for her murder. The other reason is connected to her nephew, who had been privy to confidential documents relating to the sinking of the Argentinean ship, *The General Belgrano*, during the Falklands War.[59] Another prominent anti-nuclear activist, William McRae, was found shot dead in his car a year after Murrell was murdered. Nuclear waste, he believed, 'should be stored where Guy Fawkes put his gunpowder'; a view that brought him into conflict with the authorities. He had been opposing the Dounreay nuclear power plant when he had been killed. Although the official investigation into his death ruled that McRae had shot himself, the gun was found some distance from the car, as were documents he had been carrying. Two briefcases that McRae always travelled with were missing although they were later returned to his brother by the police, who were unwilling to say how they had come into their possession. A nurse who worked at the Aberdeen Royal Infirmary later told her MP that there had been two bullets in McRae's brain, not one, a fact which ruled out suicide.[60]

Anti-nuclear activists were also harassed and beaten up for opposing the plans for dumping nuclear waste at the Nuclear Industry Radioactive Waste Executive, or NIREX, in Lincolnshire in 1986. One protester was attacked after receiving threatening phone calls warning her about stopping contractors on the site. Her head was smashed against a wall causing injuries to her face, a fractured wrist and a cracked rib. The woman recalled her attacker saying 'You have been warned.'[61]

THE ANTI-ROADS MOVEMENT

The 1990s have seen the mushrooming of grassroots and 'radical' environmental groups, who have mainly formed around the issue of road-building. Earth First! and other environmental groups appeared because individuals had become disillusioned with the 'mainstream' environmental groups such as Greenpeace, and FoE whom they saw as too big, bureaucratic, hierarchical and mainstream without any real campaigning grassroots.[62] Although some of these traditional groups worked on transport, there was still a need for grassroots groups that were prepared to undertake mass, non-violent direct action. A void caused by the inability of the mainstream groups to act on the growing problem of road-building and increasing car use needed to be filled. With this new type of environmental campaigner has come a new type of green backlash from industry, the government and from the press. 'A backlash to anti-roads protests from vested interests within the road lobby was as inevitable as Christmas,' said Brian Hanson, of the Architects and Engineers for Social Responsibility.[63]

In the late 1980s, it was still the government's policy to build as many roads as were needed to meet the demand for predicted increased growth. In 1989, the Department of Transport (DoT) published *Roads to Prosperity*, which was the official announcement of the predicted 83 to 142 per cent increase in traffic between 1988 and 2025.[64] The forecasts acted as a self-fulfilling prophecy for the DoT, who built roads to meet their predicted demand. In essence, policy dictated the forecasts, which in turn dictated the policy.[65] The result was an £18 billion road-building programme, later updated to £23 billion, which would dissect many protected areas of the countryside, destroying 160 Sites of Special Scientific Interest and 800 ancient monuments.[66]

The reason that people feel that they have no alternative but to demonstrate against road-building is what they see as the inherently undemocratic nature of the decision-making and consultative process. The primary complaint is that although there is a token public consultation process with road-building, it is inherently in favour of building roads. For example, over a five-year period, of the 146 public inquiries into trunk roads, only five were rejected by the inspectors.[67] Protesters believe that inspectors who do not rule in favour of the road in a public inquiry, are not given another inquiry to adjudicate over. Furthermore, the rules of inquiries have been strengthened in the last eighteen months, further eroding the powers of people opposing the road in question.[68]

One twenty-five year battle, incorporating four public inquiries, that was won by protesters was at Archway in North London. Crucial to the protesters' success was the concession by the Inspector to hold hearings in the evening when working people could attend. The fact that for most of the subsequent inquiries the inspectors have not granted this concession is a backlash against the Archway victory. At the Archway Inquiry the current Home Secretary, Michael Howard, QC, acted as the government's legal representative and lost. Howard labelled the Archway protesters anti-democratic. The campaigners also had to endure personal smears from the government, a dirty tricks campaign and threatening phone calls which included labelling them as violent. Ironically, the new-style protesters have caused their own backlash from the Archway campaigners, who believe that many of the protesters are career-orientated, and who have never stopped a road being built. There is open in-fighting between the two groups.[69]

TWYFORD 'RISING'

The campaign to stop a road being built through Twyford Down, an ancient historical chalk downland near the town of Winchester, is widely seen as the beginning of the anti-roads direct action movement, which has led to protests all over the country. A six-lane motorway should never have been bulldozed through the Down in the first place. It was one of the most heavily protected pieces of countryside in England, with many orders of statutory protection, including two scheduled ancient monuments, two Sites of Special Scientific Interest (SSSIs), and an Area of Outstanding Natural Beauty (AONB).[70] The hill was also shrouded in mystery; ancient Celtic field systems and trackways traversed the Down. Still the bureaucrats decided to carve a huge hollow through the middle to cut off a few minutes' driving time and to complete the 'missing link' of the M3 motorway.

EF! joined up with the Twyford Down Association which had been fighting the road for over twenty years. FoE, who had been involved until this stage,

Plate 12.2 Traffic moving through Twyford Down, showing the extent of the cutting though the chalk downlands.
Source: E.P.L./Steve Morgan

pulled out under threat of legal action from the government. This was the first time that legal intimidation was used to stop protesters in the United Kingdom, but it would not be the last on the Twyford campaign, or other anti-roads protests. EF! also merged with the Dongas, people who had set up camp on Twyford Down to oppose the road and who took their name from the ancient trackways that cross the Down. In December 1992 they would undertake what was the first of many direct actions against the road.[71]

'There was no violence until security guards came on the scene, and they were used for the first time on 9 December 1992,' recalls Rebecca Lush, one of the prominent anti-road activists at Twyford. 'That was the first time they were used and they were extremely violent on that occasion. That was our first encounter with security guards.'[72] The violence on 'Yellow Wednesday', as the day became known, because of the sea of yellow fluorescent jackets worn by the security guards, brought in by Tarmac, was intense.

That morning the protesters were due in Winchester Crown Court to be evicted from the site, but they were woken up by hundreds of security guards who had surrounded the camp under cover of darkness.[73] The activists were beaten up by the security guards, many were subjected to both sexual and physical assault. 'They were really going for women's breasts, really punching and pinching and twisting and groping,' recalls Rebecca Lush. 'They were also trying to rip clothes off and basically trying to humiliate. It just makes me really angry, being beaten up and sexually assaulted at the same time.'[74] Arms were broken and ligaments torn. People were knocked unconscious and, at one point, four people were waiting for ambulances to take them to hospital. David Bellamy, the well-known conservationist who was present, said that he had never seen violence like it in twenty years of protesting.[75]

Private security guards had arrived on the environmental scene in Britain. But the violence even sickened some of the guards, with ten resigning the first day and twelve the next.[76] However, many of the guards seemed to enjoy the violence. 'They were absolute thugs. We had never come across this type of roughness before. A lot of them were really getting into hurting people, they were really violent,' says Lush.[77] 'Obviously there were people who were very aggressive,' recalls Jason Torrance, a founder of EF! 'Some people got hurt and some people got punched'.[78] 'When we got on site we were unofficially told to make sure no cameras were watching,' recalls one guard from Group 4, 'Some of our guys went over the top.'[79] One protester was so badly beaten that when examined by a police doctor he was said to have been 'subject to a systematic beating'.[80] The violence became so abhorrent that Labour's John

Plate 12.3 Police at the Newbury by-pass with faces blacked out by balaclavas, and with no ID badges on them.
Source: E.P.L./Steve Johnson

Denham was to question the 'unacceptable force' used against protesters by Group 4 Security in a parliamentary debate.[81]

During Yellow Wednesday, women were particularly singled out for abuse by the predominantly white male security guards, a pattern that has also been repeated many times at Twyford and subsequently across the country at protest sites. 'Sexual assault on women involved in direct action happens all the time,' says Lush, two and a half years after Yellow Wednesday. 'It's not just sexual assault, it's also sexual harassment as well from security guards who think it's amusing to get a woman on the floor and call her slut and slag, and stick their pelvis in your face.'[82]

The tactics used by the security guards at Twyford, those of violence, intimidation, and sexual harassment would be repeated around the country. The individual cases of intimidation and violence are numerous, but there are certain trends that occurred. For example, when a big action or demonstration was planned, the level of violence would increase, as would violence away from the cameras. For the protesters, if they were assaulted, it was difficult to identify faces amongst the sea of yellow jackets, as guards, all in uniform, are not required to wear identification numbers.[83] At later protests, activists would complain that both police and guards would make identification difficult by hiding either their faces or ID.

SURVEILLANCE LEADS TO SLAPPS

As the protesters began to receive prominent media attention and sympathetic coverage, as well as growing support, the government took measures to identify and effectively nullify the leaders. In what was believed to be the first time that a government department rather than the security services had hired a private detective agency to spy on environmental campaigners, Brays Detective Agency of Southampton was hired by the DoT to collate evidence against the Twyford protesters. David Crocker, chairman of the Twyford Down Association and a former Conservative councillor, called the decision 'a national disgrace'. 'What we are now witnessing is actions typical of a totalitarian regime,' said Crocker, whose sentiments were backed up by Liberty, the civil liberties organisation. 'This cannot be acceptable in a democratic society,' it said.[84]

Brays were to run up a bill at Twyford of £267,000 spying on the protesters. Security firms such as Brays are completely unaccountable, and their involvement in spying on anti-roads protesters raises other issues. 'Although it may be legitimate under some circumstances to collect information on people specifically for a civil action,' argued Liberty, 'we believe that collecting information on people solely because they are protesters is a breach of their privacy under international law.'[85]

The results of Brays' surveillance soon became apparent. The government announced that it had been granted an injunction banning a number of key protesters from the site and that it was seeking an initial £1.9 million in damages and compensation from those people for delaying the road scheme. The tactic, effectively a state SLAPP against the protesters, caused outrage. Simon Fairlie, from *The Ecologist* magazine, who would later be sent to prison for breaking the injunction remarked that 'the regular use of SLAPPs to quash protest has the potential to immobilize the environmental movement in the same way that the UK trade union movement has been paralysed by the threat of sequestration of its funds'.[86]

The tactics were deplored by the protesters and politicians, alike. 'The trial is an attempt to snuff out legitimate public protest,' said Philip Pritchard, one of those arrested.[87] Labour MP, John Denham agreed, 'It has everything to do with intimidating others from taking part in any form of protest.'[88] Those being sued included a mature student of photography from Southampton who was there to take pictures of the demonstrations, and a man who went to Twyford Down once with Mr Denham and two other MPs.[89]

'It seems to be a tactic adopted by the state to suppress civil disorder,' says Michael Swartz, a solicitor from Bindman and Partners, who has represented protesters from Twyford and other road protests. Swartz goes on to explain his reasoning:

> There were two claims there, one is for damages and one for compensation. The figures that the DoT themselves have supplied, the amount of compensation they claimed started off at between £1 and £1.9 million, and at the end of the case they were claiming £3.5 million by way of compensation for disruption. Clearly, if you looked at the means of the people they were suing, many of them were either unemployed, or claiming income support, or just lived off the land, or lived off good will. The DoT must have been aware of that. They knew that they were dealing with an alternative society. Only a few of them were home-owners and they were not the major trespassers. Clearly they had no realistic prospect of getting damages, getting compensation.[90]

Because the authorities had only applied for a temporary interlocutory injunction there was no way any of the named people could actually defend themselves or argue their case in court. But the DoT had sprung another legal tactic as well. It said that not only were the named protesters liable for their own actions but also for the actions of the other defendants. The message was simple to the few affluent Conservative protesters whom the DoT wanted to isolate. 'If the seventy-five others can't afford to pay £3.5 million then we are going to go after your home,' says Swartz, who describes this move by the DoT in these circumstances as 'unprecedented and used for political means as well as practical ones'.[91]

Evidence that the government's intention was not to sue but to scare, was unveiled in June 1995, when it was announced that the DoT was dropping its £1.9 million claim against the protesters. The police tactics of mass and wrongful arrests were already beginning to backfire in other ways. In January 1995 ten protesters had actually been awarded damages totalling £50,000 for wrongful arrest, with a further forty protesters expected to receive damages of £500,000.[92]

Many of the protesters did break the injunction, an act that led to eight of them being sentenced to prison. But the tactic was also effective at whipping up support for the campaigners, and in that sense the legal action backfired. Even the sentencing judge called civil disobedience an 'honourable tradition'.[93] Those people locked away received enormous publicity and became the M3 martyrs. They were also visited by Labour MP, Chris Smith and the EC Environment Commissioner. 'It really makes you think about freedom and the kind of society you are living in,' believes Torrance, one of those sent to prison. 'Standing up for what you believe in, you can be put in prison, but the vast majority of the good things we have in this society has been born out of struggle. I think the anti-roads movement is one of those.'[94]

PHYSICAL AND LEGAL INTIMIDATION SPREADS

The roads protests began to spread as more people decided to fight the government's road-building plans. The next confrontation spot was the M11 link road in East London, another road that had been fought by the local community for two decades. Once again, the protesters were met with both legal and physical confrontation.

Just as the authorities had targeted the Twyford protesters with SLAPPs, so they did with key activists at the M11, who were also served with an injunction. This is a deliberate tactic according to Mike Swartz:

> In August 1994, the DoT and its contractors started a similar campaign against eleven campaigners at the M11. It is no accident that there were eleven. It is clear that the DoT wanted to give the campaigners some publicity. The campaigners could play their own publicity campaign out of it, after all they had something to lose from it, but the DoT wanted the message to spread far and wide, that, if you get involved in roads protests, then we are going to sue you. What easier way of spreading that message than these people being labelled the 'M11 eleven'. It's no accident.[95]

Swartz sees the use of legal intimidation as something people should be concerned about. 'It's a worrying development because it doesn't address the fundamental issues,' he says:

> A lot of people have concerns about road-building or car use or loss of resources and they express their views through direct action. Their protest is a concern, first about the environment, and secondly the failure of the political establishment to reflect those concerns. Instead of dealing with either of those points: problems with the political

establishment, problems with environmental destruction, what does the government or state do? It just clamps down on the protest, it just tries to kill the messenger, rather than address the real issues. That is very worrying, because it is a very blinkered way of dealing with issues. It is a sign of a healthy democracy that new issues are put on the agenda, and by clamping down on new issues, or those that try to put new issues on the agenda, then you are negating democracy.[96]

IN THE EYES OF THE LAW

Just as had happened at Twyford, the protesters were met with force. In the face of heightened violence by the security guards, the police seemed to do nothing. From the outset, the anti-roads protesters have complained that the police have acted in a totally prejudiced way towards them, and, on the whole, have not acted on reports of violence against them. The role of the police is to uphold the law in an impartial fashion, but in the case of protests, such as the M11, this has not been the case. Furthermore, not only have the police acted violently against the protesters, but by refusing to act against serious acts of security guard violence, they sanction that violence.

'There was total bias from the beginning, there was no question that we were criminals,' recalls one of the key organisers of the M11 protest, Paul Morotzo, 'if there was a confrontation between protesters and the security guards, it would be a protester that got arrested. If the security guards did a citizen's arrest on a protester, then the protester would be arrested straight away without even looking for evidence.'[97] On the other hand, if the protesters accused the security guards of assault, their complaints would be ignored. 'They [the police] would totally ignore you, they would just tell you to piss off,' says Lush, who recalled one occasion where a person being attacked by a security guard with a hammer was witnessed by six people as well as being recorded on video. The protesters, when they went to the police, were simply told to 'Go away'.[98]

Video footage from the M11 protest shows a protester being head-butted by a security guard in the presence of three police officers. When pressed to arrest the guard, the police refused to do so.[99] Michael Swartz supports the protesters' accusations. 'Despite some well documented claims against security guards on the M11, the police there, and I believe this happens elsewhere, have markedly failed to prosecute any security guard for assaulting a protester,' he says.[100] Tactics such as these 'do send out the wrong messages, and the main thing, it marks a breakdown in the rule of law,' argues Swartz. He adds:

> It is one thing having bad laws, it is another thing not enforcing the existing ones impartially. We are now in the position of having both – bad laws and partially enforced laws. Moreover, it sends out the messages to security guards that they can get away with violence. It give the message to the protesters that they can't trust the police to protect them.[101]

Violence was used against the protesters. 'People were trained to use violence in preference to minimal force,' alleges Paul Morotzo.[102] Violence, in the security business, saves you money. 'I think their remit was one of cost,' says Morotzo. 'It was just economics. You know if they used violence they'd have to employ less people because they'd be more effective.'[103] Morotzo's view is backed by John Windham from Liberty. Speaking about security firms, he says that 'their only driving force, of course, is profit. They are paid to do a job, and sometimes the profit motive will win out, there are no principles or codes of practices, there are no duties for security firms.'[104]

Protesters at the M11 noticed that the violent security guards seemed to get promoted, whereas the reasonable ones seemed to get sacked. The violence also increased the more successful the campaigners became. 'Basically, it was fairly straightforward,' says Morotzo. 'The tendency was for the violent security guards to get promoted and kept on and the less violent ones would get moved on, you wouldn't see them again, and so it was creating a culture of violence within the security company.'[105] This was a trait that happened at other sites, notably at Solsbury Hill near Bath.

Some security guards actually had criminal records for violent behaviour. Michael Swartz says:

> I was involved in a case some time ago when it came to light that a security guard working on the M11, who was a regular worker there, who had been promoted, had responsibilities over other men, had got criminal convictions. It started off with offensive weapons, then assault causing actual bodily harm, then assault causing grievous bodily harm. He had been in prison for offences like that and was obviously a violent man, in charge of others, in charge of evicting people from sites. He had not declared it to his employers, but his employers had not even bothered to check. He is guarding a security site.[106]

During an illegal eviction by the security guards at the M11 in June 1994, a protester was stabbed in the hand by a guard with a Stanley knife. As protesters were removed they were repeatedly kicked and punched. One male protester was grabbed by the genitals, punched and trampled on by security guards. 'When you get down you'd better run fast because I'm going to kick your head in,' the protester was told. A female was sexually assaulted, another was knocked

Plate 12.4 Protester being dragged off site by a security guard at the M11.
Source: E.P.L./Julia Guest

unconscious, after being pushed by security guards head first on to a brick.[107] 'It was just the most appalling level of violence you can imagine,' says Morotzo, who witnessed people being beaten, stamped on, spat on, groped, and verbally abused. Others were thrown to the ground, and dumped in wet cement.[108]

The protesters also allege that the police engaged in violence. At the M11, the local police were friendly to start with, but protesters recall that there were difficulties with the Territorial Support Group, a specialist back-up unit. Some of the most brutal tactics by police were used when evicting protesters from a chestnut tree on Wanstead Green in December 1993. The 'chestnut tree' was seen as a turning point in the campaign, which had really brought the environmental campaigners and the local community together. Protesters set up camp in the tree, obtaining a court order that proved it was a dwelling. Mail was even delivered by the postman to 'The tree, Wanstead Green, London'.

The police were accused of widespread brutality in evicting the protesters. Forty-nine complaints against the police were recorded by protesters including a 12-year-old girl being hit, the excessive use of violence on pressure points and the sawing of branches whilst protesters were still perched on them.[109] One activist had his foot broken by a policeman stamping on it.[110] The complaints were to have an effect on the police, who previously had seen their public order remit as a licence 'to use violence and physical force which they did round the chestnut tree', according to Morotzo. 'Because of all the legal problems that occurred because of their activities at the chestnut tree, they were much more hands off, much more careful.'[111]

Other tactics had been employed to evict the protesters. Nine members of an East London gang were paid £100 to remove them from the tree. Arriving in the dead of night, armed with crowbars and claw hammers they set fire to the tree. Two of the men were subsequently jailed for the attack. Those behind the assault have never been discovered.[112]

By June 1994, there were reports of increasing violence, assault, sexual and verbal harassment and intimidation by security guards at the M11 in London, Solsbury Hill near Bath and the M65 extension near Blackburn.[113] 'The increase in violence on British road protest sites has brought warnings that the guards are getting out of control,' wrote John Vidal in *The Guardian*, 'and it is only a matter of time before someone gets killed.'[114] Solsbury Hill protesters in particular were subject to violence. George Monbiot, a Visiting Fellow at Green College, Oxford, had his foot smashed when he was thrown on to rubble by two Reliance security guards. One woman protester claimed she had been kicked in the head by security guards. Another protester was knocked unconscious by guards, and a further one had their ankle broken. By June 1995, four protesters had been hospitalised.[115]

Asked who he thinks is responsible for the violence, Monbiot responds. 'It was clear that they could do that sort of thing if it was officially sanctioned, and it was also clear that working within that situation they felt encouraged to do it.' Monbiot continues:

> Whether it's been said or whether it's a tacit agreement, it's been made very clear to those security guards that they will not be punished for beating people like myself up. Indeed, one of the people who beat me up was promoted the next week.[116]

He also confirms a trend at Solsbury Hill which had happened at the M11. The most violent security guards seemed to be promoted, the least violent seemed to be sacked.[117]

Just as has happened at Twyford and Solsbury Hill, a worrying trend was that guards began picking out women protesters. 'They have a really nasty tactic of grabbing you by the breasts or lifting you by the shirt so you're bared,' said Paula Black, one of the women who had been assaulted.[118] The guards are also very wary of being caught on film. As one guard assaulted a protester at the hill he was told by another, 'Stop doing that, the guy's filming.'[119]. Another security guard was told to give protesters three warnings, and 'unofficially I was told I could do what I liked'.[120]

Further evidence of the violent nature of security guards was documented by the TV programme *Public Eye*. One security guard employed by Reliance Security named Larry Hunter, who had a criminal record, admitted he had punched and deliberately set out to hurt protesters. He was a member of the self-styled 'Kick Arse Squad'. Another ex-Reliance guard admitted the week before he had been hired, he was in police custody for assault. These confessions are not uncommon for the unregulated private security industry, but Reliance told the programme makers that 'its agency was used to supplying security officers to our quality standards'.[121]

At the M65 there were reports of intimidatory and dangerous tactics to get protesters from tree houses. Aerial walk-ways were reportedly cut with protesters still on them and fires were lit underneath trees with people in them.[122] One protester was knocked unconscious for twenty minutes after the excessive use of pressure on his neck by security guards. His neck was so badly damaged that he had to wear a neck brace for three weeks after the event.[123] Video footage from one demonstration at the M65 shows a protester, Caroline Hall, being manhandled by a security guard from Northern Security, subcontractors to Group 4. The guard dropped the woman seven or eight feet off the digger and, in the process of removing her, lifted up her clothes so her upper torso was exposed. Her request to pull her clothing back down was ignored. 'I felt so humiliated, I was in shock for the rest of the evening,' said Hall.[124] A security guard told the TV programme *Public Eye* that after the incident the guards laughed about the way the woman had been dropped. He resigned in disgust.[125]

The violence became so bad at the M65 that both Group 4 and Northern Security had to make an undertaking in court not to be violent.[126] The Highway Agency deputy director, James Bond, maintained that the Agency would not employ any security firm using 'excessive force'.[127] This said, no contract has been terminated due to violent behaviour of security guards.

BAIL OUT

At the M65, once again the police and courts used different legal tactics to deter protesters. Once protesters were arrested, the police imposed bail conditions that barred the protesters from approaching the site. 'They could not get involved in that campaign any more and the bail conditions stifled the campaign,' says Michael Swartz. It seems that the courts and the Criminal Prosecution Service are also acting in a political role. According to Swartz,

> They have not given court appearances for people in a matter of weeks, which would be normal for that sort of charge, but they are delaying things longer and longer. Now we have got to the ridiculous position in Stanworth Valley (M65), that there are people on bail eight months after an incident has taken place with no prospect of a trial date coming up. The Court, with the CPS's assistance, have delayed trial dates longer than lots of other routine cases so that those bail conditions bite and the campaign is stifled. Again it just misses the issues, which should be, do we want these roads to be built?[128]

Rebecca Lush argues that:

> bail conditions are the most effective legal weapon against protest. What is most dangerous, is that bail conditions totally immobilise a protest. They just make loads and loads of arrests and recommend to the magistrates that everyone be put on bail conditions not to go within two miles of the work site. It is like a pseudo-injunction, before you even go to trial and then they give you a late trial date. In most people's opinion it is the most effective tactic at stopping a protest. For months and months you are banned from protesting, and then when it finally gets to court, the police do not even bother to turn up.[129]

Bail conditions were used again with devastating effect at the protests against the Newbury by-pass with nearly 800 protesters arrested and bailed off-site. Like Twyford Down, 'the Third Battle of Newbury', as the protest became known, (the first two having been fought during the English Civil War) should never have happened. The road desecrated three SSSIs, twelve sites of archaeological importance and part of an AONB.[130]

In what must be deemed a political decision, stringent bail conditions were imposed which included protesters being bailed one kilometre off the route, ordered to go back to their home address under a resident's condition, and forced to sign on at a police station near their home. 'Two police officers admitted that these bail conditions are police force guidelines, although the police should treat people as individuals, not with a uniform policy', says Launa Johnson, a Law Society accredited legal advisor with the solicitors Douglas and Partners, who represented the protesters. Johnson adds that 'those sort of conditions are usually reserved for dangerous and persistent offenders'.[131] They were conditions designed to break the protest.

Mike Swartz says:

> Newbury was a highly political use of the law. The protest was identified from the start as a challenge to the state, to the planning and inquiry system, to law and order and to capitalistic interests. The response was a fist of iron in the velvet glove. You don't often see it.[132]

Illegal evictions were also carried out without notice, warnings of arrest were not always given by the police, protesters' homes were destroyed without notice being served and with people in them.[133] This said, Johnson maintains there were incidents where the police protected the protesters from eviction, acted on advice from protesters and reprimanded the security guards for using violence.[134] However, during the main evictions of the tree camps, the police sometimes made a tactical decision to look the other way during evictions.[135]

Although the general level of violence was felt by protesters to be less than on other road sites, in part because of the high media presence, there were isolated incidents of violence and violent rhetoric. Journalist John Vidal from *The Guardian* worked undercover as a security guard for Reliance Security for two days. His references or pieces of identification were never checked. On the second morning, a site supervisor told the guards, 'Anything in the trees today, you wack, right? . . . Thwack it with your helmet. Anything. And don't get caught.' Other security guards told Vidal that, 'I'm here for the action. It's like a hunt . . . You take out someone and you know you've got two men backing you up' and 'Remember, a kidney punch doesn't leave bruises' and 'Don't forget to say good morning as you break their fingers.' After an investigation into the affair, Reliance found no basis for the allegations concerning the site supervisor.[136]

Security guard hit squads in commando gear undertook a dawn raid on the Penn Wood site in February 1996, cutting ropeways and abseil lines from the tree dwellings that housed the protesters, putting protesters' lives at risk. Four witnesses signed statements saying that a woman had been put in a head-lock and punched by a security guard.[137] One protester was cut off a telegraph pole and dropped fifteen feet. 'I witnessed a senior security guard at the scene of this

Plate 12.5 A bloodied protester is led away by police and security at Newbury.
Source: E.P.L./Andrew Testa

incident and it was admitted to me that he was present beforehand,' says Launa Johnson.[138] Local vigilante groups shot at protesters and two vehicles were fire-bombed. One bus had a pregnant woman and 6-year-old boy inside.[139]

Dirty tricks were also used, the government accusing the protesters of spiking trees and making hoax calls to the fire brigade, although both the protesters and fire brigade denied the charges.[140] A well-reported case of a protester apparently deliberately sabotaging the brake wires of a coach carrying security guards, turned out to be neither deliberate nor sabotage.[141] The activists also allege that smoky fires were deliberately lit downwind close to the protest sites, and underneath tree-houses, causing a serious fire risk and health hazard.[142]

SECURITY FOR WHOM?

The use of unaccountable security guards on road construction sites is a serious cause for concern and for many people the mushrooming numbers of security guards *per se* are a real worry. There are now more private security guards than police and their numbers, coupled with their unaccountability and lack of training, are extremely troubling. 'The private security industry has grown significantly in recent years,' reported the Home Affairs Select Committee in May 1995. 'Alongside that growth there has been an increase in concern about the quality of the industry and the level of protection afforded to the public.'[143] The Committee received evidence of 'pitiful' standards, inadequate training and criminality within the industry. In one firm examined by the Committee, 11 of the 26 employees had convictions for 74 offences that ranged from burglary, vehicle theft, firearms offences, to rape, and threats to kill.[144] The Committee concluded that 'the current standards, particularly standards of training in much of the private security industry, are unsatisfactory and below the level the public need and have a right to expect'.[145] The manned guarding sector, which includes road construction sites, was said to be where the worst examples were to be found.

In its evidence to the Committee, Liberty expressed concerns specifically regarding road protesters. Liberty wrote:

> Under international law, the state has a positive obligation to protect the right of peaceful protest. The increased use of private security personnel raises concerns about the lack of democratic and legal control of private security firms; and the incidents of physical violence towards protesters and unlawful detention.[146]

Liberty recommended 'the employment of private security personnel in the policing of roads protests should cease immediately, because the delegation of state powers to the private sector is incompatible with international human rights standards.[147]

THE CRIMINAL (IN)JUSTICE ACT

A month after Liberty's evidence was submitted, the Criminal Justice and Public Order Act 1994 (CJA) passed on to the statute book. The new law criminalised many activities undertaken by anti-roads protesters, hunt saboteurs, squatters, ravers, and travellers. Many saw this legal intimidation as an attempt to scapegoat people on the margins of society in a torrid attempt for political credibility with the mainstream. Hailed by the government as 'the most comprehensive attack on crime', all these groups believed that, in fact, the legislation was 'the most comprehensive attack on human rights'.[148]

Under the CJA it has become a criminal offence to trespass on land with the intention of disrupting or obstructing lawful activity; an offence called aggravated trespass. Before it was criminalised, for over 800 years trespassing was a civil offence. Liberty considered

> The creation of this new offence of aggravated trespass not only places further unacceptable restrictions on the right of peaceful assembly, which the government has a positive obligation to uphold under international law, but also, by criminalising certain kinds of trespass, may in *practice*, have the effect of encouraging further violence towards peaceful protesters by private security personnel.[149]

Liberty concluded: 'The European Commission has said that the right to peaceful assembly is a fundamental right, and one of the foundations of a democratic society. Peaceful protest is a form of peaceful assembly.'[150]

The CJA is also likely to breach other internationally agreed human rights standards by eroding a right to a fair trial, the right to freedom of expression and assembly, the right to freedom of discrimination, the right to freedom from arbitrary arrest and detention and the right to privacy.[151] So fundamental were the changes in the Act that opposition came from some surprising quarters: the Police Federation, the Prison Officers Association and the Law Society were all opposed to sections of the Act.[152] 'It appears to be legislation against a certain section of the population and that is a recipe for disaster,' said the chairman of the Metropolitan Police Federation, Mike Bennet. 'The whole aim of policing is to alienate the criminal – not to make criminals of people.'[153] Despite the police's objections, this did not stop them announcing that 2,000 more officers from Special Branch would be monitoring protests.[154]

In effect, for hunt saboteurs and road protesters the CJA is the ultimate SLAPP because it penalises intent. 'A person can also be charged if the police officer believes that the person intended to commit an offence,' says Mike Swartz. He continues:

> In court it will be the policeman's word against the protesters. You have this double whammy – you penalise intent and the witness who gives evidence of that intent is not only the person in the dock, but also the copper who supplants his or her overviews of what the intent was.[155]

Across the country there have been countless demonstrations and meetings against the CJA. Many of these were by young people, disillusioned by what they saw as a government that did not seek to understand but imprison. At one such meeting on 17 November, two weeks after the CJA had become law, Jan

Clarke from the Green Party condemned the Act as a 'shabby and cynical attempt by a discredited government to find scapegoats for the problems in our society'. Caroline Ellis from Charter 88, said the Act was one of the 'worst pieces of civil rights bashing legislation ever seen in the country'. 'We know that this Act is wrong, it is a form of tyranny against civil liberties and against environmental values,' added Charles Secrett from FoE. 'The CJA has just very crudely put it to the general public that we are no longer living in a democracy,' Rebecca Lush declared.[156]

The CJA, though, has failed to stop protest, although it was the main legislation used to arrest protesters at Newbury. If anything, it has made more people protest because the Act is seen as so unjust. Activists generally believe that a whole new generation of people have been politicised by the Act, and people are out to break the Act as a point in itself.[157] But it seems that it is not just legal intimidation that the government have resorted to in order to stop the anti-roads protests.

PR AND DIRTY TRICKS

Paul Morotzo also believes that because of the rising security violence and because of the public outcry that was developing, the Highways Agency launched a campaign to label the protesters as violent. It would also make the level of violence that was occurring on road sites acceptable in the eyes of the public. Furthermore, the Highways Agency advertised for external PR consultants and launched a PR campaign against the protesters.[158] The Highways Agency is reported to have spent £200,000 a year on PR between 1992 and 1995.[159]

The security companies were also on the PR offensive, 'by continually talking about how violent the protesters were and how wonderful their guards were', alleges Rebecca Lush.[160] Roads protesters were 'no better that the IRA' one motor industry representative went on the record to say.[161] After Reclaim the Streets organised an anti-car street party in London, they were depicted as 'urban terrorists and dangerous lunatics' by *Auto Express* magazine.

The month of June 1994, the same one as the Highways Department started looking for a PR company, had seen some of the worst violence against the protesters ever. 'Green guerrillas booby-trap sites' was the headline of the article in *The Sunday Times* on 3 July 1994, written by John Harlow, the transport correspondent.[162] The article depicted a policeman holding spikes that had supposedly been dug from a pit at Batheaston. It showed a drawing of someone falling down booby-trapped stairs, which had been rigged by 'eco-terrorists', and the face of a masked man with a crossbow. 'Green extremists are carrying out a campaign of violence against road-builders,' wrote Harlow. 'Police estimate that about 150 hard-core activists have committed themselves to a "summer of hate" against construction companies.'[163] Harlow alleged that one in five security officers at Solsbury Hill had been hurt, that 'lethal mantraps straight of out the Vietnam war' had been found, as had high-tension wires strung across trees at head height. At the M65 a spring-loaded catapult had supposedly hurled barbed wire and rocks into the face of a guard. The article finished by saying that environmentalists could soon be using parcel bombs.

The article was totally inaccurate and the picture of the spikes had been running in the construction press the week before.[164] 'The spikes had actually been handed in to the security guards, who gave them to the police,' says Lush. Harlow even put Twyford Down in the wrong county – Wiltshire instead of Hampshire. The article caused outraged condemnation from the anti-road organisations who took *The Sunday Times* to the Press Complaints Commission.

The timing of the article is also damning. Coming a day after the biggest non-violent protest against a road scheme, and after a month of increased violence against the protesters. As the article alleged that the increasing use of violence was by the protesters and not security guards, it could have alienated the activists, the driving force behind the anti-roads movement and some of their middle-class supporters.

Madeleine Bunting writing in *The Guardian* reiterated many of the false arguments that had appeared in Harlow's article, asking 'how long before their middle-class supporters quit in disgust?' at the supposed violent tactics of the extremists.[165] In a letter to *The Guardian*'s editor the anti-road groups highlighted how Bunting had failed to talk to them before writing the article, and that they were 'totally committed to upholding an historical and international tradition on non-violent direct action'.[166] Less than two months later Harlow was again attacking the environmental movement, alleging that he had seen a copy of an 'eco-terror magazine' called *Terra-ist*, which 'shows that green "warriors" can blind, maim or kill road construction and security workers'. 'Special Branch,' wrote Harlow, were 'concerned about increasing levels of violence at environmental protests'.[167]

Harlow's articles were the latest in a long list of scaremongering articles that have appeared, many of them in the London *Evening Standard* that profess that the environmental movement is using or about to use violence and/or terrorist tactics to achieve their ends. 'In the beginning the *Evening Standard* tried to depict EF! along with organisations like the IRA and Blood and Honour,' recalls Jason Torrance.[168] In November 1992, Paul Charman, writing in the *Standard* warned that 'security experts are also concerned about the dramatic rise in the number of "New Age" groups on the fringes of the Green movement who are prepared to use terror in support of animal rights and ecology issues'.[169] Charman wrote that EF! expressed 'sympathy or support for terrorist activity'.[170] By July 1993 the paper warned that 'Ecologically sound "green" incendiary bombs have been designed by a new group of environmental militants, for attacks on construction giants and other big businesses'.[171]

This kind of campaign by the press, which Harlow admits is based on information from the Special Branch amongst others, is the 'background to the current state of play *vis-à-vis* radical Green activists'[172] believes Larry O'Hara, a researcher who specialises in studying the underbelly of British politics – the activities of the state and far Right. 'The state has been busy attempting to crush, distort and manipulate such initiatives,' believes O'Hara. 'The most simple form this has taken is the spreading of lies and general disinformation about the tactics of the anti-road movement.'[173]

It is widely believed that with the demise of the cold war and the potential end to terrorism in Ireland, the security services are looking for new areas of activity that are reported as organised crime, drug-trafficking, money laundering, computer-hacking, nuclear proliferation and animal rights groups.[174] There has been speculation as to whether MI5 would expand its role from animal rights to monitor the activities of the environmental movement, especially the more radical groups. Some people say that this has already happened, whilst others say it is impossible because it strays so fundamentally outside MI5's remit, which is meant to confine itself to matters of national security. This loosely defined term is suitably ambiguous, but has been described as 'the safeguarding of the State and the community against threats to their survival or well-being'.[175]

What is known is that the anti-nuclear movement was targeted by MI5, and that Special Branch already holds an intelligence index on animal rights and environmental activists.[176] It is also known that Special Branch has targeted the *Green Anarchist* and confiscated the group's files.[177] The security companies are in regular contact with Special Branch. For example, Reliance Security, whose guards protected the sites at Bath, Wymondham, the M11 and Newbury, admit having contacts with Special Branch.[178] Moreover, in the aftermath of the high media profile of the Newbury campaign, it was reported that senior police officers engaged in anti-terrorist activity would be asked to gather intelligence on leading anti-roads campaigners.[179]

There is circumstantial evidence to suggest that surveillance and bugging are already occurring.[180] Moreover, there is other evidence to suggest that phone-tapping is not as widespread as people believe. Some activists have complained of having their rubbish taken away by mysterious vans which did not belong to the council.[181] Others recall being followed or watched by vans belonging to both Brays and Group 4. Anti-road groups also believe that they are being infiltrated. Two campaigners at the M11 were thought to be police informants as well as one woman at the Solsbury Hill campaign.[182] In August 1995, campaigners were planning to squat in disused buildings in the path of the Newbury by-pass, and had discussed the issue at a local meeting and nowhere else. A week later, the houses were demolished, before they could be squatted. This had happened before the actual contract had been finalised for the road.[183]

Both Reliance Security and Group 4 have high level contacts with the Conservative Party. Between 1992 and 1994, when the two companies were awarded a number of contracts, Norman Fowler was not only the Chairman of the Conservatives but also a Director of Group 4. Lord Lane, the Chairman of the National Union of Conservative Party Associations was a Director of Reliance.[184] Brays surveillance did not stop at the M3, they were also used at the M11 (£185,000), the Batheaston by-pass (£109,000), the A11 (£450), the M65 by-pass at Blackburn (£2,000) and Newbury. By April 1995, Brays had pocketed £705,250 to spy on protesters.[185] In May 1995, the National Audit Office predicted that some £26 million could be paid for private security firms to guard Britain's roads before the building was completed. The government was spending £575,000 every month on security at the construction sites.[186]

THE DEAD-END ROAD

Anti-road protesters are part of the fastest growing grassroots movement in Europe. Continental campaigners are looking to Britain to see a way forward in activism. The new campaign is against the very culture of the car and the companies that profit from pollution, and unregulated resource extraction. Attitudes are changing against the car, albeit slowly. An ICM poll for *The*

Guardian, in August 1995, was the first to find a majority of people in favour of banning cars from city centres.[187] There is a significant difference between someone saying they believe cars should be left at home, and that person actually taking an alternative form of transport.

Other potential death knells had been sounded for the building programme. After years of denying that building roads generated traffic, the government's own sponsored research proved it wrong. Dr Phil Goodwin, a member of the government's own SACTRA (Standing Advisory Committee on Trunk Road Assessment), estimated that new roads created 10 per cent of new traffic in the short term and 20 per cent in the long term.[188] Furthermore, the government's own Royal Commission on Pollution called for a massive shift towards public transport in an in-depth report published in October 1994. After a two and a half year research project it concluded that a continuing increase in road transport was neither environmentally or socially acceptable, and called for a halving in the road-building programme, a doubling in the price of fuel and a ten-year investment programme into public transport and cycling.[189]

It is still too early to tell if there will be a serious change in direction in the government's transport policy. What is certain though is that many motorists will not give up their cars without a fight. Nor will the all-powerful road/car lobby, and a government that has traditionally received large financial support from it. The worst backlash against the growing anti-car and consumerism movement seems yet to happen.

McLIBEL

The use of SLAPPs and surveillance in Britain does not just concern the government and anti-roads protesters, though. McDonalds, one of the world's largest multinational food giants with over 14,000 outlets in seventy-two countries is currently suing two unemployed environmental activists from London Greenpeace. Despite the name, London Greenpeace does not have any affiliation to Greenpeace International. The two activists, Dave Morris and Helen Steel, questioned the company's social, health, employment and environmental record, by circulating a leaflet called 'What is Wrong with McDonalds'. Undercover agents were employed by McDonalds to spy on London Greenpeace which led to writs being issued against five campaigners. Three backed down, but Morris and Steel, now dubbed the 'McLibel Two', refused. 'It makes me angry that multinationals can use the libel laws in this country to try to silence their critics,' says Steel.[190]

The McLibel Two are not the first to be threatened with legal action or sued. McDonalds is suing animal rights campaigners in Austria and has sued the Scottish TUC.[191] It has also threatened to SLAPP the *Bournemouth Advertiser*, BBC *Nature*, *Spitting Image*, Veggies Nottingham, *The Militant*, *The Guardian*, *Today* newspaper, Channel 4 TV company, Worldwide Fund for Nature, Vegetarian Society, and the Transnational Information Centre (TIC).[192] It is ironic that the threat of legal action actually closed the TIC down, an organisation which had been formed to monitor companies such as McDonalds. It is widely understood that McDonalds believed that the threat of legal action would chill the five into silence as McDonalds had succeeded in doing with these other organisations. 'I think we are one of the first people to have stood up to McDonalds in this way, and it has really wrong-footed McDonalds,' says a spokesperson for the McLibel Support Group, 'Because they never thought the case would get this far, they thought the members would stop distributing the leaflets and apologise.'[193]

When the activists refused to apologise, McDonalds had no alternative but to continue with a case that has made British legal history by becoming its longest running civil case. Having commenced on 28 June 1994, it is not predicted to finish until sometime in 1996. The odds were stacked against the McLibel Two from day one. Up against one of the most powerful multinationals on earth, for example, McDonalds spends some £5,000 per day on legal costs, the two were not eligible to legal aid because such assistance is denied in libel cases.[194] Furthermore, McDonalds successfully argued that the link between diet and cancer was 'too complex' for an average jury and therefore it should only be held in front of a judge. The Court of Appeal refused to reverse the decision. Despite the odds, the high-profile libel action is beginning to dent the carefully constructed public image that McDonalds spends some $1.4 billion a year (or $160,000 a day) promoting to the world and, specifically, its children.

The cracks are beginning to appear in the clown's mask as McDonalds has already had some embarrassing moments. Central to the food giant's original libel case was the quote in the offending London Greenpeace leaflet that 'a diet in fat, sugar, animal products and salt and low in fibre, vitamins and minerals is linked with cancer of the breast and bowel and heart disease'. Despite this, McDonalds' expert witness on cancer has admitted in the dock that this is a 'reasonable' statement, and it has also emerged that the defendants' position is the same as that of the World Health Organisation. This has led McDonalds' QC to admit that 'We would all agree' that there is a link between high-fat, low fibre diet and certain forms of cancer. McDonalds have subsequently altered their original claim, and are now asking the defendants to prove that 'McDonalds sell meals which cause cancer and heart disease in their customers', an allegation that did not even appear in the factsheet, and something Dave Morris calls 'completely absurd'.[195]

Having sued various people for alleging that the company was in any way responsible for tropical deforestation, it has also had to acknowledge that some of its beef in the 1980s was raised on ex-rain forest land. Other embarrassing moments have come when McDonalds have had to concede that rubbish put in recycling bins was just dumped like other rubbish. Or the time when the company's Head of Training admitted that it received between 1,500 and 2,750 complaints of food poisoning a year. It has also had to concede that its label for its food, that of being 'nutritious', simply means 'containing nutrients'. Even the actor who played the character Ronald McDonald has admitted, 'I brainwashed youngsters into doing wrong. I want to say sorry to children everywhere for selling out to concerns who make millions by murdering animals.'[196] Paul Preston, President of McDonalds in the UK has acknowledged that market research showed that even the company's customers found it, 'Loud, brash, American, successful, complacent, uncaring, insensitive, disciplinarian, insincere, suspicious and arrogant.'[197]

The trial has become a public relations disaster for the company. It has failed to silence its critics or to stop the leaflets being distributed, the case has attracted worldwide publicity, and protests have been held against the company across the globe.[198]

Furthermore, in May 1994, with what has proven to be the best tactic against corporate legal intimidation worldwide, the McLibel Two SLAPPed back, suing McDonalds because the company had been distributing leaflets saying that the defendants were deliberately circulating lies against the company.[199] The trial continues: 'We thought it was important to defend free speech,' adds Morris. 'That people weren't continually silenced by McDonalds and the other multinationals who don't want the truth about their practices to be heard.'[200]

'So it becomes clear what this trial is about – the globalisation of culture and belief systems,' wrote Bryan Appleyard, in *The Independent*, a week after the trial opened. ' . . . To be in a McDonalds is to glimpse the world as reducible to a management and marketing issue. This is the globalisation of profit.'[201]

13

A FISHY TALE TO FINISH

MAGNUS THE MAVERICK

Magnus Gudmundsson, an Icelandic film-maker has been running a campaign against what he sees as animal rights and environmental extremists since the late 1980s. He is undoubtedly one of the first people to make a living out of the green backlash, although his fund-raising base has shifted from his original heartland of Iceland and Norway to the Wise Use movement in America. The main focus of Gudmundsson's crusade against environmental organisations has been Greenpeace. To achieve this end, the Icelandic film-maker has gone to some spurious lengths and started mixing with some prominent anti-green characters from within different industries and the American political Right. He has also received funding from different fisheries, whaling and governmental organisations in the Nordic countries that have wanted to see Greenpeace invalidated.

Despite Gudmundsson's repeated claims that he is solely an independent journalist interested in exposing the 'manipulation, fabrication and falsification' of the environmental movement, the reality is somewhat different.[1] Either at the invitation of various industries, such as proponents of the nuclear, fishing or whaling industries, or, increasingly, Wise Use groups, Magnus Gudmundsson now travels the world attempting to spread the anti-green gospel. And just as he accuses environmentalists of solely being in it for the money, Gudmundsson too seems to be making a healthy living out of anti-environmentalism. Furthermore, Magnus Gudmundsson seems to be guilty of the very things he accuses environmentalists of: manipulation, fabrication and falsification.

Plate 13.1 Whale meat at a quayside in Iceland.
Source: Greenpeace/Reeve

Gudmundsson's original backers were the whaling and fisheries organisations in Iceland, Norway, Greenland and the Faeroe Islands. Due to the clashes that many of these countries have had with environmental groups concerning whaling and sealing, there has been a basis of anti-environmental feeling for years. Therefore, although Magnus Gudmundsson is not widely known in Europe outside of the whaling region, his work finds a positive resonance with the public, press and politicians associated with the marine resource debate. For example, the former Minister of Fisheries in Iceland, Halldor Asgrimsson, urged other Nordic countries to 'create a united front against Greenpeace' way back in 1986.[2] Throughout the region, environmental and animal rights groups have suffered from the image of meddling imperialistic city dwellers who know nothing about traditional ways of life in the Arctic, and who care more about animals than the continuance of rural traditions, and a way of life that has remained constant for generations.

Since 1989, Magnus Gudmundsson has made three films with the purpose of attacking animal rights and environmental groups, *Survival In the High North*, *Reclaiming Paradise?* and *The Man in the Rainbow*. The first, *Survival in the High North*, made in 1989, was an attempt to highlight the devastating effect that certain environmental policies have had on the communities of the high Arctic such as Greenland and Northern Canada. He also accused Greenpeace of using false footage in their anti-sealing campaigns of the 1970s and in their campaigns against the kangaroo hunts of the 1980s. Gudmundsson's allegations about Greenpeace using false footage in their kangaroo campaign are based on the accusations of the Danish journalist Leif Blaedel, who has been attacking Greenpeace since the early 1980s, and who appeared in the film. Blaedel's accusations were later refuted by a Swedish court of arbitration as false.[3]

It seems others wanted the film made and wanted Greenpeace attacked. A proportion of the money for research for the film, as well as some footage which was donated free, were provided by the Icelandic Ministry of Fisheries, a fact that Gudmundsson has refused to acknowledge.[4] Money also came from the Vestnorden Fund, which promotes regional cooperation from the Faeroe Islands, Greenland and Iceland. Money towards Gudmundsson's second film *Reclaiming Paradise?* came from the Icelandic Film Fund.[5] On 8 June 1989, Magnus Gudmundsson gave a press conference at the National Press Club in Washington, where *Survival in the High North* was shown. The conference was paid for by *21st Century Science and Technology*, the magazine that is affiliated with political extremist Lyndon LaRouche.[6] *21st Century* and other magazines associated with LaRouche, particularly the *Executive Intelligence Review* have also widely publicised Gudmundsson's other films and his work around the world, more of which is expanded on later.

So outraged were Greenpeace about allegations in *Survival in the High North* that they had faked film footage of a seal culling and used 'terrorist' tactics, that they sued Gudmundsson in Norway. It would be the start of several legal battles that have ensued between Greenpeace and Gudmundsson as the film-maker persists in making unsubstantiated allegations against the environmental organisation. The court ordered various parts of the film to be cut and ordered Gudmundsson to pay Greenpeace 30,000 Norwegian kroner in damages.[7] Although Magnus Gudmundsson appealed against the court's decision, this was refused.[8] Gudmundsson, whose court costs were paid by the Norwegian Fisheries Association, or Fiskarlaget, an organisation that itself receives funding from the Norwegian government, took four years to pay the damages.[9]

Despite knowing that Greenpeace had successfully sued Gudmundsson, the American Wise Use group Putting People First distributed a revised version of the film to all members of the House of Representatives in 1993, saying this is the 'video Greenpeace doesn't want you to see'. In their attached letter, although PPF mentioned the Norwegian court case, they falsely stated that Magnus Gudmundsson 'was acquitted of all libel charges and compensation demands'.[10] Given its pro-whaling policy, it is not surprising that PPF, which professes to have members in Canada, Norway and Japan and outside of the United States, should support Gudmundsson's work and that of other pro-whaling individuals. It has also distributed Gudmundsson's two other films *Reclaiming Paradise?* and *The Man in the Rainbow*, as well as actively promoting funding for Gudmundsson, and supporting a boycott of Greenpeace. Wise Use groups see support for Norwegian whalers and Arctic coastal communities as a further strategy with which to attack the environmental movement.

THE HIGH NORTH ALLIANCE

PPF has made other links with key anti-environmentalists in Scandinavia. The High North Alliance (HNA), headed by Georg Blichfeldt, is a pro-whaling and anti-environmental organisation that lobbies fora such as the International Whaling Commission to lift the moratorium on commercial whaling and which also promotes Magnus Gudmundsson's films. It has an association with PPF and disseminates PPF literature to the Norwegian media.[11] Georg Blichfeldt publishes his pro-whaling views in his aptly named newsletter, the *International Harpoon: The Paper with a Point*. The relationship between PPF and the HNA goes further. In 1993, the HNA announced that it had commissioned American Attorney William Wewer, Kathleen Marquardt's husband and a key Wise Use activist, to research the maritime environmental group, the Sea Shepherd Conservation Society, and whether its anti-whaling actions were considered an act of piracy under US law.[12]

The HNA originally took its name from Gudmundsson's first film, *Survival in the High North*, and only subsequently changed it to the High North Alliance. The HNA is just one of many organisations that have sprung up in whaling, fishing and sealing communities worldwide to counter environmentalists, such as the Alaska Eskimo Whaling Commission (AEWC), the Pilot Whalers' Association of the Faeroe Islands, the Japanese Small-Type Coastal Whalers' Association, the Greenlandic Hunters' and Fishermen's Association and the Norwegian Small-Scale Whalers' Association.[13] In 1992, a coalition of pro-whaling and sealing representatives from the governments of Iceland, Norway, Greenland and the Faeroe Islands was also formed, calling itself the North Atlantic Marine Mammal Commission (NAMMCO).[14]

Like many of the above organisations, the HNA tries to portray itself as a small-scale subsistence organisation:

> working for the future of coastal cultures and the sustainable use of marine resources. Most major organisations representing the interests of fishermen, whalers and sealers in Greenland, Iceland, the Faeroes and Norway, together with a number of local authorities, are members of the organisation.[15]

However, the HNA receives other funding, much of which comes from two sources in the Norwegian government – the Norwegian Foreign Department and the Norwegian Ministry of Fisheries.[16]

On the one hand, the HNA argues for the debate on whaling to be based on science and common sense, on the other, it disseminates anti-Greenpeace press articles. Georg Blichfeldt repeats many of Gudmundsson's allegations as well as those that have appeared in other press attacks against Greenpeace. For example, in 1992, a journal called *Eurofish Report* ran a story entitled 'Greenpeace Accused of Buying Anti-Whaling Votes', in which the review stated that 'Norwegian environmentalist Georg Blichfeldt claimed that he has documentary evidence that Greenpeace has "bought" at least 12 members of the International Whaling Commission.'[17] These claims have been refuted by Francisco Palacio, the supposed source of the allegations.[18]

The HNA paid for Magnus Gudmundsson to attend the International Whaling Commission (IWC) meeting in Mexico in 1994 in order to hold a press conference to discredit Greenpeace.[19] Throughout the IWC meeting, Blichfeldt worked closely with the Norwegian government to promote the country's pro-whaling position. Indeed, BBC press were referred to Blichfeldt by the Norwegian Embassy in England when they required an interview on Norway's position. Magnus Gudmundsson's press conference somewhat backfired, as he was asked about his relationship with right-wing extremists, and what his attack on Greenpeace had to do with saving whales. 'I have no relationship with any right-wing organisation or extremists in the United

States,' said Gudmundsson during his press conference. 'This is a total fabrication, and it is nothing new, it is something Greenpeace has been trying to spread since 1989.'[20]

LINKING UP WITH WISE USE AND INDUSTRY

Since 1989 Gudmundsson has been joining forces and networking with other prominent anti-environmental activists and organisations, some of which could be classified as being on the extreme of the Right. In 1989 Gudmundsson claims to have attended the Wilderness Impact Research Foundation's (WIRF) conference in Utah where he told the audience there that what was happening in the countries of the far north would also happen in the USA.[21] The following year, 1990, he was back in Salt Lake City again at the WIRF's conference.[22] Also in attendance were two of Gudmundsson's closest companions: Kristjan Loftsson, an Icelandic whaler and Steinar Bastesen, President of the Norwegian Small-Type Whaling Association, who believes Greenpeace are a bunch of 'parasites'. During his speech Gudmundsson alleged that Greenpeace was out to destroy the economy of Norway, that it wanted to forcefully move the people of the far north and that it bought influence with countries by investing tens of millions of dollars in the country.[23]

So impressed with Gudmundsson were the conference organisers at WIRF, that he was invited back the following year. 'This is a true environmental protection conference even though utilisers of nature are behind it,' said Gudmundsson to the delegates at the conference in 1991.[24] After the conference expressed support for Icelandic whaling, the Icelandic Foreign Minister, Jon Baldvin Hannibalsson, also sent a note of moral support to the participants.

But Gudmundsson was beginning to look for other areas to discuss his anti-environmental work and he found an eager audience within the nuclear industry. On 11 June 1991 he addressed the Canadian Nuclear Association on anti-nuclear 'advocacy groups'.[25] Since then, Gudmundsson's films have been shown by nuclear industry organisations in Belgium and Canada. Allegations from his films have been used by a nuclear industry representative in Brazil.[26] Within the next couple of years Gudmundsson would speak to audiences in Japan at various fisheries conferences and in New Zealand, at the invitation of the New Zealand Fishing Industry Association (NZFIA). The NZFIA stated that

Gudmundsson had been invited to 'balance the environmental debate'.[27] Gudmundsson repeated the *Survival in the High North* allegations during his New Zealand tour at a speech to the Fishing Industry Association, as well as on television.[28] TV NZ, when told of the facts by Greenpeace, offered an apology, whereas Gudmundsson refused to retract his allegations.[29] Once again he was sued by Greenpeace.

In Japan in 1992, he said of animal rights 'this movement seems to have adopted, either coincidentally or on purpose, some of the ideas Herr Hitler publicised in his *Mein Kampf*'.[30] During a lecture at the Greater Japan Fisheries Conference in February 1994, Gudmundsson openly accused Greenpeace and WWF of using 'more than five million dollars to bribe and buy support of delegates at the IWC'.[31]

In 1993 Magnus Gudmundsson and Steinar Bastesen again made a trip to another anti-environmental conference in the US, this time to the Wise Use conference at Reno organised by Alan Gottlieb's and Ron Arnold's Center for the Defense of Free Enterprise. The conference passed a resolution supporting Norwegian whaling.[32] Bastesen has had a close relationship with the Wise Use conference for years, attending in 1990, 1991, 1993 and 1994.[33] Arnold sees Gudmundsson and Bastesen as his two key contact people in Scandinavia. 'We pay for their travels and they attend our conferences,' Arnold told the Norwegian *Verdens Gang* newspaper in 1994.[34] Bastesen, in turn, reiterates Arnold's rhetoric and quotes him as the source of the fact that the environmental movement is a 'front' organisation for the communists. 'Arnold', Bastesen believes, is 'foresighted'.[35]

OUT TO SINK THE RAINBOW

It is no coincidence that Arnold, given his history of vehement anti-environmental rhetoric, would appear as a character witness against Greenpeace in Gudmundsson's third film, entitled *The Man in the Rainbow* in Danish or *The Rainbow Man* in English. Made by the Danish company Nordisk with Gudmundsson as a consultant for the Danish TV channel TV2, the film was originally shown in Denmark on 14 November 1993. It would later be shown in Denmark, Sweden, Finland, Malaysia and Iceland. The accusations made in the film, however, circumnavigated the world.

Magnus Gudmundsson tells different stories in different places, as to whose idea the film was. At meetings in the USA and Mexico, such as at the International Whaling Commission (IWC) and Convention on International Trade in Endangered Species (CITES), Gudmundsson recalls that he was persuaded to make a film about Greenpeace after having been approached by Nordisk, although he did not particularly want to. However, during a lecture at the Greater Japan Fisheries Conference in February 1994, Gudmundsson stated that the film was entirely his idea and he approached the film company.[36]

The film was an attack on Greenpeace and, in particular, its Honorary Chairman, David McTaggart, based mainly on old allegations that had appeared in press reports and some new exaggerated and unsubstantiated research. Appearing alongside Ron Arnold, who appeared as 'a writer', was the private investigator and Wise Use activist Barry Clausen, who works closely with *21st Century Science and Technology*'s associate editor Roger Maduro to produce *Ecoterrorism Watch*. Maduro himself has met Gudmundsson on numerous occasions and professes to have the same beliefs on the environmental movement as his Icelandic counterpart.[37]

Within twenty-four hours of *The Rainbow Man* being shown, *21st Century Science and Technology* was distributing a press release, which was also posted on the Internet. Information that could have only been known by the producers was contained in the release, which suggested some degree of cooperation between *21st Century* and the film-makers.[38] A week after the film was shown, members from the LaRouche group, Patriots for Germany, including LaRouche's wife Helga, interrupted a Greenpeace meeting in Dusseldorf by shouting allegations from the film.[39] *21st Century Science and Technology* also carried a four-page article on the film in its Winter edition, written by a Dane called Poul Rasmussen, who is Chairman of LaRouche's Schiller Institute in Copenhagen. The magazine also published a fax number where one could order the film.[40]

The film accused Greenpeace of bribing the IWC, of having secret bank accounts and of working in collaboration with the 'terrorist organisation' Earth First!. Arnold also accused Greenpeace of bribing politicians. It was a cleverly put together piece of questionable journalism. However, in their desire to discredit Greenpeace, the film-makers were prepared to go to some pretty disreputable lengths themselves. After the German TV channel NDR had bought the film, it was due to be shown in Germany. However, having been warned by Greenpeace, one of their reporters, Christoph Luetgert was told to examine the way it had been made. Luetgert found out, among other things, 'that statements made in interviews with the marine biologist Francisco Palacio had been cut and taken out of context in such a way that he is presented – wrongly – as a key witness against Greenpeace'. Luetgert considered the film 'a nasty piece of work, the nastiest I've come across to date'.[41]

Francisco Palacio had been promised by Gudmundsson and the producer of the film, Michael Klint, that he would see his interview before broadcast, but they never kept their promise. In Palacio's injunction against the film being shown in Germany, it stated that it was:

> not until May 1994, that Palacio learnt that his interview with the two Danish journalists [Gudmundsson and Klint] had been cut and combined with other statements and questions so that not only was the impression given, but it was directly alleged, that the plaintiff conceded that Government representatives were bribed with millions by Greenpeace and the WWF.[42]

The German court ordered that, as Palacio's interview had been manipulated, *The Man in the Rainbow* could not be shown publicly.[43]

There was victory too for David McTaggart, the subject of much personal vilification in the film. McTaggart's application to the court said that various statements made by interviewees in the film were 'false', 'incorrect', and 'insulting'. Once again the court ruled that the film could not be shown with the offending sequences.[44] Greenpeace also alleged that the film contained 'unfounded allegations of fact'. Once again the court ruled in their favour.[45] TV2 also had to apologise to David McTaggart over allegations in the press release which accused McTaggart of hash smuggling, bribery and taking money from the CIA, which were neither true, nor appeared in the film.

Despite NDR's reservations, and the rulings of the German court, Gudmundsson continues to publicise and sell the film. In 1994 he attended the Fly-in for Freedom with his old associates Bastesen and Icelandic whaler Kristjan Loftsson. During the conference Gudmundsson professed to be happy at seeing how the Alliance For America was growing, saying that, 'I am dedicated to the cause and to the truth. If you guys were up to something bad, I wouldn't hesitate in letting people know.' Also at the AFA was Eugene LaPointe, a former CITES Secretary-General, who paid tribute to Gudmundsson.[46] LaPointe had been dismissed from his CITES office following a strong lobbying campaign by NGOs who contended that he was favouring the interests of wildlife traders. Gudmundsson not only showed *The Rainbow Man* but also Greenpeace footage of seal clubbing that he claimed was staged. This is despite the court ruling in Norway that proved otherwise.

Introducing *The Rainbow Man*, Gudmundsson told his audience that 'Everything in this film is true. It is based on fact.' He then added that:

> Greenpeace's latest claim is that the interviews in this film were taken out of context . . . We have invited international journalists to our studios to look at the raw footage to make their minds up. It was obvious that nothing was taken out of context.

As people watched the film, they applauded environmentalists being beaten up. 'Shoot 'em' rung around the room.[47] Gudmundsson told his audience:

> Greenpeace is only 1,000 individuals, not 5 million individuals as they say. You can deal with them. You can win. They are not even that clever. They cannot even deal with me and I am alone, and I am not that clever.[48]

THE IWMC AT THE IWC

Later in 1994 Gudmundsson teamed up with Eugene LaPointe from the International Wildlife Management Consortium or IWMC at the CITES meeting in November in Fort Lauderdale, Florida. The IWMC, based in Geneva, is a new organisation wrapped in Wise Use type language that is committed to 'wildlife management' and 'sustainable use' based on 'science'. In essence, it has copied the language of NAMMCO, which supports 'conservation and sustainable utilisation of living resources'.[49] At the IWC meeting at Dublin in 1995, the IWMC supported the 'philosophy of sustainable use and the value of reason' against the people who 'remain blind to the realities of mankind's role in the environment.'[50]

In the IWMC's brochure, it calls itself a worldwide coalition of wildlife managers 'who believe and consider the rational utilisation of wildlife resources as the most dynamic incentive for encouraging human beings to support conservation efforts . . . This union of all wildlife hunters represents the last hope for the future of wildlife species.'[51] The brochure was published with the assistance of the Safari Club International, whose motto is 'Conservation of wildlife and protection of the hunter', and which has links to the Parsons Group, a lobby group retained by the High North Alliance to push for the resumption of whaling.[52]

The hunters, fishers, trappers, whalers, sealers have found a voice in the IWMC. 'Wildlife conservation policy', says the IWMC, must be 'of the people, by the people and for the people'.[53] 'It took us a long time to realise that rights granted to animals are rights taken away from humans,' says LaPointe. In this regard its message is the same as those of Wise Use or Share groups – that environmental 'extremists' have left mankind out of the ecological debate. The IWMC, which professes to a membership of over sixty 'conservation' organisations, is clearly associated with people who want to lift restrictions on the trade in endangered species, who advocate the resumption of commercial whaling and sealing, and who are paid by or represent industries and governments that would benefit from such activities. An ambiguous name like the International Wildlife Management Consortium only serves to confuse, and even by wrapping itself in green rhetoric cannot disguise what is an international anti-environmental organisation. Publishing a magazine called *Conservation Tribune* is an additional smokescreen.

During the CITES meeting at Fort Lauderdale, the IWMC organised Gudmundsson's press conference entitled 'Facts or Fiction: Human Rights Versus Animal Rights' and showed his three films. Gudmundsson was also a registered delegate for the IWMC.[54] The flyer for the event labelled Greenpeace as an international protectionist organisation whose 'propaganda' was often 'disguised as environmental concerns'.[55] The IWMC also published articles by Georg Blichfeldt and Gudmundsson. The latter's article was entitled 'Inspiration for an Environmental Film-maker,' in which he accused greens of being 'high priests' of a 'new cultism' who when unmasked would be shown to be 'irrational, impractical and inhumane'. He finished his article with the cryptic quote 'Beware of evil, my son, for it has many faces but the face it most often will wear is the face of good.'[56]

The 'Issue Specialists' that the IWMC paraded to 'discuss sustainable use' give an insight into IWMC's agenda. These included a lawyer who represents a pro-hunting caucus in Washington, a representative from the Wise Use groups, the Alliance for America and National Wilderness Institute. Also included was someone from the leading anti-green think-tank, the Competitive Enterprise Institute, and people with links to the fishing and whaling industries in Venezuela and Japan.

Steve Boynton was described by the IWMC as 'a Washington DC-based attorney specialising in wildlife and marine resource law'. He is Vice President of an organisation called Henke and Associates, which is run by Janice Scott Henke who was a member of the IWMC delegation to the IWC in 1994 and 1995 and a member of the 'Norwegian Small-Type Whalers' Association' delegation to CITES in 1994. Boynton has also worked for the Japanese

Institute of Cetacean Research, and is also general counsel to the Congressional Sportsmen's Foundation. The latter organisation is the non-profit affiliate to the Congressional Sportsmen's Caucus, which represents hunting and fishing interests in Congress, listing 218 US senators and representatives as members.[57] The Caucus supports the resumption of commercial minke whaling by Norway and believes that the Endangered Species Act should be altered to increase the cost of listing fragile species and habitat.[58]

Boynton himself argues that the 'United States is fostering an illegal policy to obstruct whaling' in a legal briefing paper for the Washington Legal Foundation, another Coors family-backed right-wing think-tank.[59] Another bill promoted by the Congressional Sportsmen's Caucus advocates captive breeding of wildlife species instead of habitat protection, a measure that the National Wildlife Federation has described 'as a blueprint for extinction'. Peter Kelley, communications director for the League of Conservation Voters calls the Caucus a 'conduit for the oil and gas lobby and developers to influence members on outings'.[60]

The Congressional Sportsmen's Foundation has ties to right-wing funding sources and the gun lobby. Its staff work rent-free out of an office of the Olin Corporation, the ammunition manufacturer. On the board are Ray Arnett, former executive Vice-President of the National Rifle Association and Peter Coors, chief executive officer of ultra-conservative Coors family brewing company and G. Bersett, president of Olin's Winchester Division.[61] The Foundation also hosted a lunch for Gudmundsson in May 1995, which was attended by both Barry Clausen and Roger Maduro.

Both Boynton and John Barthel, from the Wise Use group the American Trappers Association, and who is also a member of the Alliance for America and another IWMC specialist, stopped off in Iceland on their way to the IWC in Dublin in 1995. They informed the press that attitudes in America were changing towards whaling and against environmental organisations like Greenpeace.[62] In November 1995, the Icelandic press reported that Boynton had been hired as a consultant for the Icelandic Embassy in Washington for a period of three months, earlier in the year.[63] In December, in a move showing increasing solidarity between the Wise Use movement and its European counterparts, Bruce Vincent, from the Alliance for America, also visited Iceland for the formal launch of a new organisation called Marine Utilization, which will lobby for the resumption of commercial whaling.

Another IWMC Specialist is Ike Sugg from the right-wing Competitive Enterprise Institute, whose book *Elephants and Ivory* was published by the

Institute of Economic Affairs in London.[64] Sugg is a vocal critic of the Endangered Species Act and is a member of the Property Rights Task Force, along with other prominent anti-green activists.[65] Yosho Kaneko, from the Global Guardian Trust (GGT) from Japan is also an IWMC Specialist.[66] The GGT's rhetoric fits its congenial name, and its aims are to 'promote the sustainable use of natural resources and to use the best available scientific information as the basis for conserving and ensuring the rational uses to natural resources'.[67] Whilst the GGT promotes 'sustainable use', it accuses environmental organisations, particularly Greenpeace of 'pursuing their cause in disregard of the principle of' sustainable development.[68] The GGT's main objective is 'the revision of the extremist environmental protection movement', which is in evidence at discussions at both the IWC and at CITES.[69] 'Having formed the GGT, we absolutely must confront this environmental protection movement that has gone too far,' says Kunio Yonezawa, the chairman of the board of GGT.[70]

Yonezawa, like other anti-green activists, contends that:

> Now that the world's communist movement has gone, I think the only one circulating around the world is the extremist environmental movement of Greenpeace and the like. It's a world political party which has encroached into the political core of Anglo-Saxon-led countries such as the UK, Australia etc.[71]

In order to confront Greenpeace, the GGT maintains that it will 'foster cooperation with countries and NGOs that share the same aim as the GGT, advocating sustainable development based on scientific foundations'.[72] Widely believed to be little more than a front for the Japanese fishing and whaling industry, the GGT is also looking to extend its membership to include the farming, forestry and the energy industries.[73]

Dr Francisco Herrera-Teran is another IWMC Specialist, and former Minister for Fisheries in Venezuela. He appeared with Gudmundsson at his press conference at the IWC to castigate a Venezuelan biologist, Aldemero Romero Diaz, who has published over 300 articles and four books on marine mammals in the country. In a charge that Gudmundsson has repeated in various newspaper articles and conferences, the two men accused Romero of inciting fishermen to kill dolphins for campaigning purposes and of falsifying footage of dolphins being killed. Romero is adamant that the footage is not doctored by him but by the Venezuelan authorities, who want to extradite him to stand trial.[74] The US State Department calls the Venezuelan legal system 'corrupt', and there is no doubt that Romero would not receive a fair trial in Venezuela.[75] There is also evidence that the authorities have threatened other environmental organisations to publicly discredit Romero and distance themselves from him. If this is the case, it would not be unlikely that the authorities had doctored the film to discredit Romero. On Icelandic radio, Magnus Gudmundsson remarked that Romero had been sentenced for his crime and that he was being sought by Interpol, accusations which turned out to be false.[76]

What we are seeing in the IWMC is a well-funded, articulate organisation that has been set up to counter environmental and animal welfare organisations at events such as CITES and IWC. Although it only appeared on the map two years ago, it has already built up a formidable network between key anti-environmental organisations and individuals that looks set to continue.

As if to emphasise the fact, at the IWC meeting in Dublin in 1995, the IWMC was again well represented by a large delegation. It also helped organise an anti-environmental demonstration by Norwegian coastal women.[77] Given the Norwegian presence, it was interesting to see that Steinar Bastesen had joined the IWMC's ranks, as well as a representative from Iceland's fishermen.[78] Gudmundsson gave a talk entitled 'Save the whale: the propaganda war' and showed *The Rainbow Man* on behalf of the IWMC.[79] Other articles published by the IWMC in Dublin were written by Yoshio Kaneko, Francisco Herrera-Teran, Steve Boynton and Michael De Alessi from the Competitive Enterprise Institute, who has also criticised Greenpeace over its Brent Spar campaign.[80]

Three months later Magnus Gudmundsson was back in the USA, this time addressing the right-wing Heritage Foundation. Supposedly talking about 'Exposing the Left's Environmental Agenda in the 104th Congress', Gudmundsson once again lambasted Greenpeace.[81] Gudmundsson's speech reiterated many old Wise Use and LaRouchite accusations with convincing clarity. According to the film-maker, Greenpeace and other environmental organisations had become 'the new religion'. Greenpeace were a politically motivated group who hated humans, but Americans most of all, said Gudmundsson. Greenpeace had huge revenue sources, including funds from industry, which it strived to conceal from the public. Greenpeace, alleged Gudmundsson, had received some two and a half million dollars from Shell and, when the company stopped donating money, it prompted Greenpeace to attack the Brent Spar. [Greenpeace does not receive corporate funding.]

Gudmundsson also claimed that Greenpeace has no scientific evidence to back up its positions and there was no scientific evidence to prove ozone depletion. [The latter is a favourite topic of Roger Maduro from *21st Century Science and Technology*.] Furthermore, Greenpeace disliked technology, free enterprise, capitalism, and 'cheap and abundant energy'. [Other Wise Use favourites – Greenpeace has been promoting alternative technologies such as the Greenfreeze refrigerator and solar power, but Gudmundsson's last comment was a plug for the nuclear industry.] Most federal agencies had been infiltrated by Greenpeace, said Gudmundsson.[82]

Magnus Gudmundsson has been actively associated with the Icelandic and Norwegian governments, whalers, sealers, fishermen, the nuclear industry, the Wise Use movement, associates of Lyndon LaRouche, the IWMC, and various right-wing think-tanks such as the Heritage Foundation. Icelandic environmentalists believe that Gudmundsson's association with the US anti-greens puts him in a difficult position. On the one hand, his credibility with the Wise Use movement stems from the fact that he is Icelandic. Indeed, Gudmundsson and his associates represent an effort by the Wise Use movement to internationalise its campaign against regulation of resource utilisation. On the other, Gudmundsson's increasing reliance on anti-environmentalists damages his credibility in Iceland, where Wise Use policy on ozone depletion, global warming and other topical ecological issues, meets no sympathy at all.

This said, it seems the networking between key resource-dependent industries, anti-environmentalists, right-wing commentators and the Wise Use movement is becoming more coordinated. For example, in December 1995, following the International Conference on the Sustainable Contribution of Fisheries to Food Security in Kyoto, the Norwegian Whalers Union signed a 'joint declaration of non-governmental organizations interested in responsible aquatic resource utilization', along with fishing and whaling groups from Canada, Russia, Japan, Korea, Latin America, Switzerland, Zimbabwe and the United States. Over eighty American Wise Use groups signed, including the Alliance for America, BlueRibbon Coalition, People for the West! and Putting People First. The organising group was the Fishermen's Coalition, which had been set up after a demonstration against Greenpeace's tuna campaign in 1992.[83]

Two months later, in February 1996, Teresa Platt from the Fishermen's Coalition and Bruce Vincent from the Alliance for America wrote to President Bill Clinton arguing that:

> With the completion of the revised management procedure, the IWC should finally allow, recognise and support the humane and sustainable use of abundant cetaceans as practised for thousands of years by citizens of Norway, Iceland, Canada, the Faeroe Islands, Japan, the Caribbean nations, South America, the United States and many other countries.[84]

In June, further international networking was reflected in a panel discussion at the Alliance for America's Fly-in for Freedom, entitled 'Going Global: Making a Difference in the International Arena', chaired by Teresa Platt. Attending a Wise Use conference for the first time was Georg Blichfeldt from the High North Alliance, speaking on the 'Conflict Between US Preservationists Values and the People of the High North'. Other speakers on the panel included Judy Mashinya, from the Africa Resources Trust, talking about 'Elephants and Africans: Is Coexistence Possible?' and Ike Sugg, from the Competitive Enterprise Institute, speaking on 'Economic Value and Wildlife: Impossibility or Necessity'. Gudmundsson also held a workshop on Icelandic films.

There is no doubt that the IWMC and their Wise Use allies will have an increasing influence on debates at the IWC and at CITES, and very likely at other conventions such as those on biodiversity. The whales that the world thought had been saved in the 1980s by the moratorium on commercial whaling are far from safe, and their future looks increasingly bleak, as anti-environmental forces coalesce across the globe. The international battle over the whales, and increasingly now over the rapidly dwindling fish stocks, looks set to become one of the most controversial resource conflicts of the coming decade. In parallel with America, many people embroiled in restarting commercial whaling are out to gut the Endangered Species Act, the domestic protection for American wildlife.

There will be strenuous efforts to focus the agenda on to 'sustainable utilisation', and 'balance' which will be no more than privatisation of wildlife in disguise. Or extinction by another name.

14

CONCLUSION

Beating the backlash

We can do things about what man has done to the Earth. We can redesign human systems, and we can make money by doing it right. You can run an economy, you can run a civilization, on respect for your home planet. And we will develop that respect, because we have no alternative. But it's a little bit hard to get people to understand that at the moment. We will. We must do that.

David Brower[1]

The green backlash, born out of both the success of the environmental movement and its failure, is still to run its course. Many more activists will be intimidated, beaten up, vilified and killed for working on ecological issues. The backlash is now an intricate part of working on, writing on, speaking on, campaigning on, or even teaching ecological issues. The paradigm shift that is occurring across the globe looks set to continue.

The lessons learnt from the stories in this book are that with the collapse of communism, environmentalists are now increasingly being identified as a global scapegoat for threatening the vested interest of power: the triple engines of unrestricted corporate capitalism, right-wing political ideology and the nation state's protection of the *status quo*. In all probability the backlash will get worse, as the resource wars of the coming decades intensify, as more people fight over less. We have already had the fish wars, but conflicts over water, wood, whales, metals, minerals, energy, cars and even consumerism will all happen and all inevitably create backlash.

The green backlash, though, has to be seen in a positive light, because if the environmental movement absorbs the right lessons it will start the next millennium in a much better shape than it finished this one. The backlash must force the movement to reevaluate itself and realise that it has not run out of ideas, it has not run its course, but it is still in danger of running out of time.

It must also face up to some painful realities: many people were corrupted by the very system that they set out to change. Other people with legitimate concerns felt that the environmental movement had failed them, too.

To beat the backlash, there are three distinct areas that need to be addressed.

1. The environmental movement has to rediscover its roots.
2. It has to broaden out to work closely with other groups.
3. It has to start putting forward solutions and a positive alternative coherent vision for the future.

Environmentalism must rediscover its roots, grow from its roots, because that is where the strength and the ideas will come from. Most people who have written about the future of the American environmental movement believe that the best direction for activism lies, not with the mainstream groups from Washington, but in the environmental justice movement.[2] The mainstream groups need to reanalyse their agenda. As Peter Montague warned in *Rachel's Environment & Health Weekly* in 1995:

> The big environmental groups – the traditional environmental 'movement' as they like to call themselves – now lack viable political ideas that can reshape and regain power by appealing to the American people. They are too timid or too indentured to speak of the nation's real problems: the raw power that the global corporations wield over our jobs, our quality of life, our mass media, our elections, our legislature, our schools, our courts, and indeed our minds.[3]

The backlash in America will hopefully make some US environmental groups change direction and, in part, this will mean going back to the grassroots and start organising, campaigning and talking face to face, door to door, street to street, community to community. Because that is the only way to start rebuilding a definite agenda and a positive message. Grassroots organising is definitely an area where the anti-environmental movement has beaten the environmentalists at their own game over the last few years. There is no doubt either that they have been able to exploit the weaknesses of the mainstream groups. One of the most glaring of these is its apparent neglect of people and social concerns. It is undeniable that some mainstream environmental groups in the USA have advocated a position of wilderness preservation without evaluating the human consequences of such policy decisions.

This said, environmentalists also have to demonstrate that the anti-green policies of polarisation serve neither the people who promote such tactics, nor workers or the community in which they live. The only beneficiaries of such actions are the corporations whose primary interest is profit maximisation. It also has to highlight that many of the corporations backing anti-environmentalism have a history of working against the public interest and not for it.

Environmentalists have to start working with other campaign groups, with workers, with women, unions and other progressive organisations to address social, cultural and development issues, as well as ones of ecological equity. Vandana Shiva, a leading environmentalist from India warns that if the environmental movement does not change there will be problems. 'Otherwise they won't even need a backlash to become irrelevant, and forgotten,' she says. 'The environmental movement that does not link up to social justice and equity issues will be made irrelevant, just by the changing times and there will be more pressing problems that society will face'.[4]

Vandana Shiva continues:

> If environmental activists try and act alone, without connecting up with movements for justice, movements for human rights, movements for democracy, they can be contained very easily, not just by backlash, but also by polarisation, by constantly making it look like the environmental interest is a secondary interest, whereas jobs and the survival interest are the primary interest. This will be exacerbated by people refusing to recognise the environmental base is also a livelihood base, and environmental issues are tied very closely to economic survival. I think what is needed now very rapidly is broad-basing of our environmental work. Building into trying to create large citizens' alliances and trying to deal with deregulated commerce, and uncontrolled power of capital.[5]

There are some positive examples where coalitions have formed and common ground has been found. A coalition of environmental, human rights and development groups formed after the death of Ken Saro-Wiwa to fight for justice for the people of the Niger Delta. In Britain a disparate group of people came together to fight the Criminal Justice Act. So squatters and anti-road activists demonstrated with ravers and hunt saboteurs against a law that they felt was unjust. Through the common ground, the feeling of political alienation has sparked other protest.

The environmental justice movement in the USA and the movement against the Criminal Justice Act in Britain, are examples of grassroots community-based action that have influenced the political process. They are examples of where the empowerment of activism overcame political disillusionment. But groups working together need to broaden out even further, building coalitions across continents. In essence, the world is getting smaller. With the globalisation of the market place, the internationalisation of industry and trade and the threat of global pollution problems, the environmental movement needs to respond to these challenges with other groups where common interest can be found.

Environmentalists should be working and assisting the 'seed satyragraha' movement which rose up to oppose GATT in India. Other groups are forming to fight free trade, and to question the plundering of their biodiversity and natural resources by multinationals, and they will need support too.

Finally, environmentalists are always being criticised for being negative, always pointing out the problems without offering solutions. In putting forward positive solutions the movement can prove the scaremongers and scapegoaters wrong. Working with other groups, environmentalists have to undertake the hardest challenge, to find the best solution to the most pressing problem of all. It is no longer good enough to just protest at the inequality of world trade, or highlight the global environmental, social and health problems associated with our current economic system, the environmental movement has to come up with positive solutions to the problem.

Take the debate of sustainable development, for example. The current business-as-usual corporate definition of sustainability is irreconcilable with the goals of eco-justice. We may have to ask some difficult questions, and hopefully make business face up to some awkward realities, about the fact that their business may never be sustainable. Environmentalists also have to work with others to somehow make multinationals accountable for their ecological, social and cultural impact.

Whilst the majority of business remains unsustainable, the corruption of the political system continues. So too does the revolving door between business and politicians. The relationship has to be challenged. As we reach the millennium, there will be a lot of debate about society and civilisation itself. If we are to have ecologically sound development based on social justice and equity, then we need a debate about whether we need a new economic system. It could be argued that the eco-justice and equity part of sustainable development cannot be achieved under the current economic model. If we are to have a greater degree of equity, there needs to be a system where people and planet are prioritised before profit. We need to tip the scales towards balance. The new millennium is a watershed for us to realise that it could be time to try something new. Mark Dowie writes:

With prevailing economic systems in various degrees of disarray, the worldwide environmental movement seems uniquely positioned to serve as a vehicle for a civilisation ordered on a new basis. It is clear now that both capitalism and socialism, in all their experimental forms, have failed to create ecologically sustainable economics.[6]

Economically, communism has failed, and many commentators argue that capitalism is also failing.[7] Environmentally, both have been pretty disastrous. Maybe the best reason for change is that capitalism is not working, in that it is not providing long-term secure employment for the masses, on which individuals and communities can base a sustainable future.[8]

That could be the ultimate challenge of the global environmental movement. To find a global economic system that provides work without destroying the world. To achieve a balance between the need for jobs, and security and the need for environmental protection. To formulate an ecologically sustainable world. In which community democracy, equality and justice feature. Local empowerment rather than global repression has to be the way forward. We do not want an unregulated anarchic global sweatshop, run by all-powerful companies, in bed with corrupt politicians, who do not care about worker and environmental protection, financed by speculators addicted to the global gamble.

The backlash has given the environmental movement the opportunity to change for the better, it should not blow that chance.

NOTES

INTRODUCTION

1. V. Shiva, interview with author, 21 July 1995

1 ROLL BACK THE MILLENNIUM

1. R. Arnold, quoted by 'Fossil Bill' Kramer, *Building Material Retailer*, September 1995.
2. L. Neergaard, 'Environmentalists mark 25th Earth Day protesting legislation', *AP Worldstream*, 22 April 1995.
3. P. Montague, 'A "movement" in disarray', *Rachel's Environment & Health Weekly*, 19 January 1995, No. 425.
4. M. Dowie, *Losing Ground: American Environmentalism at the Close of the Twentieth Century*, Massachusetts Institute of Technology, 1995, p.177.
5. G. Lean, 'Assault on Green Laws endangers newt brigade', *The Independent on Sunday*, 23 April 1995, p.15.
6. *UPI*, 'Environmental groups flunk Congress', Washington, 10 October 1994.
7. A. Reilly Dowd, 'Environmentalists are on the run', *Fortune*, 19 September 1994, p.91.
8. M. Dowie, *Losing Ground: American Environmentalism at the Close of the Twentieth Century*, Massachusetts Institute of Technology, 1995, p.xiii.
9. G. Lee, 'Environmentalists try to regroup: on Earth Day, groups to launch fight against setbacks on Hill', *Washington Post*, 22 April 1995.
10. Ibid.
11. Associated Press, reported in 'Private enterprise also has its champion', *The Tribune*, 26 August 1979.
12. Ibid.

13. M. Megalli and A. Friedman, Pacific Legal Foundation, *Masks of Deception: Corporate Front Groups in America*, Essential Information, December 1991.
14. Ibid.
15. B. Wood and T. Barry, *Power Brokers in the Rockies: Privately-Minded in the Public Interest*, A New Mexico People and Energy Power Structure Report, Albuquerque, date unknown.
16. R. Bellant, *The Coors Connection: How Coors Family Philanthropy Undermines Democratic Pluralism*, South End Press, Boston, 1991, p.85.
17. R. Bellant, interview with author, 25 November 1994.
18. R. Bellant, *The Coors Connection: How Coors Family Philanthropy Undermines Democratic Pluralism*, South End Press, Boston, 1991, p.84.
19. B. Wood and T. Barry, *Power Brokers in the Rockies: Privately-Minded in the Public Interest*, A New Mexico People and Energy Power Structure Report, Albuquerque, date unknown.
20. M. Megalli and A. Friedman, National Legal Center for the Public Interest, *Masks of Deception: Corporate Front Groups in America*, Essential Information, December 1991; R. Bellant, *The Coors Connection: How Coors Family Philanthropy Undermines Democratic Pluralism*, South End Press, Boston, 1991, pp.85, 86.
21. R. Bellant, *The Coors Connection: How Coors Family Philanthropy Undermines Democratic Pluralism*, South End Press, Boston, 1991, p.84; quoting *Rocky Mountain Magazine*, 1981, March/April, p.29.
22. M. Dowie, *Losing Ground: American Environmentalism at the Close of the Twentieth Century*, Massachusetts Institute of Technology, 1995, p.72.
23. J. Prodisjackson, 'Another Look: James Watt', *Associated Press*, 28 April 1991.
24. A. Cockburn and J. Ridgeway, 'James Watt: The apostle of pillage', *Village Voice*, 26 January to 3 February 1981.
25. Senate Congressional Record, *Watt-ism*, 20 October 1981, S11722.
26. B. Stall and B. Hand, 'How Watt is helping the environmentalists', *San Francisco Chronicle*, 28 September 1981; D. Russakoff, 'Watt and his foes love their mutual hate', *Washington Post*, 23 March 1982.
27. M. Donnelly, 'Dominion theology and the 'Wise' Use movement', reprinted from *Wild Oregon*, the Journal of the Oregon Natural Resource Council, no date.
28. R. Maughan and D. Nilson, *What's Old and What's New About the Wise Use Movement*, Department of Political Science, Idaho State University, 1993; a version of the paper was presented at the Western Social Science Association Convention, 23 April 1993.
29. M. Satchell, 'Any colour but green', *US News & World Report*, 21 October 1991, p.75.
30. R. Maughan and D. Nilson, *What's Old and What's New About the Wise Use Movement*, Department of Political Science, Idaho State University, 1993.
31. M. Satchell, 'Any colour but green', *US News & World Report*, 21 October 1991, p.75.
32. R. Bellant, *The Coors Connection: How Coors Family Philanthropy Undermines Democratic Pluralism*, South End Press, Boston, 1991, pp.15, 88.

33. R. Arnold, *At the Eye of the Storm: James Watt and the Environmentalists*, Regnery Gateway, Chicago, 1982, p.248.
34. Ibid., pp. 27, 35–8.
35. W. Kronholm, 'Conservative Book Portrays "Real Jim Watt",' *Associated Press*, Washington, 9 November 1982.
36. R. Sangeorge, 'Environmental leader blasts Watt biography', *United Press International*, 9 November 1982.
37. S. L. Udall and W. K. Olson, 'Perspective on environmentalism: me first, God and Nature second; With Communism gone, the Far Right is seeing red in the green movement; public land preservation is under attack, *Los Angeles Times*, 27 July, Part B, p.5.
38. J. Hamburg, 'The Lone Ranger', *California Magazine*, November 1990, p.92.
39. C. Williams, 'The park rebellion, Charles Cushman, James Watt and the attack on the National Parks', *Not Man Apart*, A Friends of the Earth Reprint, June 1982.
40. T. Manjikian, 'Watt wafts West, Former Secretary of the Interior James Watt generates new controversy', *California Business*, September 1986, p.12.
41. J. Halpin and P. de Armond, 'The Merchant of Fear', *Eastsideweek*, 26 October 1994.
42. D. Junas, *Rising Moon: The Unification Church's Japan Connection*, Institute For Global Security Studies, Seattle, Washington, 1989; United Press International, *Parkersburg News*, 8 June 1989.
43. D. Junas, interview with author, 22 November 1994.
44. D. Junas, 1991, 'Rev. Moon Goes to College', *CovertAction Information Bulletin*, Fall 1991, Number 38.
45. M. Knox, 'Meet the anti-greens: the 'Wise Use' Movement fronts for industry', *The Progressive*, October 1991, p.21.
46. W. P. Pendley, *It Takes A Hero: The Grassroots Battle Against Environmental Oppression*, a project of the Mountain States Legal Foundation, Free Enterprise Press, Bellevue, Washington, p.271. The Free Enterprise Press is a division of the Center of the Defense of Free Enterprise, of which Gottlieb is the founder and Director.
47. R. Arnold, *At the Eye of the Storm: James Watt and the Environmentalists*, Regnery Gateway, Chicago, 1982, pp. 21, 52, 56.
48. *Western Horizons*, 'CDFE: Out front, and pulling strings behind the scenes', newsletter of the Wise Use Public Exposure Project, Western States Center, Portland, Oregon, September 1993, p.14.
49. *Eastsideweek*, 'Is this the most dangerous man in America?', 26 October 1994; J. Halpin and P. de Armond, 'The merchant of fear', *Eastsideweek*, 26 October 1994.
50. J. Halpin and P. de Armond, ibid.
51. T. Ramos, 'The case of the Northwest timber industry', in *Let the People Judge: Wise Use and the Private Property Rights Movement*, ed. J. Echeverria and R. Booth Eby, Island Press, 1995, p.86.
52. J. Hamburg, 'The Lone Ranger', *California Magazine*, November 1990, p.90; L. Callaghan, 'The high priest of property rights', *The Columbian*, 17 May 1992; D. Helvarg, *The War Against the Greens: The 'Wise-Use' Movement, the New Right, and Anti-Environmental Violence*, Sierra Club Books, San Francisco, 1994, p.149.

53. L. P. Gerlach and V. H. Hine, *People, Power, Change: Movements of Social Transformation*, Bobbs-Merrill Company, Indianapolis, New York, 1970.
54. T. McKegney, 'Only A Movement Can Combat A Movement', Environmental Campaigners Say', report of the Atlantic Vegetation Management Association 'Education Seminar' quoting Ron Arnold, Natural Resources: Forest Extension Service, 25 October 1984.
55. R. Arnold, *Ecology Wars: Environmentalism As if People Mattered*, Free Enterprise Press, Bellevue, Washington, 1993, p. 145.
56. Ibid.
57. R. Arnold, 'Loggerheads over landuse', *Logging and Sawmilling Journal*, April 1988, reprinting paper that was presented to the Ontario Forest Industries Association in Toronto in February.
58. R. Stapleton, *On the Western Front: Wise Use Movement*, Cover Story, National Parks and Conservation Association, 1993, p.32; K. Long, 'A grinch who loathes green groups', *Toronto Star*, 21 December 1991.
59. T. Egan, 'Fund-raisers tap anti-environmentalism', *New York Times*, 19 December 1991.
60. R. F. Nash, *The Rights of Nature: A History of Environmental Ethics*, The University of Wisconsin Press, 1989, p.9.
61. J. Krakauer, 'Brown fellas', *Outside*, December 1991, p.71.
62. R. Arnold, *Ecology Wars*, remarks given at the Maine Conservation Rights Institute, Second Annual Congress, 20 April 1992.
63. Ibid.
64. R. L. Barry, *The Wise Use Movement: A Briefing Paper for Montana (and the Northern Rockies)*, Montana Alliance for Progressive Policy, 1992, p.3.
65. K. O'Callaghan, 'Whose agenda for America?', *Audubon*, September/October 1992.
66. P. de Armond, personal communication to author, 24 November 1994.
67. R. Arnold, *Ecology Wars: Environmentalism As if People Mattered*, Free Enterprise Press, Bellevue, Washington, 1993, p.125.
68. R. Maughan and D. Nilson, *What's Old and What's New About the Wise Use Movement*, Department of Political Science, Idaho State University, 1993.
69. T. Ramos, interview with author, 22 December 1994.
70. Ibid.
71. D. Mazza, *God, Land and Politics: The Wise Use and Christian Right Connection in 1992 Oregon Politics*, Western States Center and Montana AFL/CIO, 1993, p.3.
72. R. Arnold, *The Wise Use Agenda: The Citizen's Policy Guide to Environmental Resource Issues – A Task Force to the Bush Administration by the Wise Use Movement*, (ed.) A.M Gottlieb, The Free Enterprise Press, 1989, p.ix.
73. Ibid., pp.157–66.
74. Ibid., p.xv; R. Rÿser, *Anti-Indian Movement on the Tribal Frontier*, Center for World Indigenous Studies, June 1992. p.48.
75. Ibid., p.xx.
76. Ibid., pp.5–18.

77. *U.S. Newswire*, 'AFC announces national rallies to support Operation Desert Storm', 1 February 1991.
78. G. Spohn, 'Moon Church move to college examined', *Chicago Tribune*, 2 October 1992, p.8.
79. R. Grant, 'The American Freedom Coalition and Rev. Moon', *The Washington Post*, 29 October 1989, p.B7.
80. American Freedom Coalition, *Environmental Task Force: Serving Mankind and the Environment*, no date.
81. American Freedom Coalition of Washington, *Non Profit Corporation Annual Report*, C/O Prentice-Hall Corp. System, Seattle, 1 March 1988; American Freedom Coalition of Washington, *Non Profit Corporation Annual Report*, Bellevue, Washington, 28 February 1989; American Freedom Coalition of Washington, *Non Profit Corporation Annual Report*, Bellevue, Washington, 28 February 1990; American Freedom Coalition of Washington, *Non Profit Corporation Annual Report*, Bellevue, Washington, 12 February 1991.
82. M. Hume, 'Resource-use conference had links to Moonie cult', *The Vancouver Sun*, 8 July 1989, p.A6.
83. D. Junas, interview with author, 22 November 1994.
84. T. Ramos, interview with author, 22 December 1994.
85. M. Hume, 'Resource-use conference had links to Moonie cult', *The Vancouver Sun*, 8 July 1989, p.A6.
86. T. Ramos, interview with author, 22 December 1994.
87. Ibid.
88. T. Ramos, 'The case of the Northwest timber industry', in *Let the People Judge: Wise Use and the Private Property Rights Movement*, (eds) J. Echeverria and R. Booth Eby, Island Press, 1995, pp. 88–94.
89. T. H. Watkins, 'Discouragements and clarifications', in *Let the People Judge: Wise Use and the Private Property Rights Movement*, (eds) J. Echeverria and R. Booth Eby, Island Press, 1995, p.52.
90. T. Ramos, 'The case of the Northwest timber industry', in *Let the People Judge: Wise Use and the Private Property Rights Movement*, (eds) J. Echeverria and R. Booth Eby, Island Press, 1995, p.89.
91. T. Ramos, *The Wise Use Movement: An Overview*, draft, Western States Center, December 1994.
92. J. Krakauer, 'Brown fellas', *Outside*, December 1991, p.71.
93. D. Hupp, 'The Wise Use Movement', in the *Western States Center Newsletter*, Summer 1992, p.6.
94. T. Ramos, 'The case of the Northwest timber industry', in *Let the People Judge: Wise Use and the Private Property Rights Movement*, (eds) J. Echeverria and R. Booth Eby, Island Press, 1995, p.95.
95. R. L. Barry, *The Wise Use Movement: A Briefing Paper for Montana (and the Northern Rockies)*, Montana Alliance for Progressive Policy, 1992, p.3.
96. T. Ramos, *The Wise Use Movement: An Overview*, draft, Western States Center, December 1994.

97. R. Maughan and D. Nilson, *What's Old and What's New About the Wise Use Movement*, Department of Political Science, Idaho State University, 1993.
98. D. Kuipers, 'Putting themselves first', *The Bay Guardian*, 15 July 1992, p.15.
99. C. P. Alexander, 'Gunning for the Greens', *Time*, 3 February 1992, p.51.
100. A. E. Ladd, 'The environmental backlash and the retreat of the state', *Blueprint for Social Justice*, January 1993, Vol. XLVI, No. 5.
101. C. Berlet, interview with author, 4 November 1994.
102. T. Ramos, interview with author, 22 December 1994.
103. P. Brick, 'Taking back the rural West', in *Let the People Judge: Wise Use and the Private Property Rights Movement*, (eds) J. Echeverria and R. Booth Eby, Island Press, 1995, p.62.
104. K. O'Callaghan, 'Whose agenda for America?', *Audubon*, 1992, Vol. 94, No. 5, p.80.
105. T. Ramos, interview with author, 22 December 1994.
106. A. E. Ladd, 'The environmental backlash and the retreat of the state', *Blueprint for Social Justice*, January 1993, Vol. XLVI, No. 5.
107. J. Stauber, interview with author, 22 March 1995.
108. M. Dowie, *Losing Ground: American Environmentalism at the Close of the Twentieth Century*, Massachusetts Institute of Technology, 1995, pp. 98, 124.
109. Ibid., p. xiii.
110. Ibid., p. 65.
111. Center for the Defense of Free Enterprise, *Programme for the Wise Use Leadership Conference*, John Ascuaga's Nugget Hotel, Reno, Nevada, 5, 6 and 7 June 1992.
112. D. Kuipers, 'Putting themselves first', *The Bay Guardian*, 15 July 1992; Center for the Defense of Free Enterprise, *Programme for the Wise Use Leadership Conference*, John Ascuaga's Nugget Hotel, Reno, Nevada, 5, 6 and 7 June 1992; A. M. Gottlieb (ed.), *The Wise Use Agenda: The Citizen's Policy Guide to Environmental Resource Issues – A Task Force to the Bush Administration by the Wise Use Movement*, The Free Enterprise Press, 1989, p.16.
113. A. L. Rawe and R. Field, 'Tug-o-war with the Wise Use movement', *Z Magazine*, October 1992, p.63.
114. J. Basl, 'Grassroots organisations unite!', *Alliance News*, February 1992, Vol. 1, Issue 1.
115. Ibid.
116. B. Ruben, 'Root rot', *Environmental Action*, Spring 1992, p.26.
117. T. Harding, 'Mocking the turtle: backlash against the environmental movement', *New Statesman and Society*, 24 September 1993, Vol. 6, No. 271, p.45.
118. B. Ruben, 'Root rot', *Environmental Action*, Spring 1992, p.26.
119. *Alliance News*, Alliance for America: Mission Statement, February 1992, Vol. 1, Issue 1.
120. Alliance for America, *Fly-In for Freedom Participants List*; R. Arnold, *The Wise Use Agenda: The Citizen's Policy Guide to Environmental Resource Issues – A Task Force to the Bush Administration by the Wise Use Movement,* (ed.) A. M Gottlieb, The Free Enterprise Press, 1989, pp.157–66.
121. B. Ruben, 'Root rot', *Environmental Action*, Spring 1992, p.26.
122. M. Kriz, 'Land mine', *The National Journal*, 23 October 1993, Vol. 25, No. 43, p.2531.

123. B. Ruben, 'Root rot', *Environmental Action*, Spring 1992, p.26.
124. W. P. Pendley, *It Takes A Hero: The Grassroots Battle Against Environmental Oppression*, Free Enterprise Press, 1994, p.vii.
125. Ibid.
126. Ibid.
127. Ibid., p.xii.
128. S. O'Donnell, *Report of the Sixth Annual Wise Use Conference*, 15–17 July 1994.
129. Ibid.
130. D. Helvarg, *The War Against the Greens: The 'Wise-Use' Movement, the New Right, and Anti-Environmental Violence*, Sierra Club Books, San Francisco, 1994, pp.290, 300.
131. T. Ramos, interview with author, 22 December 1994.
132. S. Diamond, interview with author, 7 February 1995.
133. D. Junas, interview with author, 22 November 1994.
134. R. Ross, 'Biodiversity bashing: how a LaRouche follower orchestrated the death of an environmental treaty', *Washington Post*, 23 April 1995.
135. Ibid.; S. O'Donnell, *Report of the Sixth Annual Wise Use Conference*, 15–17 July 1994.
136. R. Ross, 'Biodiversity bashing: how a LaRouche follower orchestrated the death of an environmental treaty', *Washington Post*, 23 April 1995.
137. J. Margolis, 'Odd trio could kill nature pact', *Chicago Tribune*, 30 September 1994; D. J. Barry and K. A. Cook, *How the Biodiversity Treaty Went Down: The Intersecting Worlds of Wise Use and Lyndon LaRouche*, CLEAR, Environmental Working Group, Washington, draft, 13 October 1994.
138. R. Ross, 'Biodiversity bashing: how a LaRouche follower orchestrated the death of an environmental treaty', *Washington Post*, 23 April 1995.
139. S. O'Donnell, *Report of the Sixth Annual Wise Use Conference*, 15–17 July 1994.
140. Women's Environment and Development Organisation, 'Women's Leaders Combating "Contract on America"', *News and Views*, June 1995.
141. J. Looney, 'The "Wise Use" Triple Threat: Unfunded Mandates * Private Property Takings * Cost Benefit/Risk Assessment', *Gaining Ground*, Global Action and Information Network, Autumn, 1994, Vol. 2, No. 3.
142. NRDC, *NRDC Report Documents Sweeping Retreat On Environment*, Press Release, 21 February 1995.
143. J. J. Supon, 'GOP waging war on environment', Washington, *United Press International*, 10 February 1995.
144. A. Lewis, 'Abroad at home: the one and the many', *The New York Times*, 30 December 1994, Section A, p.31.
145. J. Mathews, 'Green sweep', *The Washington Post*, 18 December 1994.
146. Sierra Club, *A Status Report On The War On The Environment*, August 1995.
147. National Audubon Society, *Private Property Rights and the Environment*, Washington, no date.
148. J. Echeverria, 'The takings issue', in *Let the People Judge: Wise Use and the Private Property Rights Movement*, (eds) J. Echeverria and R. Booth Eby, Island Press, 1995, p.147.
149. N. D. Hamilton, 'The value of land, seeking property rights solutions to public environmental concerns', in *Let the People Judge: Wise Use and the Private Property*

Rights Movement, (eds) J. Echeverria and R. Booth Eby, Island Press, 1995, pp.154–5.
150. M. Dowie, *Losing Ground: American Environmentalism at the Close of the Twentieth Century*, Massachusetts Institute of Technology, 1995, p.98.
151. Friends of the Earth, *Unfunded Mandates, Risk, Takings: Deceit in Washington*, May/June 1994.
152. J. Echeverria, 'The takings issue', in *Let the People Judge: Wise Use and the Private Property Rights Movement*, (eds) J. Echeverria and R. Booth Eby, Island Press, 1995, p.148.
153. W. K. Burke, 'The Wise Use movement: right-wing anti-environmentalism', *The Public Eye*, a publication of the Political Research Associates, June 1993, p.5; National Audubon Society, *Private Property Rights and the Environment*, Washington, no date.
154. M. Lavelle, 'The "Property Rights" Revolt', in *Let the People Judge: Wise Use and the Private Property Rights Movement*, (eds) J. Echeverria and R. Booth Eby, Island Press, 1995, p.40.
155. W. K. Burke, 'The Wise Use movement: right-wing anti-environmentalism', *The Public Eye*, a publication of the Political Research Associates, June 1993, p.5 quoting Tarso Ramos.
156. N. E. Roman, 'Fed-up states seize on the 10th Amendment', *Washington Times*, 7 July 1994.
157. C. W. LaGrasse, *The Return to the Stone Age of Government: Positions on Property*, The Property Rights Foundation of America, March/April 1994, Vol. I, No. 1.
158. Friends of the Earth, *Unfunded Mandates, Risk, Takings: Deceit in Washington*, May/June 1994; quoting American Federation of State, County and Municipal Employees, AFL-CIO, Congressional Testimony, 28 April 1994.
159. OMB Watch, *Unfunded Mandates Crisis – Alert*, December 1994.
160. Sierra Club, *A Status Report On The War On The Environment*, August 1995.
161. J. Looney, 'The "Wise Use" Triple Threat: Unfunded Mandates * Private Property Takings * Cost Benefit/Risk Assessment', *Gaining Ground*, Global Action and Information Network, Autumn 1994, Vol. 2, No. 3.
162. J. Herbert, *Environmental Gridlock*, *Associated Press*, Washington, 7 October 1994.
163. C. Safina and S. Iudicello, 'Wise Use below the high-tide line: threats and opportunities', in *Let the People Judge: Wise Use and the Private Property Rights Movement*, (eds) J. Echeverria and R. Booth Eby, Island Press, 1995, p.120.
164. *UPI*, 'Endangered species report card', 30 October 1995.
165. Figures from the Nature Conservancy.
166. *Business Wire*, statement by Secretary of the Interior, Washington, 10 May 1995.
167. *PR Newswire*, 'Hundreds rally for "Sensible Reform" and property rights', Californians for Sensible Environmental Reform, 26 April 1995.
168. H. Josef Hebert, 'Gingrich-species', *Associated Press*, 26 May 1995.
169. T. Kenworthy, 'Panel supports stronger Species Act: effect of study on upcoming hill environmental debate seen as questionable', *Washington Post*, 25 May 1995; R. Snodgrass, 'The Endangered Species Act – a commitment worth keeping', in *Let the*

People Judge: Wise Use and the Private Property Rights Movement, (eds) J. Echeverria and R. Booth Eby, Island Press, 1995, p.278.

170. *UPI*, 'Environmentalists hail High Court ruling', Washington, 29 June 1995; T. Reichhardt, 'Environmental groups get Supreme Court boost on Endangered Species', *Nature*, Vol. 376, 6 July 1995.
171. V. Allen, 'Environment policies hinge on budget battle', *Reuters*, 26 August 1995; *Christian Science Monitor*, title unknown, 29 June 1995.
172. D. Morgan, 'House conservatives step up assault on regulations', *Washington Post*, 19 July 1995.
173. T. Kenworthy and D. Morgan, 'Budget axe chops landmarks of environmentalist legislation', *Washington Post*, 16 March 1995.
174. H. Josef Hebert, 'Federal fire sale', *Associated Press*, Washington, 25 May 1995.
175. H. Josef Hebert, 'Budget-Environment', Washington, *Associated Press*, 16 May 1995; M. Walker, 'Licence to pollute the free world', *The Guardian*, 8 September 1995, Society, pp.4–5.
176. J. St. Clair, 'Stopping on green: why environmentalists are dumping Democrats in the West', *Washington Post*, 23 October 1994.
177. T. Kenworthy and D. Morgan, 'Unit doubles timber cut allowed on federal land', *Washington Post*, 3 March 1995; *Reuter*, 'Republicans move to chop logging restrictions', Washington, 10 February 1995; D. DellaSala and D. Olsen, 'Deregulation in the USA: a comment', *Arborvitae*, The IUCN/WWF Forest Conservation Newsletter, September 1995, p.4.
178. *Associated Press*, 'GOP offers mining reform bill', 7 March 1995.
179. S. Sonner, 'Logging-NAFTA', *Associated Press*, 29 August 1995.
180. *Reuters*, 'Anchorage', 30 August 1995.
181. D. Morgan, 'Republicans defect to kill curbs on EPA: House rejects provisions limiting Agency's power to enforce Clean Air and Water standards', *Washington Post*, 29 July 1995.
182. *BNA*, 'Bjerregaard calls anti-environment drive in U.S. Congress blow to global leadership', Brussels, 27 July 1995.
183. M. Walker, 'Licence to pollute the free world', *The Guardian*, 8 September 1995, Society, pp.4–5.
184. Sierra Club, *A Status Report On The War On The Environment*, August 1995.
185. S. Sonner, 'Congress-Environment', Washington, *Associated Press*, 26 February 1996.
186. H. Josef Hebert, 'Environmental Offensive', Washington, *Associated Press*, 5 April 1995.
187. *Reuters*, 'Groups urge Congress to stop rollback of green laws', 1 November 1995.
188. D. E. Kalish, 'Green rebound', *Associated Press*, 8 June 1995.
189. J. Yang, 'Petroleum industry assists backers of Alaska drilling', *Washington Post*, 25 October 1995.
190. M. Walker, 'Licence to pollute the free world', *The Guardian*, 8 September 1995, Society, pp.4–5.
191. Ibid.

2 CULTURE WARS AND CONSPIRACY TALES

1. N. Gibbs, 'The blood of innocents', *Time*, 1 May 1995, p.41.
2. The Blue Mountain Working Group, *A Call to Defend Democracy and Pluralism*, 17 November 1994.
3. S. Pharr, *The Right's Agenda*, Women's Project, Little Rock, Arkansas, 1994.
4. C. Berlet, 'The Right rides high', *The Progressive*, October 1994, p.26.
5. The Blue Mountain Working Group, *A Call to Defend Democracy and Pluralism*, 17 November 1994.
6. C. Berlet, interview with author, 4 November 1994.
7. S. Diamond, *Spiritual Warfare: The Politics of the Christian Right*, South End Press, 1989, p.57.
8. Dr S. Diamond, *Right-Wing Movements in the United States, 1945–1992*, dissertation, submitted for the degree of Doctor of Philosophy in Sociology in the graduate division of the University of California, pp.11–12.
9. Rep. William Dannemeyer, 'Environmental party: a major threat to our national security', *Inside Washington*, 1 September 1990.
10. *Campus Report*, 'Environmentalism Becomes Radical', Accuracy in Academia, Washington, April 1990, Vol. V, No. 4.
11. *Human Events: The National Conservative Weekly*, 'Radical environmentalists fuel Earth Day', 1990, Vol. L, No. 17.
12. L. H. Rockwell Jr, 'An anti-environmentalist manifesto', *Buchanan from the Right*, 1990, Vol. 1, No. 6.
13. People for the American Way, 'The "Greening" of the Right', *Right Wing Watch*, September 1992, p.3.
14. Dr J. Hardisty, 'Constructing homophobia', *The Public Eye*, March 1993, p.6.
15. J. Bleifuss, 'The God of Mammon: Christian coalition makes corporate allies', *PR Watch*, 1994, Fourth Quarter, p.8.
16. S. Diamond, interview with author, 7 February 1995.
17. J. Bleifuss, 'The God of Mammon: Christian coalition makes corporate allies', *PR Watch*, 1994, Fourth Quarter, p.9.
18. W. J. Lanouette, 'The New Right – "revolutionaries" out after the "Lunch-Pail" vote', *The National Journal*, 21 January 1978, Vol. 10, No. 3, p.88.
19. M. Quigley and C. Berlet, 'Traditional values, racism, and Christian theocracy: the right-wing revolt against the modern age', *The Public Eye*, December 1992, p.2.
20. *The Public Eye*, 'Free Congress foundation in Rio', Political Research Associates, December 1992, p.10; R. Bellant, *The Coors Connection: How Coors Family Philanthropy Undermines Democratic Pluralism*, South End Press, Boston, 1991, p.21.
21. T. Ramos, interview with author, 22 December 1994.
22. NET, *One if By Land . . .* , Washington, 1994.
23. Ibid.
24. W. P. Pendley, *It Takes A Hero: The Grassroots Battle Against Environmental Oppression*, Free Enterprise Press, 1994.

25. R. Bellant, *The Coors Connection: How Coors Family Philanthropy Undermines Democratic Pluralism*, South End Press, Boston, 1991, p.16.
26. L. Hatfield and D. Waugh, 'Right wing's smart bombs', *San Francisco Examiner*, 24 May 1992; quoting Burton Yale Pines.
27. Ibid.
28. Ibid.
29. Economics America, *The Right Guide: A Guide to Conservative and Right-of-Center Organisations*, Ann Arbor, Michigan, 1993.
30. S. Diamond, interview with author, 7 February 1995.
31. C. Deal, *The Greenpeace Guide to Anti-Environmental Organisations*, Odonian Press, Berkeley, 1993, p.79.
32. *A CLEAR View*, The Competitive Enterprise Institute, CLEAR, Vol. 3, No. 3, 21 February 1996.
33. CLEAR, *'Defund the Left' Campaign*, Memorandum, 14 June 1995.
34. T. Roszak, 'Green guilt and ecological overload', *The New York Times*, 9 June 1992, p.27.
35. The EDA includes the following American think-tanks: The American Council on Science and Health, New York; The Goldwater Institute, Arizona; the Center for Individual Rights, Washington; the Claremont Institute, California; Committee for A Constructive Tomorrow, Washington; Competitive Enterprise Institute, Washington; Heritage Foundation, Washington; Independence Institute, Colorado; Institute of Political Economy, Utah; James Madison Institute, Florida; National Center for Policy Analysis, Texas; Pacific Research Institute, San Francisco; Political Economy Research Center, Massachusetts and the Reason Foundation in California. Non-American think-tanks include the Institute of Public Affairs, Melbourne, Australia; Institute Euro 92 in Paris, France, The Fraser Institute in Vancouver, Canada and the Institute of Economic Affairs in London, England.
36. *Organisation Trends*, 'Free market environmentalism: an interview with Fred Smith', Capital Research Centre, 1990, April, p.3.
37. Earth Day Alternatives, *The Free Market Manifesto*, Competitive Enterprise Institute, Washington, 1990.
38. P. Hawken, *The Ecology of Commerce: A Declaration of Sustainability*, HarperBusiness, 1994, p.66.
39. *Reuter*, 'Support unravels for California pollution trading', Los Angeles, 14 August 1995.
40. *Organisation Trends*, 'Free market environmentalism: an interview with Fred Smith', Capital Research Centre, 1990, April, p.3.
41. C. Berlet, interview with author, 4 November 1994.
42. Environmental Working Group, *Earth Day Information*, 20 April 1996.
43. D. Helvarg, *The War Against the Greens: The 'Wise-Use' Movement, the New Right, and Anti-Environmental Violence*, Sierra Club Books, San Francisco, 1994, pp.19, 20.
44. C. Deal, *The Greenpeace Guide to Anti-Environmental Organisations*, Odonian Press, Berkeley, 1993, p.58.
45. M. Dowie, *Losing Ground: American Environmentalism at the Close of the Twentieth Century*, MIT Press, 1995, pp.83–4.

46. J. K. Andrews Jr, D. Armey, F. Barnes, G. L. Bauer, T. Bethell, D. Boaz, T. J. Bray, S. Brunelli, A. Carlson, A. L. Chickering, M. E. Daniels Jr, D. J. Devine, P. du Pont, M. Eberstadt, L. Edwards, J. A. Eisenach, P. J. Ferrera, E. J. Feulner Jr, P. Gramm, J. Helms, H. Hyde, F. C. Ikle, R. Kirk, C. Mack, A. Meyerson, J. C. Miller III, G. Norquist, W. Olson, D. J. Popeo, V. I. Postrel, R. W. Rahn, P. Robertson, P. Schlafly, W. E. Simon, K. Tomlinson, M. Wallop, G. Weigel, P. M. Weyrich, K. Zinsmeister, The Vision Thing, 'Conservatives Take Aim at the '90s', 1990, *Policy Review*, Spring, 1990, No. 52, p.4. Among the authors were Paul Weyrich and Pat Robertson. The other signatories are the president of the Independence Institute, president of the Family Research Council, a senior editor of *The New Republic*, Washington editor of *The America Spectator*, executive vice president of the Cato Institute, executive director of the American Legislative Exchange Council (ALEC), president of the Rockford Institute, vice president of the Institute for Contemporary Studies, president and CEO of the Hudson Institute, chairman of Citizens for America, executive editor of the *National Interest*, president of the Washington Policy Group, senior fellow at the Cato Institute, president of The Heritage Foundation, distinguished scholar at the Center for Strategic and International Studies, editor of *Policy Review*, chairman of Citizens for a Sound Economy, president of Americans for Tax Reform, senior fellow at the Manhattan Institute, General Counsel of the Washington Legal Center, *Reason Magazine*, vice president and chief economist of the US Chamber of Commerce, president of Eagle Forum in Alton, Illinois, president of the John M. Olin Foundation, executive editor of *Reader's Digest*, president of the Ethics and Public Policy Center in Washington, and a research associate with the American Enterprise Institute.
47. J. Sugarmann, *NRA: National Rifle Association, Money, Firepower, Fear*, National Press Books, Washington DC, 1992, p.131.
48. R. Rÿser, *Anti-Indian Movement on the Tribal Frontier*, Center for World Indigenous Studies, June 1992, p.44.
49. P. de Armond, interview with author, 24 November 1995.
50. *Who's Who in America*, Alan Gottlieb, galley proof, Marquis Who's Who, New Providence, 1995, 49th Edition.
51. S. O'Donnell, interview with author, 9 November 1994.
52. T. Ramos, interview with author, 22 December 1994.
53. P. de Armond, interview with author, 24 November 1994.
54. R. Arnold, *Ecology Wars: Environmentalism As if People Mattered*, Free Enterprise Press, Bellevue, Washington, 1993, p.30.
55. *New Gun Week* is published by the Second Amendment Foundation, Publisher: Alan M. Gottlieb, Editor: Joseph P. Tartaro: Terrence P. Duffy, Contributing editors: Ron Arnold and others.
56. D. Helvarg, *The War Against the Greens: The 'Wise-Use' Movement, the New Right, and Anti-Environmental Violence*, Sierra Club Books, San Francisco, 1994, p.66.
57. J. Sugarmann, *NRA: National Rifle Association, Money, Firepower, Fear*, National Press Books, Washington DC, 1992, p.155.
58. T. Egan, 'Fund-raisers tap anti-environmentalism', *New York Times*, 19 December 1991.

59. J. Krakauer, 'Brown fellas', *Outside*, December 1991, p.114.
60. D. Helvarg, *The War Against the Greens: The 'Wise-Use' Movement, the New Right, and Anti-Environmental Violence*, Sierra Club Books, San Francisco, 1994, p.137.
61. P. de Armond, interview with author, 24 November 1994.
62. J. Halpin and P. de Armond, 'The merchant of fear', *Eastsideweek*, 26 October 1994.
63. D. Helvarg, *The War Against the Greens: The 'Wise-Use' Movement, the New Right, and Anti-Environmental Violence*, Sierra Club Books, San Francisco, 1994, p.66.
64. Ibid., p.129.
65. M. Burdman, 'Greenpeace: shock troops of the New Dark Age', *Executive Intelligence Review*, 21 April 1989.
66. D. King, *Lyndon LaRouche and the New American Fascism*, Doubleday, 1989, p.ix.
67. D. King and P. Lynch, 'The empire of Lyndon LaRouche', *Wall Street Journal*, 27 May 1986.
68. D. King, *Lyndon LaRouche and the New American Fascism*, Doubleday, 1989, p.xiv.
69. Anti-Defamation League of B'nai B'rith, Extremism on the Right, A Handbook, date unknown; J. Sommer, *Briefing Paper on LaRouche for Greenpeace Germany*, 1994; LaRouche's European organisation is centred in the Schiller Institute in Weisbaden in Germany, but there are also offices in Dusseldorf, Copenhagen, Milan, Paris, Rome and Stockholm. An office has also recently been opened in Russia. In Central and Latin America, countries where LaRouche organisations or publications have an office include Mexico, Colombia and Peru, but they are also active in Venezuela and recently Panama. Australia too has been a new target country for LaRouche. International Bureaus of EIR are listed in Bangkok, Bogota, Bonn, Copenhagen, Houston, Lima, Mexico City, Milan, New Delhi, Paris, Rio de Janeiro, Rome, Stockholm, Washington, and Weisbaden.
70. J. Sommer, *Briefing Paper on LaRouche for Greenpeace Germany*, 1994.
71. D. King, *Lyndon LaRouche and the New American Fascism*, Doubleday, 1989, p.89.
72. M. Burdman, 'Greenpeace: shock troops of the New Dark Age', *Executive Intelligence Review*, 21 April 1989, p.24.
73. Ibid.
74. Ibid., pp.25, 26.
75. *21st Century Science and Technology*, 'The world needs nuclear energy', March–April 1989, p.2.
76. M. Burdman, 'Greenpeace: shock troops of the New Dark Age', *Executive Intelligence Review*, 21 April 1989, p.27.
77. L. LaRouche, 'How all my enemies will die', Wiesbaden, no date, 7 January.
78. Ibid.
79. J. Sommer, *Briefing Paper on Lyndon LaRouche*, Greenpeace Germany, 1994.
80. D. King, *Lyndon LaRouche and the New American Fascism*, Doubleday, 1989, p.270; C. Berlet, interview with author, 2 February 1995.
81. C. Berlet, interview with author, 2 February 1995.
82. D. King, *Lyndon LaRouche and the New American Fascism*, Doubleday, 1989, p.44.
83. K. Kalimtgis, D. Goldman and J. Steinberg, *Dope Inc: Britain's Opium War Against the US*, New Benjamin Franklin House Publishing, New York, 1978.

84. *New Federalist*, 'Never again! stop the United Nations Genocide Conference', May 1994.
85. N. Hamerman, From the Editor, *Executive Intelligence Review*, Vol. 19, No. 25, 19 June 1992, p.1.
86. C. Berlet, interview with author, 2 February 1995.
87. D. Barry and K. Cook, *How the Biodiversity Treaty Went Down: The Intersecting Worlds of 'Wise Use' and Lyndon LaRouche*, Environmental Working Group, Washington, 1994, p.14.
88. Ibid., pp.14–5.
89. W. P. Pendley, 'Do grizzlies have the same rights as humans?', *21st Century Science and Technology*, Fall 1993, p.69; K. Marquardt, 'Extremists attack Norway for the resumption of whaling', *21st Century Science and Technology*, Spring 1993, pp.65–6; M. Coffman, 'The pagan roots of environmentalism', *21st Century Science and Technology*, Fall 1994, pp.55–64.
90. D. Barry and K. Cook, *How the Biodiversity Treaty Went Down: The Intersecting Worlds of 'Wise Use' and Lyndon LaRouche*, Environmental Working Group, Washington, 1994, p.18; indexed in Spring 1992; *Executive Intelligence Review*, 'Pacific Tuna fishermen take on Greenpeace', 1 October 1993.
91. D. King, *Lyndon LaRouche and the New American Fascism*, Doubleday, 1989, p.160.
92. W. Welch, 'NSA investigated a LaRouche-linked group', *Associated Press*, 13 August 1987.
93. D. King, *Lyndon LaRouche and the New American Fascism*, Doubleday, 1989, p.167.
94. N. Starcevic, 'Few LaRouche supporters in Europe have never elected candidate', *Associated Press*, 22 March 1986.
95. J. Sommer, *Briefing Paper on LaRouche for Greenpeace Germany*, 1994.
96. Ibid.
97. D. King, *Lyndon LaRouche and the New American Fascism*, Doubleday, 1989, p.65.
98. C. Berlet, interview with author, 2 February 1995.
99. C. Berlet and M. Lyons, 'Militia nation', *The Progressive*, June 1995, p.22.
100. J. Cohen and N. Solomon, 'Knee-jerk coverage of bombing should not be forgotten', *Alternet*, 4 May 1995.
101. D. Junas, 'The rise of the militias', *CovertAction*, Spring 1995, p.25.
102. K. Stern, *Militias: A Growing Danger*, an American Jewish Committee background report, April 1995.
103. R. Hathaway, *The Disenfranchised Americans*, Eagle Constitutional Militia, 20 April 1995.
104. N. Toiczek, 'Make-believe world inspires US terror, *The Independent on Sunday*, 6 August 1995, p.17.
105. T. Egan, 'Federal uniforms become target of wave of threats and violence', *The New York Times*, 25 April 1995, p.A10.
106. Stormfront BBS, 23 August 1995.
107. M. Cooper, 'A visit with MOM', *The Nation*, 22 May 1995.
108. C. Berlet and M. Lyons, 'Militia nation', *The Progressive*, June 1995, p.25.
109. Ibid., p.24.
110. From the Patriot Archives site on the Internet, published in the *San Francisco Bay Guardian*, 17 May 1995.

111. C. Berlet, interview with author, 2 February 1995.
112. M. Cooper, 'A visit with MOM', *The Nation*, 22 May 1995.
113. B. Hawkins, 'Patriot games', *Detroit Metro Times*, 12 October 1994.
114. D. Junas, 'The rise of the militias', *CovertAction*, Spring 1995, p.25.
115. *Spotlight*, 'Russian choppers confirmed', 5 September 1994, pp.1, 5.
116. R. Shelton, 'The new Minutemen', *Kansas City News Times*, 22 February 1995.
117. P. Weiss, 'Outcasts digging in for the apocalypse', *Time*, 1 May 1995, pp.34–5.
118. C. Berlet, interview with author, 2 February 1995.
119. P. Weiss, 'Outcasts digging in for the apocalypse', *Time*, 1 May 1995, pp.34–5.
120. *Western Horizons*, 'Inside the 1993 Wise Use Leadership Conference', September 1993, Vol. 1, No. 3, p.2.
121. M. Cole, 'Video used by militia features Chenoweth', *The Idaho Statesman*, 28 April 1995.
122. S. O'Donnell, interview with author, 9 November 1994.
123. D. Junas, interview with author, 22 November 1994.
124. T. Ramos, interview with author, 22 December 1994.
125. P. de Armond, interview with author, 24 November 1995.
126. Ibid.
127. B. Clark, 'John Birch meets John Wayne', *The StranGer*, 17 May 1995.
128. Ibid.
129. K. Durbin, 'Environmental terrorism in Washington State', *Seattle Weekly*, 11 January 1995; *A CLEAR View*, Vol. 3, No. 2, 29 January 1996.
130. R. Downes and G. Foster, 'On the front lines with Northern Michigan's militia', *Northern Express*, 22 August 1994.
131. B. Knickerbocker, 'The radical element: violent conflict over resources', *The Christian Science Monitor*, 2 May 1995, p.10.
132. J. Ridgeway and L. Zeskind, 'Revolution USA', *Village Voice*, 25 April 1995.
133. K. Sneider, 'Bomb echoes extremists' tactics', *The New York Times*, 26 April 1994, p.A12.
134. D. Helvarg, 'The anti-enviro connection', *The Nation*, 22 May 1995, p.722.
135. P. de Armond and J. Halpin, 'Steal This State: One Man's Journey into the Secession Movement Underground', *Eastsideweek* 1994, 17 August, p.14.
136. S. O'Donnell, interview with author, 9 November 1994.
137. County Commissioners, *A Brief Description of the County Government Movement*, 3 August 1993.
138. P. de Armond and J. Halpin, *Steal This State: One Man's Journey into the Secession Movement Underground*, Eastside Week 1994, 17 August, p.14.
139. F. Williams, 'Sagebrush Rebellion 11', in *Let the People Judge: Wise Use and the Private Property Rights Movement*, (eds) J. Echeverria, R. Booth Eby, Island Press, 1995, p.130.
140. C. McCoy, 'Cattle prod: Catron County leads a nasty revolt over eco-protection', *The Wall Street Journal*, 3 January 1994.
141. W. Perry Pendley, *It Takes A Hero: The Grassroots Battle Against Environmental Oppression*, Free Enterprise Press, 1994, p.141.

142. F. Williams, 'Sagebrush Rebellion 11', in *Let the People Judge: Wise Use and the Private Property Rights Movement*, (eds) J. Echeverria, R. Booth Eby, Island Press, 1995, p.132.
143. J. Snyder, *States Rights : County Movement Rides Again*, People for the West!, 1994, date unknown.
144. S. W. Reed, 'The County Supremacy Movement: mendacious myth marketing', *Idaho Law Review*, 1994, Vol. 30, p.527; T. Kenworthy, 'U.S. enters range war, suing Nevada County', *The Wall Street Journal*, 9 March 1995.
145. T. Egan, 'Court puts down rebellion over control of federal land', *The New York Times*, 16 March 1996.
146. *New Scientist*, 'Arizona fights for the right to stay cool', 29 April 1995.
147. NFLC, 'Why there is a need for the militia in America', *Federal Lands Update*, October 1994.
148. Ibid.
149. R. Kaiser, 'United grassroots efforts continue to succeed', *Federal Lands UpDate*, November 1994.
150. Ibid.
151. D. Helvarg, 'The anti-enviro connection', *The Nation*, 22 May 1995, p.724.
152. Ibid.
153. Ibid.
154. K. Sneider, 'Bomb echoes extremists' tactics', *The New York Times*, 26 April 1994, p.A12.
155. R. Crawford, S. L. Gardner, J. Mozzochi, and R. L. Taylor, '*The Northwest imperative*', Coalition for Human Dignity, Northwest Coalition Against Malicious Harassment, 1994, pp.1.18–1.20.
156. M. Hecht, '"Wise Use" and environmentalists both played by same forces', *21st Century Science and Technology*, Summer 1995, p.9.
157. Ibid.
158. Ibid.
159. A. Chaitkin, 'A Warning on the "Wise Use" movement', *21st Century Science and Technology*, Summer 1995, p.11.

3 THE DEATH OF DEMOCRACY

1. A. Carey, *Taking the Risk out of Democracy: Propaganda in the US and Australia*, University of New South Wales Press, Ltd, Sydney, 1995.
2. P. Hawken, *The Ecology of Commerce: A Declaration of Sustainability*, HarperBusiness, 1994, p.120.
3. M. Dowie, *Losing Ground: American Environmentalism at the Close of the Twentieth Century*, MIT Press, 1995, p.86.
4. V. Cable, 'What price integrity today?', *The Independent*, 18 April 1995, p.15.

5. C. Berlet, interview with author, 4 November 1994.
6. R. Kazis and R. L. Grossman, *Fear At Work: Job Blackmail, Labour and the Environment*, The Pilgrim Press, 1982, p.77.
7. D. Helvarg, *The War Against the Greens: The 'Wise-Use' Movement, the New Right, and Anti-Environmental Violence*, Sierra Club Books, San Francisco, 1994, p.61–2.
8. B. Williams, *US Petroleum Strategies in the Decade of the Environment*, PennWell Books, Oklahoma, 1991, p.305.
9. M. Useem, *The Inner Circle: Large Corporations and The Rise of Business Political Activity in the U.S. and the U.K.*, Oxford University Press, 1984, pp.17–18.
10. Ibid., pp.36–40.
11. M. Useem, interview with author, 28 March 1995.
12. *The Guardian*, 'Captains keep on running into each other', 15 July 1995; companies and banks linked in some way through their directors include ICI, Zeneca, Unilever, Boots, RTZ, The Prudential, Marks and Spencer, Whitbread, British Airways, National Westminster Bank, Barclays, BP, BOC, Gmet, and British Gas.
13. J. Nelson, *Sultans of Sleaze: Public Relations and the Media*, Between the Lines, 1989, p.15.
14. Dr S. Diamond, *Right-Wing Movements in the United States, 1945–1992*, dissertation, submitted for the degree of Doctor of Philosophy in sociology in the graduate division of the University of California, p.267.
15. W. Greider, *Who Will Tell the People?: The Betrayal of American Democracy*, Touchstone, Simon and Schuster,1991, p.35.
16. P. Hawken, *The Ecology of Commerce: A Declaration of Sustainability*, HarperBusiness, 1994, p.109.
17. W. Greider, *Who Will Tell the People?: The Betrayal of American Democracy*, Touchstone, Simon and Schuster, pp.25, 39.
18. P. Hawken, *The Ecology of Commerce: A Declaration of Sustainability*, HarperBusiness, 1994, p.91.
19. K. Watkins, 'The foxes take over the hen house', *The Guardian*, 17 July 1992.
20. T. Hines and C. Hines, *The New Protectionism, Protecting Against Free Trade*, Earthscan, 1993, p.34.
21. *World Bank*, World Bank Atlas 25th Anniversary Edition, 1993; *Fortune*, The Fortune Global 500, 1992, p.55 (Do not include excise taxes).
22. *Corporate Crime Reporter*, 'Multinational corporations are growing beyond nations' ability to control them', OTA report Finds, 1993, Vol. 7, No. 35, 13 September, p.8.
23. *Corporate Crime Reporter*, 'Shareholders association reports that most American corporations undermine or deny "One Share, One Vote", Other Forms of Shareholder Rights', 1990, Vol. 4, No. 14, 9 April.
24. P. Donovan and C. Barrie, 'Mr 75 per cent suffers ordeal by gas shareholders', *The Guardian*, 1 June 1995, p.1; M. Gagan and W. Gleeson, 'Fury fails to burst gas bubble', *The Independent*, 1 June 1995, p.3.
25. *The Guardian*, 'Tell Sid he's not wanted: British Gas has made a mockery of people's capitalism', Editorial, 1 June 1995.

26. *The Sunday Telegraph*, 'RTZ seeks to thwart protests', 4 June 1995.
27. L. Buckingham and R. Cowe, 'Spar showdown is no surrender', *The Guardian*, 24 June 1995, p.38.
28. N. Chomsky, 'The free market myth', *Open Eye*, 1995, No. 3, p.10.
29. W. Morehouse, *Accountability, Regulation and Control of Multinational Corporations*, testimony prepared for the Permanent People's Tribunal on Industrial and Environmental Hazards and Human Rights, London, 28–30 November 1994, corrected draft.
30. Ibid.
31. M. Useem, *The Inner Circle: Large Corporations and The Rise of Business Political Activity in the U.S and the U.K.*, Oxford University Press, 1984, p.133.
32. Ibid., p.150.
33. M. Alperson, A. Tepper Marlin, J. Schorsch and R. Will, *The Better World Investment Guide*, Council on Economic Priorities, Prentice Hall Press, 1991, p.34.
34. W. Greider, *Who Will Tell the People?: The Betrayal of American Democracy*, Touchstone, Simon and Schuster, p.48.
35. M. Useem, *The Inner Circle: Large Corporations and The Rise of Business Political Activity in the U.S and the U.K.*, Oxford University Press, 1984, p.134.
36. D. Rogers, 'How oil money helped to change the face of Congress', *The Boston Globe*, 10 September 1981.
37. Ibid.
38. R. Kazis and R. L. Grossman, *Fear At Work: Job Blackmail, Labour and the Environment*, The Pilgrim Press, 1982, p.92.
39. Ibid.
40. M. Alperson, A. Tepper Marlin, J. Schorsch and R. Will, *The Better World Investment Guide*, Council on Economic Priorities, Prentice Hall Press, 1991, p.35.
41. W. Greider, *Who Will Tell the People: The Betrayal of American Democracy*, Touchstone, Simon and Schuster, pp.48, 259.
42. M. Dowie, *Losing Ground: American Environmentalism at the Close of the Twentieth Century*, MIT Press, 1995, p.85.
43. D. Duston, 'Congress-investments', *Associated Press*, 20 August 1995.
44. M. Walker, 'Licence to pollute the free world', *The Guardian*, 8 September 1995, Society, pp.4–5.
45. V. Cable, 'What price integrity today?', *The Independent*, 18 April 1995, p.15.
46. D. Alton, 'Standards in public life', *The Independent on Sunday*, 30 October 1994, Register of MPs' Interests.
47. Ibid.
48. Ibid.
49. *The Guardian*, 'Watchdog urges checks on MP's income', 12 May 1995, p.6.
50. *The Independent on Sunday*, 'Parliament, who lobbies whom – Register of Members' interests', 30 October 1994.
51. P. Hosking, 'Why companies say Yes Minister', *The Independent on Sunday*, 10 September 1995, Business, p.1.
52. Ibid.
53. *Labour Research*, 'Earth slips under Tories' feet', December 1994, p.9.

54. Ibid., p.10.
55. Ibid., pp.9–10.
56. R. Smith, 'Lorry ban "Ended by Tory backers"', *The Guardian*, 8 June 1994.
57. D. Alton, 'Standards in public life', *The Independent on Sunday*, 30 October 1994, Register of MPs' Interests.
58. *Labour Research*, 'Earth slips under Tories' feet', December 1994, p.11.
59. Published in *The Guardian*, 12 May 1995, p.1.
60. *Corporate Crime Reporter*, 1991, 'Two reports indicate public corruption may be endemic to current political system', 1991, Vol. 5, No. 27, 8 July, p.9.
61. Ibid.
62. Ibid.
63. *Corporate Crime Report*, 'Consumer groups blast auto industry efforts to defeat fuel efficiency measure, say Transportation Department is illegally lobbying for industry', 1991, Vol. 5, No. 27, 8 July, p.7. CCR quotes Clarence Ditlow, executive director of the Center for Auto Safety (CAS), who 'said that more fuel efficient automobiles can be produced without reducing vehicle size. A CAS analysis determined that the nation's cars could attain average fuel economy standards of 40–45 miles per gallon, while reducing the death rate by 20 percent, by the year 2001'.
64. W. Greider, *Who Will Tell the People?: The Betrayal of American Democracy*, Touchstone, Simon and Schuster, pp.37, 38–9.
65. Ibid.
66. D. Nicolson-Lord, 'Facing up to the factoids', *The Independent on Sunday*, 10 July 1994, p.19.
67. R. Nixon, 'Science for sale: truth to the highest bidder', *CovertAction*, Spring 1995, pp.49–50.
68. Ibid., p.50.
69. Ibid., p.49.
70. Ibid., p.52.
71. J. Senker, *Biotechnology at SPRU*, conference proceedings, Welcome Institute, London, 15 March 1995.
72. N. Woodcock, 'Fossil fuels: a crisis of resources or effluents?', *Geology Today*, July–August 1993.
73. *New Scientist*, 'Who kidnapped science?', 15 July 1995, p.3.
74. *Nature*, 'British science sent back to dog-house', 13 July 1995, p.101.
75. R. Nixon, 'Science for sale: truth to the highest bidder', *CovertAction*, Spring 1995, p.49.
76. C. Crossen, *Tainted Truth: The Manipulation of Fact in America*, Simon and Schuster, New York, 1994, p.19.
77. M. Megalli and A. Friedman, Pacific Legal Foundation, *Masks of Deception: Corporate Front Groups in America*, Essential Information, December 1991, pp.2–3.
78. C. Berlet and W. K. Burke, 'Corporate fronts: inside the anti-environmental movement', *Greenpeace Magazine*, January/February/March 1992.
79. J. O'Dwyer, 'Citizens' Groups a Front, Says Green Org', *Jack O'Dwyer's Newsletter*, Vol. 25, No. 3, 15 January 1992, p.7.

80. *Corporate Crime Report*, 'Consumer groups blast auto industry efforts to defeat fuel efficiency measure, say Transportation Department is illegally lobbying for industry', 1991, Vol. 5, No. 27, 8 July, p.7.
81. *Corporate Crime Reporter*, 'Public interest groups name worst advertising of the year'. Stroh's Swedish Bikini Team makes list', Vol. 5, No. 47, 9 December 1991, p.9.
82. M. Megalli and A. Friedman, Pacific Legal Foundation, *Masks of Deception: Corporate Front Groups in America*, Essential Information, December 1991, p.56.
83. Ibid., pp.175–6; D. Levy, 'Talking a green game: corporate environmentalism shows its true colors', *Dollars and Sense*, May/June 1994, p.16.
84. M. Megalli and A. Friedman, Pacific Legal Foundation, *Masks of Deception: Corporate Front Groups in America*, Essential Information, December 1991, pp.18–22.
85. C. Deal, *The Greenpeace Guide to Anti-Environmental Organisations*, Odonian Press, 1993, p.56.
86. *IPS*, 'U.S. big business infiltrates United Nations', 8 February 1995; P. Scupholme, *Response to Questionnaire from Earth Resources Research*, HSE Policy Unit, London, 21 May 1992; Amoco, *Response to Questionnaire from Earth Resources Research*, 1992; Global Climate Coalition, *Statement of Purpose and Principles*, Washington, no date; Texaco Inc., 'Ours to protect, environment', *Health & Safety Review*, White Plains, New York, 1992, p.47; FDCH, *Congressional Testimony*, 16 November 1993, Attachments; Global Climate Coalition, *Energy Efficiency in American Industry*, Membership List and Background Information, October 1993; Global Climate Coalition, *The United States Versus The European Community: Environmental Performance*, GCC Report, August 1993; Global Climate Coalition, *Leadership in Energy Efficiency*, GCC Report, March 1993.
87. J. C. Stauber, 'Going . . . going . . . green!', *PR Watch*, 1994, Vol. 1, No. 3, Second Quarter, p.2.
88. *Business Wire*, 'Scientists warn of disruption from global warming', 18 September 1995.
89. *IPS*, 10 June 1992.
90. *IPS*, 'U.S. big business infiltrates United Nations', 8 February 1995.
91. Global Climate Coalition, *Developing Countries Escape Climate Treaty Negotiation With No New Obligation*, press release, 7 April 1995.
92. F. Pearce, 'Fiddling while the Earth warms', *The New Scientist*, 25 March 1995, p.14.
93. Ibid.
94. *Der Spiegel*, 'Hohepriester im Kohlenstoff-Klub', 1995, 14, pp.36–8.
95. *PR Newswire*, 'ICCP commends results of international climate talks: urges business leadership on technology assessment', 12 April 1995; ICCP's members include Allied Signal, American Standard/Trane, AT & T, BP America, Carrier/United Technologies, Celotex/Jim Walter Corporation, Dow, Du Pont, Enron, AB Electrolux/ White Consolidated/ Frigidaire, Elf Atochem, ICI America, 3M, York International, Air Conditioning and Refrigeration Institute, Alliance for Responsible Atmospheric Policy, Association Home Appliance Manufacturers, Japan Flon Gas Association, and the Polysocyanurate Insulation Manufacturers Association.
96. *PR Newswire*, 'ICCP commends results of international climate talks: urges business leadership on technology assessment', 12 April 1995.

97. F. Pearce, 'Fiddling while the Earth warms', *The New Scientist*, 25 March 1995, p.15.
98. K. O'Callaghan, 'Whose agenda for America?', *Audubon*, National Audubon Society, September/October 1992, Vol. 94, No. 5.
99. J. Skow, 'Earth Day blues', *Time*, 24 April 1995, p.75; *A CLEAR View*, National Wetlands Coalition, CLEAR, Vol. 3, No. 2, 29 January 1996.
100. E. Newlin Carney, 'Industry plays the grass-roots card', *The National Journal*, 1 February 1992, Vol. 24, No. 5, p.281.
101. C. Berlet and W. K. Burke, 'Corporate fronts: inside the anti-environmental movement', *Greenpeace Magazine*, January/February/March 1992.
102. D. Barry, interview with author, 7 November 1994.
103. *O'Dwyer's PR Services*, 'Links with activist groups get results in environmental PR', Vol. 8, No. 2, February 1994, p.1.
104. T. Harding, 'Mocking the turtle: backlash against environmental movement', *New Statesman and Society*, 1993, Vol. 6, No. 271, p.45.
105. Alliance for America, *Democracy is Not A Spectator Sport*, Fly-In for Freedom, Washington DC, 17–21 September 1994.
106. *Eye Witness Report*, Alliance for America Conference, Missouri, 5–7 May 1992.
107. *Eye Witness Report*, 'Wise use leadership conference, John Ascuaga's nugget', Reno, Nevada, 5–7 June 1992.
108. T. Ramos, interview with author, 22 December 1994.
109. C. Deal, *The Greenpeace Guide to Anti-Environmental Organisations*, Odonian Press, 1993, p.78; *High Country News*, 1 July 1991.
110. D. Helvarg, *The War Against the Greens: The 'Wise-Use' Movement, the New Right, and Anti-Environmental Violence*, Sierra Club Books, San Francisco, 1994, p.453.
111. R. Ekey, 'Wise Use and the Greater Yellowstone Vision Document: lesson learned', *Let the People Judge: Wise Use and the Private Property Rights Movement*, (eds) J. Echeverria and R. Booth Eby, Island Press, 1995, p.344.
112. *A CLEAR View*, Western State Coalition Summit V, Conference Review, 17 April 1996, Vol. 3, No. 6.
113. BlueRibbon Coalition, 'BlueRibbon Support', *BlueRibbon Magazine*, January 1992, p.19; BlueRibbon Coalition, BlueRibbon Support', *BlueRibbon Magazine*, March 1992, p.19.
114. M. Knox, 'Meet the anti-greens: the "Wise Use" movement fronts for industry', *The Progressive*, October 1991.
115. *BlueRibbon Magazine*, List of 1994 Board of Directors, April 1994, p.3.
116. T. Williams, 'Greenscam', *Harrowsmith Country Life*, May/June 1992, p.32.
117. BlueRibbon Coalition, 'Special thanks to Senator Symms', *BlueRibbon Magazine*, January 1992, p.3.
118. C. Collins, 'Modern day "Minute Men",' May/June 1992, *BlueRibbon Magazine*, September 1994, p.3; T. Williams, 'Greenscam', *Harrowsmith Country Life*, May/June 1992 p.3; C. Collins, *Why All the Fuss?* no date; T. Egan, 'Fund-raisers tap anti-environmentalism', *New York Times*, 18 December 1991, p.18; S. O'Donnell, *Report of the Sixth Annual Wise Use Conference*, 15–17 July 1994.

119. T. Walker, 'Conservative network launches on cable TV', *BlueRibbon Magazine*, January 1994, p.15.
120. A. Cook, 'Hands across the ocean', *BlueRibbon Magazine*, April 1994, p.3.
121. S. Diamond, interview with author, 7 February 1995.
122. L. Hatfield and D. Waugh, 'Where think tanks get money', *San Francisco Examiner*, 26 May 1992; C. Deal, *The Greenpeace Guide to Anti-Environmental Organisations*, Odonian Press, 1993, pp.18–19.
123. M. Megalli and A. Friedman, Pacific Legal Foundation, *Masks of Deception: Corporate Front Groups in America*, Essential Information, December 1991, p.185; R. Bellant, *The Coors Connection: How Coors Family Philanthropy Undermines Democratic Pluralism*, South End Press, Boston, 1991.
124. Competitive Enterprise Institute, List of CEI Contributors, no date.
125. C. Deal, *The Greenpeace Guide to Anti-Environmental Organisations*, Odonian Press, 1993, p.79; S. Diamond, 'Free market environmentalism', *Z Magazine*, December 1991, p.54.
126. Heritage Foundation, List of Donors, 1992.
127. J. Karliner, 'The Bhopal tragedy: ten years after', *Global Pesticide Campaigner*, December 1994, Vol. 4, No. 4, p.1.
128. J. Nelson, 'Pulp and propaganda', *Canadian Forum*, July/August 1994, p.16.
129. P. Hawken, *The Ecology of Commerce: A Declaration of Sustainability*, HarperBusiness, 1994, pp.197–8.
130. Dr V. Shiva, *GATT and Free Trade: A Prescription for Environmental Apartheid*, evidence submitted to the Permanent People's Tribunal on Industrial and Environmental Hazards and Human Rights, 28 November–2 December 1994.
131. J. Nelson, *Sultans of Sleaze: Public Relations and the Media*, Between the Lines, 1989, p.102.
132. M. Iba, 'The same old trends', *Third World Guide 91/92*, New Internationalist, 1990, p.105.
133. *WHIN*, 'Sri Lanka: occupational diseases among workers in the Free Trade Zones', *Workers' Health International Newsletter*, Spring 1994, No. 39, p.11.
134. J. Nelson, *Sultans of Sleaze: Public Relations and the Media*, Between the Lines, 1989, pp. 97, 106.
135. Ibid., p.106.
136. R. Hindmarsh, 'Corporate biotechnological hegemony and the seed', *Pesticide Monitor*, July 1993, Vol. 2, No. 1, January 1993, p.1.
137. *Seedling*, 'Threats from the test tubes', December 1994, p.7; V. Shiva, 'Biotech hazards transferred south', *WHIN*, Summer 1994, p.4.
138. *Seedling*, 'Threats from the test tubes', December 1994, p.8.
139. *Pesticide News*, '10 years after Bhopal . . . ', December 1994, No. 26, p.8.
140. K. Bruno, *Greenpeace Submission to the Fourth Session of the London Tribunal on Industrial Hazards and Human Rights on the Occasion of the Tenth Anniversary of the Bhopal Tragedy*, November 1994.
141. B. Barclay and J. Steggall, 'Obsolete pesticide crisis', *Global Pesticide Campaigner*, February 1992, Vol. 2, No. 1, p.1.

142. *Swiss Review of World Affairs*, 'Bhopal: ten years later', 1 February 1995; S. Elsworth, *A Dictionary of the Environment*, Paladin, 1990, p.61; J. Karliner, 'The Bhopal tragedy: ten years after', *Global Pesticide Campaigner*, December 1994, Vol. 4, No. 4, pp.1, 6–8; C. Urquhart and S. J. Benbow, 'The Bhopal legacy lingers on', *Pesticide News*, December 1994, p.4.
143. J. Karliner, 'The Bhopal tragedy: ten years after', *Global Pesticide Campaigner*, December 1994, Vol. 4, No. 4, p.8.
144. C. Urquhart and S. J. Benbow, 'The Bhopal legacy lingers on', *Pesticide News*, December 1994, pp.5–6; D. Dembo, 'Bhopal, settlement or sellout?', *Global Pesticide Monitor*, 1989, Vol. 1, No. 1, p.4.
145. *IPS*, 'Lawyers from Malaysia to Mongolia', Eugene, Oregon, 8 March 1995.
146. K. Bruno, *Greenpeace Submission to the Fourth Session of the London Tribunal on Industrial Hazards and Human Rights on the Occasion of the Tenth Anniversary of the Bhopal Tragedy*, November 1994.
147. B. Dinham, 'Industrial hazards and human rights', *WHIN*, Autumn 1994, p.20.
148. *Corporate Crime Reporter*, 'Study reveals polluters have special access to U.S. trade negotiators', 1992, Vol. 6, No. 1, 6 January, p.7; quoting study by Public Citizen, *Trade Advisory Committees: Preferential Access for Polluters*; P. Hawken, *The Ecology of Commerce: A Declaration of Sustainability*, HarperBusiness, 1994, p.97.
149. Dr V. Shiva, *GATT and Free Trade: A Prescription for Environmental Apartheid*, evidence submitted to the Permanent People's Tribunal on Industrial and Environmental Hazards and Human Rights, 28 November–2 December 1994.
150. W. Morehouse, *Accountability, Regulation and Control of Multinational Corporations*, testimony prepared for the Permanent People's Tribunal on Industrial and Environmental Hazards and Human Rights, London, 28–30 November 1994, corrected draft.
151. T. Hines and C. Hines, *The New Protectionism: Protecting Against Free Trade*, Earthscan, 1993, p.52.
152. *Open Eye*, 'GATT: the end of the citizen?', 1995, No. 3, p.14.
153. Ibid.
154. H. Gleckman and R. Krut, *Towards a New International System to Regulate International Companies*, paper for the Permanent People's Tribunal on Industrial and Environmental Hazards and Human Rights, London, 28 November–2 December 1994.
155. Ibid.
156. Ibid.

4 GET ON THE GLOBAL GREEN

1. J. Bleifuss, 'Flack attack', *Utne Reader*, January/February 1994, p.77.
2. J. Nelson, *Sultans of Sleaze: Public Relations and the Media*, Between the Lines, 1989, pp.130-1.
3. K. Bruno, *Greenpeace Guide to Greenwash*, Greenpeace, 1992.

4. J. Lowe and H. Hanson, 'A look behind the advertising', *Earth Island Journal*, Winter 1990, pp.26–7.
5. J. Bruggers, 'Chevron drilling fought by environmentalists', *Contra Costa Times*, 27 November 1988; D. Baum, 'Indians oppose drilling on Montana land', *San Francisco Examiner*, 11 November 1990; J. Lowe and H. Hanson, 'A look behind the advertising', *Earth Island Journal*, Winter 1990, pp.26–7.
6. Council on Economic Priorities, *Mobil Oil Corporation: A Report on the Company's Environmental Policies and Practices*, Corporate Environmental Data Clearinghouse, The Council on Economic Priorities, 1991, New York, p.8.
7. *Chemical and Engineering News*, 'Mobil Chemical is sued for making false claims about its degradable plastics', 1990, 25 September, p.14; Council on Economic Priorities, *Mobil Oil Corporation: A Report on the Company's Environmental Policies and Practices*, Corporate Environmental Data Clearinghouse, The Council on Economic Priorities, 1991, New York, p.8.
8. P. Hawken, *The Ecology of Commerce: A Declaration of Sustainability*, HarperBusiness, 1994, p.130.
9. *Corporate Crime Reporter*, 'Environmentalists charge Earth Tech '90 companies are "Wolves dressed in sheep's clothing"', Vol. 4, No. 14, 9 April 1990, pp.7–8.
10. *Corporate Crime Reporter*, 'Public interest groups name worst advertising of the year', Vol. 5, No. 47, 9 December 1991, p.9.
11. *Corporate Crime Reporter*, 'Consumer groups blast companies for the most "misleading" ads of 1992', Vol. 5, No. 47, 9 December, p.9; J. Stauber and S. Rampton, *Toxic Sludge is Good for You: Lies Damn Lies and the Public Relations Industry*, Common Courage Press, 1995, p.38.
12. *Corporate Crime Reporter*, 'Nuclear power industry attempting to overcome public opposition to waste problems through high-powered public relations campaigns', Vol. 5, No. 47, 9 December 1991.
13. N. Verlander, 'Pressure Group Perspective – Friends of the Earth "Green Con of the Year Award"', *Greener Marketing: A Responsible Approach to Business*, Greenleaf, (ed.) Martin Chater, date unknown; *Earth Matters*, 'The Green Con Award 1989', Spring 1990, p.20; *Earth Matters*, 'Green Con of the Year 1990', Spring 1991.
14. N. Verlander, 'Green cons and eco labels', *Earth Matters*, 1992, month unknown.
15. P. Stevenson, 'Eight years later, industry advertising still violates FAO Code', *Global Pesticide Campaigner*, Vol. 3, No. 4, November 1993, pp.3–5.
16. Ibid.; *Pesticide Monitor*, 'Corporate greenwash', 1993, Vol. 2, No. 3, p.12.
17. P. Stevenson, 'Eight years later, industry advertising still violates FAO Code', *Global Pesticide Campaigner*, Vol. 3, No. 4, November 1993, p.5.
18. *PR Watch*, 'Plutonium is our friend', 1994, Fourth Quarter, p.11.
19. European Nuclear Society, International Workshop on Nuclear Public Information, Lucerne, Switzerland, 30 January–2 February 1994.
20. *Tomorrow*, 'Mitsubishi refuses to roll over', October–December 1993, p.33.
21. *Pesticide News*, 'Rhone-Poulenc: the image and the reality', September 1993, p.14.
22. K. Bruno, *Greenpeace Guide to Greenwash*, Greenpeace, 1992.

23. W. Morehouse, *Accountability, Regulation and Control of Multinational Corporations*, testimony prepared for the Permanent People's Tribunal on Industrial and Environmental Hazards and Human Rights, London, 28–30 November 1994, corrected draft.
24. J. Stauber, interview with author, 22 March 1995.
25. M. Dowie, *Losing Ground: American Environmentalism at the Close of the Twentieth Century*, MIT Press, 1995, p.85.
26. J. Bleifuss, 'Science in the private interest: hiring flacks to attack the facts', *PR Watch*, Vol. 2, No. 1, First Quarter, 1995, p.12.
27. K. McCauley, 'Going "Green" blossoms as PR trend of the 90's', *O'Dwyer's PR Services*, January 1991, p.1.
28. J. Stauber, interview with author, 22 March 1995.
29. Ibid.
30. *O'Dywer's Directory of PR Firms*, Jack Bonner Associates, 1993.
31. L. Buckingham, 'Advertising chief to get £8m a year', *The Guardian*, 3 September 1994.
32. *PR Newswire*, 22 March 1994.
33. *PR Watch*, 'The PR industry's top 15 greenwashers, based on O'Dwyer's Directory of PR Firms and interview with Hill and Knowlton', 1995, Vol. 2, No. 1, First Quarter, p.4.
34. J. Stauber and S. Rampton, *Toxic Sludge is Good for You: Lies Damn Lies and the Public Relations Industry*, Common Courage Press, 1995, p.125.
35. J. C. Stauber, 'Going . . . going . . . green!', *PR Watch*, 1994, Vol. 1, No. 3, Second Quarter, p.2; *O'Dwyer's PR Services Report*, 'Profiles of top environmental PR firms: E. Bruce Harrison', February 1994, p.30; C. Deal, *The Greenpeace Guide to Anti-Environmental Organisations*, Odonian Press, 1993, p.16.
36. J. C. Stauber, 'Going . . . going . . . green!', *PR Watch*, 1994, Vol. 1, No. 3, Second Quarter, p.2; E. B. Harrison, *Going Green: How to Communicate your Company's Environmental Commitment*, Business One Irwin, 1993, pp.xiv–xv.
37. F. Graham Jr, *Since Silent Spring*, Hamish Hamilton, London, 1970, p.48.
38. J. C. Stauber, 'Going . . . going . . . green!', *PR Watch*, 1994, Vol. 1, No. 3, Second Quarter, p.2.
39. J. Stauber, interview with author, 22 March 1995.
40. E. B. Harrison, *Going Green: How to Communicate your Company's Environmental Commitment*, Business One Irwin, 1993, pp.8,9,14.
41. Ibid., pp.xii, 15, 31, 44, 131.
42. Ibid., p.189.
43. *O'Dwyer's PR Services*, 'Links with activist groups get results in environmental PR', Vol. 8, No. 2, February 1994.
44. J. Bleifuss, 'Covering the Earth with "Green PR"', *PR Watch*, 1995, Vol. 2, No. 1, First Quarter, p.3.
45. *O'Dwyer's PR Services*, 'Links with activist groups get results in environmental PR', Vol. 8, No. 2, February 1994, p.20.
46. J. Stauber and S. Rampton, *Toxic Sludge is Good for You: Lies Damn Lies and the Public Relations Industry*, Common Courage Press, 1995, p.66.

47. *O'Dwyer's PR Services*, 'Links with activist groups get results in environmental PR', Vol. 8, No. 2, February 1994, p.22.
48. S. Bennett, R. Frierman and S. George, *Corporate Realities and Environmental Truths: Strategies for Leading your Business in the Environmental Era*, John Wiley and Sons, 1993, p.140.
49. J. Stauber and S. Rampton, *Toxic Sludge is Good for You: Lies Damn Lies and the Public Relations Industry*, Common Courage Press, 1995, p.138.
50. *PR Watch*, 'MBD's divide-and-conquer strategy to defeat activists', October–December 1993, Vol. 1, No. 1, p.5.
51. J. C. Stauber, 'Strange bedfellows at PR Conference on activism', *PR Watch*, 1994, Vol. 1, No. 2, First Quarter, p.2.
52. J. Stauber and S. Rampton, *Toxic Sludge is Good for You: Lies Damn Lies and the Public Relations Industry*, Common Courage Press, 1995, p.54.
53. D. Alters, 'Shhhhh . . . some firms are busy spying on the nation's social activists', *Boston Globe*, July, precise date unknown.
54. J. C. Stauber, 'Spies for hire – Mongoven, Biscoe and Duchin, Inc', *PR Watch*, Vol. 1, No. 1, October–December 1993.
55. *Threshold*, Welcome to the Terrordome: corporations begin spying on SEAC', January–February 1991, p.37.
56. Pagan International, *Greenpeace: A Special Report*, October 1985.
57. *Labor Notes*, 'Shell's "Neptune Strategy" aims at countering anti-apartheid boycott', January 1988, pp.1, 10.
58. J. Nelson, *Sultans of Sleaze: Public Relations and the Media*, Between the Lines, 1989, p.14.
59. Ibid.
60. *Campaign Magazine*, title unknown, 16 February 1990.
61. A. Garrett, 'Sponsorship: business buys its way to a greener image', *The Independent on Sunday*, 10 November 1991, p.26.
62. J. Stauber and S. Rampton, *Toxic Sludge is Good for You: Lies Damn Lies and the Public Relations Industry*, Common Courage Press, 1995, p.132.
63. S. Epstein, 'BST: the public health hazard', *The Ecologist*, Vol. 19, No. 5, September/October 1989, p.191.
64. J. Dillon, 'Poisoning the grass-roots: PR giant Burson-Marsteller thinks global, acts local', *CovertAction Quarterly*, Spring 1993, No. 44.
65. J. C. Stauber, 'Shut up and eat your "Frankenfoods"', *PR Watch*, 1994, First Quarter, Vol. 1, No. 2, p.8.
66. Internal B-M documents.
67. S. Epstein, 'BST: the public health hazard', *The Ecologist*, Vol. 19, No. 5, September/October 1989, p.193.
68. J. C. Stauber, 'Shut up and eat your "Frankenfoods"', *PR Watch*, 1994, First Quarter, Vol. 1, No. 2, p.8.
69. *PR Watch*, 'Sound bites back', 1994, Third Quarter, Vol. 1, No. 4, p.12.
70. *O'Dwyer's Directory of PR Firms*, 'Burson-Marsteller', Spring 1993; C. Nevin, 'From ideals to images', *The Independent on Sunday*, 17 April 1994, p.26; *PR Week*, 9 January 1992.

71. J. Motavalli, 'Dog soldier', *7 Days*, 4 October 1989.
72. J. Nelson, 'Pulp and propaganda', *Canadian Forum*, July/August 1994, p.16.
73. J. Bleifuss, 'Covering the Earth with "Green PR"', *PR Watch*, Vol. 2, No. 1, 1995, First Quarter, p.6.
74. Burson-Marsteller, *Counselling and Communications Worldwide*, no date, New York.
75. J. Nelson, *Sultans of Sleaze: Public Relations and the Media*, Between the Lines, 1989, p.22.
76. S. Anderson and J. L. Anderson, *Inside the League: The Shocking Exposé of How Terrorists, Nazis, and Latin American Death Squads Have Infiltrated the World Anti-Communist League*, Dodd, Mead and Company, New York, 1986, pp.204,272.
77. J. Nelson, *Sultans of Sleaze: Public Relations and the Media*, Between the Lines, 1989, pp.21–2,25.
78. Ibid., p.26.
79. S. Anderson and J. L. Anderson, *Inside the League: The Shocking Exposé of How Terrorists, Nazis, and Latin American Death Squads Have Infiltrated the World Anti-Communist League*, Dodd, Mead and Company, New York, 1986, p.274.
80. J. Nelson, 'Burson-Marsteller, Pax Trilateral, and the Brundtland Gang vs. the environment', *The New Catalyst*, Summer 1993, No. 26, p.2.
81. C. Urquhart and S. J. Benbow, 'The Bhopal legacy lingers on', *Pesticide News*, December 1994, p.6.
82. S. Elsworth, *A Dictionary of the Environment*, Paladin, 1990, p.441; J. Nelson, 'Burson-Marsteller, Pax Trilateral, and the Brundtland Gang vs. the environment', *The New Catalyst*, Summer 1993, No. 26, p.2.
83. J. Reed, 'Interview with the vampire: PR helps the PRI drain Mexico dry', *PR Watch*, 1994, Fourth Quarter, p.7.
84. *AFP*, 'Five Malaysian states found to be overlogging', Kuala Lumpur, 3 October 1994.
85. J. Stauber and S. Rampton, *Toxic Sludge is Good for You: Lies Damn Lies and the Public Relations Industry*, Common Courage Press, 1995, p.150.
86. J. Nelson, 'Burson-Marsteller, Pax Trilateral, and the Brundtland Gang vs. the environment', *The New Catalyst*, Summer 1993, No. 26, p.2.
87. S. Schmidheiny, *Changing Course: A Global Business Perspective on Development and the Environment*, with the Business Council for Sustainable Development, MIT Press, 1992, pp.xii–xvii.
88. J. Nelson, 'Burson-Marsteller, Pax Trilateral, and the Brundtland Gang vs. the environment', *The New Catalyst*, Summer 1993, No. 26, p.2.
89. *The Ecologist*, 'Power: the central issue', 1992, Vol. 22, No. 4, pp.163–4.
90. Dr H. Gleckman, *Transnational Corporations and Sustainable Development: Reflections from Inside the Debate*, 21 August 1992.
91. P. Hawken, *The Ecology of Commerce: A Declaration of Sustainability*, HarperBusiness, 1994, p.168.
92. S. Schmidheiny, *Changing Course: A Global Business Perspective on Development and the Environment*, with the Business Council for Sustainable Development, MIT Press, 1992, p.xi.
93. *Tomorrow*, Business Council for Sustainable Development, October–November 1993, No. 4.

94. *Corporate Crime Reporter*, 1991, 'Bush administration lobbies against code of conduct for transnational corporations', Vol. 5, No. 27, 8 July 1991, p.7.
95. *The Financial Times*, 'Green groups merge', 30 November 1994.
96. The Alliance for Beverage Cartons and the Environment, *Dioxins: 'No Need to Worry' Say Representatives of the Beverage Carton Industry*, 7 August 1990.
97. R. Nixon, 'Science for sale: truth to the highest bidder', *CovertAction*, Spring 1995, p.48.
98. Internal B-M documents.
99. Ibid.
100. C. Peter and H-J. Kursawa Stucke, *Deckmantel Okologie*, Knaur, 1995; City of Hamburg Press Release, date unknown.
101. *City of Hamburg Press Release*, date unknown.
102. *PR Week*, 28 October 1993.
103. C. Nevin, 'From ideals to images', *The Independent on Sunday*, 17 April 1994, p.26; *Marketing*, 20 May 1993, p.23.
104. Anonymous, Personal communication with author after private function with British Department of Environment officials, April 1995.
105. J. Dillon, 'Poisoning the grass-roots: PR giant Burson-Marsteller thinks global, acts local', *CovertAction Quarterly*, Spring 1993, No. 44.
106. *O'Dwyer's Directory of PR Firms*, 'Hill and Knowlton', Spring 1993; Record at the Library of Registration.
107. J. Steed, 'The power of PR', *The Toronto Star*, 1 November 1992.
108. S. B. Trento, *The Power House: Robert Keith Gray and the Selling of Access and Influence in Washington*, St. Martin's Press, 1992 in pictures after page 238.
109. Ibid., p.70.
110. Ibid.
111. W. Greider, *Who Will Tell the People?: The Betrayal of American Democracy*, Touchstone, Simon and Schuster, p.35.
112. J. Stauber and S. Rampton, *Toxic Sludge is Good for You: Lies Damn Lies and the Public Relations Industry*, Common Courage Press, 1995, p.75.
113. Hill and Knowlton, *Letter to Mr Ed Franklin*, Fidelity Tire Company, 1989.
114. J. Carlisle, 'Public relationships: Hill & Knowlton, Robert Gray, and the CIA', *CovertAction*, Spring 1993; J. Stauber and S. Rampton, *Toxic Sludge is Good for You: Lies Damn Lies and the Public Relations Industry*, Common Courage Press, 1995, p.150.
115. S. B. Trento, *The Power House: Robert Keith Gray and the Selling of Access and Influence in Washington*, St. Martin's Press, 1992, pp.200, 315.
116. Ibid., p.viii; J. Carlisle, 'Public Relationships: Hill & Knowlton, Robert Gray, and the CIA', *CovertAction*, Spring 1993, p.20.
117. J. Steed, 'The power of PR', *The Toronto Star*, 1 November 1992.
118. J. Carlisle, 'Public relationships: Hill & Knowlton, Robert Gray, and the CIA', *CovertAction*, Spring 1993, p.20; J. Steed, 'The power of PR', *The Toronto Star*, 1 November 1992.
119. *O'Dwyer's PR Services Report*, 'Sustainable PR holds key to good environmental image', February 1994, p.24.

120. D. Ip, 'Sustainable development: beyond today's fashion', *Sustainable Development*, 1993, Vol. 1, No. 2, p.4.
121. S. Lele, 'Sustainable development: a critical review', *World Development*, 1991, Vol. 19, No. 6, p.613.
122. *O'Dwyer's PR Services Report*, 'Sustainable PR holds key to good environmental image', February 1994, p.24.
123. Business Council for Sustainable Development, 'Free trade is essential for sustainable development', *Tomorrow Magazine*, April–June 1994.
124. L. C. Van Wachem, *The Three-Cornered Challenge: Energy, Environment and Population*, The Cadman Memorial Lecture, London, 14 September 1992.
125. R. Gray and J. Bebbington, *Sustainable Development and Accounting: Incentives and Disincentives for the Adoption of Sustainability by Transnational Corporations*, a research investigation by United Nations Conference on Trade and Development and the Centre for Social and Environmental Accounting Research, University of Dundee, 1994, p.3.
126. J. Nelson, *Sultans of Sleaze: Public Relations and the Media*, Between the Lines, 1989, p.147.
127. E. B. Harrison, *Going Green: How to Communicate your Company's Environmental Commitment*, Business One Irwin, 1993, p.9.
128. M. Dowie, *Losing Ground: American Environmentalism at the Close of the Twentieth Century*, Massachusetts Institute of Technology, 1995, p.235.
129. P. Hawken, *The Ecology of Commerce: A Declaration of Sustainability*, HarperBusiness, 1994, pp.30–1.
130. R. Gray and J. Bebbington, *Sustainable Development and Accounting: Incentives and Disincentives for the Adoption of Sustainability by Transnational Corporations*, a research investigation by United Nations Conference on Trade and Development and the Centre for Social and Environmental Accounting Research, University of Dundee, 1994, p.7 quoting the Body Shop.
131. C. Flavin and J. Young, 'Shaping the next industrial revolution: environmentally sustainable development', *USA Today*, March 1994, Vol. 122, No. 2586, p.78.

5 THE PARADIGM SHIFT

1. M. Hagler, 'Contrarians revisited', *Tomorrow Magazine*, January–March 1995, p.54.
2. F. Graham Jr, *Since Silent Spring*, Hamish Hamilton, London, 1970, p.48.
3. *The Ecologist*, 'The threat of environmentalism', Vol. 22, No. 4, July 1992, p.162; J. Bleifuss, 'Journalist, watch thyself: keeping tabs on the messengers', *PR Watch*, 1995, First Quarter, Vol. 2, No. 1, p.10; J.Bleifuss, 'Covering the Earth with "Green PR"', *PR Watch*, 1995, Vol. 2, No. 1, First Quarter, p.5.
4. C. Berlet, 'Hunting the "Green Menace"', *The Humanist*, July/August 1991, p.26.
5. C. Berlet, 'Re-framing dissent as criminal subversion', *CovertAction*, Summer 1992, No. 41, p.40.

6. Ibid.
7. Ibid., p.35.
8. R. Arnold, *Ecology Wars*, remarks given at the Maine Conservation Rights Institute, Second Annual Congress, 20 April 1992.
9. W. P. Pendley, *It Takes A Hero: the Grassroots Battle Against Environmental Oppression*', Free Enterprise Press, 1994, pp.274–316.
10. Ibid., p.vii.
11. A. M. Gottlieb (ed.), *The Wise Use Agenda: The Citizen's Policy Guide to Environmental Resource Issues – A Task Force to the Bush Administration by the Wise Use Movement*, The Free Enterprise Press, 1989, p.xx.
12. T. Egan, 'Fund-raisers tap anti-environmentalism', *The New York Times*, 19 December 1991, p.A18.
13. K. Long, 'Washington man, his group set goal of ending environmental movement', *Oregonian*, 10 December 1991.
14. M. Hager, 'Enter the contrarians', *Tomorrow Magazine*, October–December 1993, No. 4, p.10.
15. M. Knox, 'Meet the anti-greens: the 'Wise Use' movement fronts for industry', *The Progressive*, October 1991, p.22.
16. J. Hamburg, 'The Lone Ranger', *California Magazine*, November 1990, p.92.
17. *Western Horizons*, 'Wise Users renounce balance and "middle ground", align themselves with extreme right wing', September 1993, Vol. 1, No. 3, p.4.
18. Ibid.
19. S. O'Donnell, *Report of the Sixth Annual Wise Use Conference*, 15–7 July 1994.
20. D. Helvarg, *The War Against the Greens: The 'Wise-Use' Movement, the New Right, and Anti-Environmental Violence*, Sierra Club Books, San Francisco, 1994, p.235.
21. Ibid., p.252.
22. T. R. Mader, *The Enemy Within*, Abundant Wildlife Society of North America, 1991, pp.3,11.
23. M. Donnelly, 'Dominion theology and the Wise Use movement', *Wild Oregon*, the journal of the Oregon Natural Resource Council, date unknown.
24. S. O'Donnell, *Report of the Sixth Annual Wise Use Conference*, 15–17 July 1994.
25. S. L. Udall and W. K. Olson, 'Me first!ers', *The Phoenix Gazette*, date unknown.
26. M. Coffman, *Fly-In For Freedom*, Alliance for America, Washington DC, 17–21 September 1994; P. Bradburn, *Fly-In For Freedom*, Alliance for America, Washington DC, 17–21 September 1994.
27. R. Mann, 'Rally draws thousands', *Humboldt Beacon*, 7 June 1990.
28. M. Knox, 'Meet the anti-greens: the 'Wise Use' movement fronts for industry', *The Progressive*, October 1991, p.21.
29. J. R. Luoma, 'Backlash', *The National Times*, January 1993.
30. M. Coffman, *Fly-In For Freedom*, Alliance for America, Washington DC, 17–21 September 1994.
31. W. Kramer, 'The freedom fighters are back battling eco-fanaticism!', *BlueRibbon Magazine*, October 1992.

32. M. M. Hecht, 'Dixy Lee Ray: in memoriam', *21st Century Science and Technology*, Spring 1994, p.28.
33. R. Kazis and R. L. Grossman, *Fear At Work: Job Blackmail, Labor and the Environment*, The Pilgrim Press, 1982, pp.65–6.
34. Ibid.
35. D. Howard, *Fly-In For Freedom*, Alliance for America, Washington DC, 17–21 September 1994.
36. S. O'Donnell, *Report of the Sixth Annual Wise Use Conference*, 15–17 July 1994.
37. D. Helvarg, *The War Against the Greens: The 'Wise-Use' Movement, the New Right, and Anti-Environmental Violence*, Sierra Club Books, San Francisco, 1994, p.161.
38. S. O'Donnell, *Report of the Sixth Annual Wise Use Conference*, 15–17 July 1994.
39. *Western Horizons*, 'Wise Users renounce balance & "middle ground", align themselves with extreme right wing', September 1993, Vol. 1, No. 3, p.8.
40. J. Krakauer, 'Brown fellas', *Outside*, December 1991, p.70.
41. D. Howard, *Fly-In For Freedom*, Alliance for America, Washington DC, 17–21 September 1994.
42. W. P. Pendley, *Fly-In For Freedom*, Alliance for America, Washington DC, 17–21 September 1994.
43. *American Timberman and Trucker*, Yellow Ribbon Coalition, June 1989, p.16.
44. M. Donnelly, 'Dominion theology and the Wise Use movement', *Wild Oregon*, the journal of the Oregon Natural Resource Council, date unknown.
45. M. L. Knox, 'The Wise Use guys', *Buzzworm: The Environmental Journal*, November/ December 1990, p.35.
46. *Fly-In For Freedom*, Alliance for America, Washington DC, 17–21 September 1994.
47. K. Long, 'Washington man, his group set goal of ending environmental movement', *Oregonian*, 10 December 1991.
48. K. O'Callaghan, 'Whose agenda for America?' *Audubon*, 1992, p.84.
49. S. O'Donnell, *Report of the sixth Wise Use leadership conference*, Reno, Nevada, 15–17 July 1994.
50. A. Icenogle, 'Rushville man coordinates fight against environmentalists', *Rushville Times*, 1 April 1992.
51. D. King, *Lyndon LaRouche and the New American Fascism*, Doubleday, 1989, p.236.
52. G. Ball, 'Wise Use nuts & bolts, *Mendocino Environmental Center Newsletter*, Summer/ Fall 1992, Issue 12, p.19.
53. *The Litigator*, 'MSLF confronts terrorists', Mountain States Legal Foundation, Summer 1990, p.1.
54. M. L. Knox, 'The Wise Use guys', *Buzzworm: The Environmental Journal*, November/ December 1990, p.33.
55. *EIR Talks*, 'Interviewer: Mel Klenetsky', 20 April 1994.
56. S. O'Donnell, *Report of the Sixth Annual Wise Use Conference*, 15–17 July 1994.
57. Ibid.
58. *Executive Intelligence Review*, 'Save the planet's humans: lift the ban on DDT', 19 June 1992, Vol.19, No. 25.

59. M. M. Hecht, 'Population control lobby banned DDT to kill more people', *Executive Intelligence Review*, 19 June 1992, Vol.19, No. 25, p.38.
60. Ibid.
61. T. H. Jukes, 'Silent Spring and the betrayal of environmentalism', *21st Century Science and Technology*, Fall 1994, p.54.
62. *21st Century Science and Technology*, 'The world needs nuclear energy', March–April 1989, p.2.
63. R. A. Maduro, 'New evidence shows "ozone depletion" just a scare', *21st Century Science and Technology*, Winter 1990, pp.38–41.
64. R. A. Maduro, 'The greenhouse effect is a fraud', *21st Century Science and Technology*, March–April 1989, p.14.
65. *21st Century Science and Technology*, advert for EIR: 'The "greenhouse effect" is a hoax', Winter 1990, p.7.
66. J. Bleifuss, 'Science in the private interest: hiring flacks to attack the facts', *PR Watch*, 1995, First Quarter, Vol. 2, No. 1, p.11.
67. ABC News, *Nightline Transcript*, 24 February 1994.
68. W. Nixon, 'Environmental overkill', *Earth Action Network*, December 1993, Vol. 4, No. 6, p.54.
69. S. Leiper, 'Trashing environmentalism: the story of Dixy Lee Ray', *Propaganda Review*, Spring 1994, p.11.
70. *The Litigator*, 'U.S. senator and governor join MSLF board', date unknown.
71. W. P. Pendley, *It Takes A Hero: The Grassroots Battle Against Environmental Oppression*, Free Enterprise Press, 1994, pp.95–7.
72. S. Leiper, 'Trashing environmentalism: the story of Dixy Lee Ray', *Propaganda Review*, Spring 1994, p.12.
73. Petr. Beckmann's obituary on the Fort Freedom BBS; B. Lyons, personal communication with author, 20 May 1995.
74. D. L. Ray and L. Guzzo, *Trashing the Planet: How Science Can Help Us Deal With Acid Rain, Depletion of the Ozone, and Nuclear Waste (Among Other Things)*, HarperPerennial, 1992, pp.123, 126.
75. Ukrainian figures announced on the ninth anniversary of the Chernobyl disaster were higher than previous estimates.
76. D. L. Ray and L. Guzzo, *Trashing the Planet: How Science Can Help Us Deal With Acid Rain, Depletion of the Ozone, and Nuclear Waste (Among Other Things)*, HarperPerennial, 1992, pp.xi, 5.
77. Ibid., p.7.
78. S. Leiper, 'Trashing environmentalism: the story of Dixy Lee Ray', *Propaganda Review*, Spring 1994, p.13.
79. D. Helvarg, *The War Against the Greens: The 'Wise-Use' Movement, the New Right, and Anti-Environmental Violence*, Sierra Club Books, San Francisco, 1994, p.228.
80. M. M. Hecht, 'Dixy Lee Ray: in memoriam', *21st Century Science and Technology*, Spring 1994, p.28.
81. G. Taubes, 'The ozone backlash', *Science*, Vol. 260, 11 June 1993, p.1582.

82. J. Margolis, 'Facts-schmacts: the twisting logic of pseudo-science', *Chicago Tribune*, 16 November 1993, p.23.
83. *Global Environmental Change Report*, 'Gore tries to discredit sceptics, but strategy backfires', Cutter Information Corp, 11 March 1994.
84. Quotes taken from *Rational Readings on Environmental Concerns*, (ed.) Jay H. Lehr, Van Nostrand Reinhold, 1992, pp.125, 140, 244, 247, 279, 291, 303, 326, 343, 369, 387, 805, 819, 822, 834.
85. E. C. Krug, 'The great acid rain flimflam', *Rational Readings on Environmental Concerns*, (ed.) Jay H. Lehr, Van Nostrand Reinhold, 1992, p.42.
86. B. Amers and L. Swirsky Gold, 'Environmental pollution and cancer: some misconceptions', *Rational Readings on Environmental Concerns*, (ed.) Jay H. Lehr, Van Nostrand Reinhold, 1992, p.162.
87. N. P. Robinson Sirkin and G. Sirkin, 'Taking the die out of dioxin', *Rational Readings on Environmental Concerns*, (ed.) Jay H. Lehr, Van Nostrand Reinhold, 1992, p.247.
88. E. M. Whelan, 'Deadly Dioxin?', *Rational Readings on Environmental Concerns*, (ed.) Jay H. Lehr, Van Nostrand Reinhold, 1992, p.225.
89. H. W. Ellsaesser, 'The credibility gap between science and the environment', *Rational Readings on Environmental Concerns*, (ed.) Jay H. Lehr, Van Nostrand Reinhold, 1992, p.695.
90. E. C. Krug, 'Just maybe . . . the sky isn't falling', *Rational Readings on Environmental Concerns*, (ed.) Jay H. Lehr, Van Nostrand Reinhold, 1992, pp.355, 356.
91. S. F. Singer, 'Global climate change: facts and fiction', *Rational Readings on Environmental Concerns*, (ed.) Jay H. Lehr, Van Nostrand Reinhold, 1992, p.402.
92. H. W. Ellsaesser, 'The credibility gap between science and the environment', *Rational Readings on Environmental Concerns*, (ed.) Jay H. Lehr, Van Nostrand Reinhold, 1992, p.695.
93. F. Singer, 'Scientific shallows of whale sanctuary idea', *The Washington Times*, 5 May 1994; *PR Newswire*, 'Proposed acid rain controls will cost consumers billions', 23 April 1990; The National Science Foundation, *A Re-Examination of Costs and Benefits of Automobile Emission Control Strategies*, 22 March 1976.
94. F. Singer, 'The latest scare: energy policy by press release', *Eco-Logic*, May 1992, p.6.
95. *PR Newswire*, 'Proposed acid rain controls will cost consumers billions', 23 April 1990.
96. E. Krug, 'Save the planet, sacrifice the people: the Environmental Party's bid for power', *Imprimis*, 1991, Vol. 20, No. 7.
97, Dr E. Krug, 'Greenhouse catapults environmentalists' agenda', *Citizen Outlook*, Committee For A Constructive Tomorrow, July/August 1991, Vol. 6, No. 2.
98. C. Deal, *The Greenpeace Guide to Anti-Environmental Organisations*, Odonian Press, Berkeley, April 1993, p.48.
99. George C. Marshall Institute, *Scientific Perspectives on the Greenhouse Problem*, Washington DC, December 1989.
100. R. J. Samuelson, 'And now the good news about the environment', *WP*, 5 April 1995.
101. R. S. Lindzen, 'The politics of global warming', *Eco-Logic*, May 1992, pp.16–18.

102. *Energy Report*, 'Near-record temperatures for 1994 consistent with warming, officials say', 23 January 1995; B. Ruben, 'Back talk: environmental problems are being misrepresented in the media', *Environmental Action Magazine*, January 1994.
103. H. W. Ellsaesser, 'The great greenhouse debate', *Rational Readings on Environmental Concerns*, (ed.) Jay H. Lehr, Van Nostrand Reinhold, 1992, p.404; S. Idso, 'Carbon dioxide and global change: end of nature or rebirth of the biosphere?', *Rational Readings on Environmental Concerns*, (ed.) Jay H. Lehr, Van Nostrand Reinhold, 1992, p.415.
104. Consumer Alert, *Briefing to be Held on 'Global Warming: Dissecting the Theory'*, 20 April 1990; D. Helvarg, *The War Against the Greens: The 'Wise-Use' Movement, the New Right, and Anti-Environmental Violence*, Sierra Club Books, San Francisco, 1994, p.211.
105. Newswire, *Global Climate Coalition Press Release*, 18 February 1992.
106. ICE internal packet, *Strategies*, p.3.
107. R.L. Lawson, *Memo to 'Coal Producer Members'*, 15 May 1991 (companies pledged support as of 15 May 1991) are AMAX Coal Industries, Anker Energy, ARCO Coal Company, Berwind Natural Resources Corp, Cyprus Coal Company, Drummond Company Inc., Island Creek Coal Company, Jim Walter Resources, Ohio Valley Coal Company, Peabody Holding Company, Pittsburgh and Midway Coal Mining, Pittston Coal Management Company, Stanley Industries, United Company and the Zeigler Coal Holding Company.
108. *Coal & Synthfuels Technology*, 'If there is global warming, it could be good, some scientists say', No. 12, 16 December 1994, pp.6–7 from *Global Warming Network On-line Today*; Competitive Enterprise Institute, *List of CEI Contributors*, no date.
109. ABC News, *Nightline Transcript*, 24 February 1994.
110. *Coal & Synthfuels Technology*, 'If there is global warming, it could be good, some scientists say', No. 12, 16 December 1994, pp.6–7 from *Global Warming Network On-line Today*; B. Ruben, 'Back talk; environmental problems are being misrepresented in the media', *Environmental Action Magazine*, January 1994.
111. ABC News, *Nightline Transcript*, 24 February 1994; C. Deal, *The Greenpeace Guide to Anti-Environmental Organisations*, Odonian Press, 1993, p.89.
112. D. Helvarg, *The War Against the Greens: The 'Wise-Use' Movement, the New Right, and Anti-Environmental Violence*, Sierra Club Books, San Francisco, 1994, p.239.
113. M. Hager, 'Enter the contrarians', *Tomorrow Magazine*, October - December 1993, No. 4, p.11.
114. H. Kurtz, 'Dr Whelan's media operations', *CJR*, March/April 1990, p.44; J. Bleifuss, 'Science in the private interest: hiring flacks to attack the facts', *PR Watch*, 1995, p.11.
115. W. P. Pendley, *It Takes A Hero: The Grassroots Battle Against Environmental Oppression*, Free Enterprise Press, 1994, p.13.
116. M. Megalli and A. Friedman, Pacific Legal Foundations, *Masks of Deception: Corporate Front Groups in America*, Essential Information, December 1991; H. Kurtz, Dr Whelan's media operations', *CJR*, March/April 1990, p.43. Funding comes from Adolph Coors Foundation, ALCOA Foundation, American Cyanamid Company, Amoco Foundation, Ashland Oil Foundation, Burger King Corporation, Carnation Company, Chevron,

Ciba-Geigy, Con Edison, Coca-Cola, Dow Chemical, Du Pont, Exxon, Ford, ICI America, General Electric, General Mills, General Motors, Johnson & Johnson, John M. Olin Foundation, Kellogg Company, Mobil Foundation, Monsanto Fund, NutraSweet Company, Pepsi-Cola, Pfizer Inc, Proctor & Gamble, the Sarah Scaife Foundation, Shell Oil, the Warner-Lambert Foundation and Union Carbide.

117. H.Kurtz, 'Dr Whelan's media operations', *CJR*, March/April 1990, p.45.
118. *Rolling Stone*, 'Hall of shame: who's the foulest of them all?', 3 May 1990.
119. M. Megalli and A. Friedman, Pacific Legal Foundation, *Masks of Deception: Corporate Front Groups in America*, Essential Information, December 1991.
120. H. Kurtz, 'Dr Whelan's media operations', *CJR*, March/April 1990, p.47.
121. J. C. Stauber, 'Burning books before they're printed', *PR Watch*, 1994, First Quarter, Vol. 1, No. 5, pp.3–4.
122. Ibid., p.2.
123. M. Hager, 'Enter the contrarians', *Tomorrow Magazine*, October–December 1993, No. 4, p.13.
124. C. Berlet, interview with author, 4 November 1994.
125. R. H. Limbaugh III, *The Way Things Ought To Be*, Pocket Books, Simon and Schuster, 1992, p.xiii.
126. Ibid., p.155.
127. Ibid., p.156.
128. World Meteorological Organisation, *Scientific Assessment of Ozone Depletion: 1994*, in Montreal Protocol on substances that deplete the ozone layer, UNEP, 1994, p.xxix.
129. R. H. Limbaugh III, *The Way Things Ought To Be*, Pocket Books, Simon and Schuster, 1992, p.162.
130. Ibid., pp.301,161,167.
131. R. H. Limbaugh III, *See, I Told You So*, Pocket Books, Simon and Schuster, 1993, pp.171–2, 177.
132. L. Haimson, M. Oppenheimer and D. Wilcove, *The Way Things Really Are: Debunking Rush Limbaugh on the Environment*, Environmental Defense Fund, 21 December 1994.
133. *EXTRA!*, 'The way things aren't: Rush Limbaugh debates reality', July/August 1994, pp.10–17.
134. D. Helvarg, *The War Against the Greens: The 'Wise-Use' Movement, the New Right, and Anti-Environmental Violence*, Sierra Club Books, San Francisco, 1994, p.284.
135. J. Passacantando and A. Carothers, 'Crisis? What crisis? The ozone backlash', *The Ecologist*, 1995, Vol. 25, No. 1, pp.5–7.
136. G. Taubes, 'The ozone backlash', *Science*, Vol. 260, 11 June 1993, p.1580.
137. ABC News, *Nightline Transcript*, 24 February 1994.
138. D. Helvarg, *The War Against the Greens: The 'Wise-Use' Movement, the New Right, and Anti-Environmental Violence*, Sierra Club Books, San Francisco, 1994, p.288.
139. Ibid., p.290.
140. M. Hager, 'Enter the contrarians', *Tomorrow Magazine*, October–December 1993, No. 4, p.19.
141. J. Mathews, 'The feelgood future', *The Guardian*, 26 July 1995, Society, p.5.
142. R. Braile, 'What the hell are we fighting for?', *Garbage*, Fall 1994, p.28.

143. *Rachel's Environment and Health Weekly*, 'The state of humanity', No. 485, 14 March 1996.
144. A. Eilly Dowd, 'Environmentalists are on the run', *Fortune*, 19 September 1994, p.96.
145. Ibid., p.96.
146. J. Passacantando and A. Carrothers, 'The ozone backlash', *The Ecologist*, 1995, Vol. 25, No.1, p.5; F. Pearce, 'Fiddling while the Earth warms', *New Scientist*, 1995, p.14.
147. B. Bolin, *Report to the Eleventh Session of the Intergovernmental Negotiating Committee for a Framework Convention on Climate Change (INC/FCCC)*, Chair of the Intergovernmental Panel on Climate Change, New York, 6 February 1995.
148. ABC News, *Nightline Transcript*, 24 February 1994.
149. World Meteorological Organisation, *Scientific Assessment of Ozone Depletion: 1994*, in Montreal Protocol on substances that deplete the ozone layer, UNEP, 1994, p.xxv.
150. NASA News, *NASA's UARS Confirms CFCs Caused Antarctic Ozone Hole*, National Aeronautics and Space Administration, Washington, 19 December 1994.
151. R. Evans, 'Ozone destruction at record in September, UN says', *Reuter*, Geneva, 4 October 1994.
152. Department of the Environment, *International Research Shows Large Ozone Reduction Over Arctic*, Environment News Release, 30 March 1995.
153. *Reuter*, 'Antarctic ozone hole gets worse, scientists say', 3 August 1995.
154. *Europe Environment*, 'Sharp thinning of the ozone layer in 1994', 13 June 1995.
155. S. Nebehay, 'Record ozone depletion reported over northern zone', *Reuter*, 12 March 1996.
156. UN, *Strengthening of International Cooperation and Coordination of Efforts to Study, Mitigate and Minimize the Consequences of the Chernobyl Disaster*, Report of the Secretary General, 1995.
157. *BNA*, 'IPCC Working Group Report documents "Discernable Human Influence" on climate', 4 December 1995.
158. C. Berlet, 'Hunting the "Green Menace"', *The Humanist*, July/August 1991, p.31.
159. Ibid.
160. Ketchum, *Crisis Management Plan for the Clorox Company*, 1991, pp.3,18; Greenpeace, *Clorox Company's Public Relations "Crisis Management Plan" Leaked to Greenpeace*, press release, 10 May 1991.
161. Fake press release claiming to be from Northern Californian Earth First!, although it does not exist.
162. *Arbitration Document between Association of Western Pulp and Paper Workers, and Louisiana-Pacific Corporation Western Division*, 5 September 1991.
163. J. Franklin, 'First they kill your dog', Muckraker, Fall 1992, pp.7–9.
164. R. Shaw, 'Bucking the tide of ecological correctness', *Insight*, 2 May 1994, p.18; *San Francisco Chronicle*, 'Sahara Club targets "eco-freaks"', 14 December 1990; Sahara Club, *Newsletter*, no date, No. 10, pp.2, 4.
165. Sahara Club, *Newsletter Number 24*, 6 July 1994.
166. D. Kuipers, interview with Rick Sieman, Sahara Club, 12 February 1992.
167. Sahara Club, *Newsletter*, No. 7, no date; Sahara Club, *Newsletter*, No. 8, Winter 1991.

168. Sahara Club, *Newsletter*, no date.
169. Ibid.
170. M. Dowie, *Losing Ground: American Environmentalism at the Close of the Twentieth Century*, MIT Press, 1995, p.210.
171. B. Lyons, interview with author, 15 July 1994.
172. J. Bari, *Timber Wars*, Common Courage Press, 1994, p.266.
173. Ibid., pp.264–700.
174. B. Clausen, *Walking on the Edge: How I Infiltrated Earth First*, Washington Contract Loggers Association, 1994.
175. J. Margolis, 'Fringe groups find niches in colourful political spectrum', *Chicago Tribune*, 3 March 1994; B. Clausen, *Ecoterrorism Watch*, November 1994, p.14.
176. J. Todd Foster, 'Fighting ecoterrorism', *Spokesman Review*, 10 April 1994.
177. K. Olsen, 'Activists say Earth First! meeting designed to incite hatred', *Moscow-Pullman Daily News*, 4 April 1994.
178. KUOI News, *Unofficial Transcript of Interview with Barry Clausen*, 11 April 1994.
179. Notes of a meeting where Barry Clausen spoke at Republic High School, 16 March 1995.
180. BCTV News Hour, *Transcript*, 27 April 1995.
181. *The Associated Press*, 'Unabomber – "hit list" link-eyed', 3 August 1995.
182. ABC News, *Evening News*, 3 August 1995.
183. *Earth First Journal*, open letter to ABC, 7 April 1996.
184. P. Terzian and R. Emmett Tyrrell, 'Faulty connections . . . and cliches', *The Washington Times*, 12 April 1996.
185. *EnviroScan*, 'The terror and violence of environmentalists', Public Relations Management Ltd, 1995, Issue No. 145.
186. CDFE, 'Ecoterror Response Network established', *The Private Sector*, *The Wise Use Men*, Spring 1996, p.2.
187. Sahara Club, *Newsletter*, No. 8, Winter 1991; R. Arnold and A. Gottlieb, *Trashing the Economy: How Runaway Environmentalism is Wrecking America*, Free Enterprise Press, 1993, p.179.
188. S. Allis, N. Burleigh, J. Carney and D. Waller, 'A moment of silence', *Time*, 8 May 1995, p.46.
189. K. Durbin, 'Environmental terrorism in Washington State', *Seattle Weekly*, 11 January 1995.
190. C. Berlet, *Clinic Violence, The Religious Right, Scapegoating, Armed Militias, and the Freemason Conspiracy Theory*, Political Research Associates, 19 January 1995.
191. H. Halpern, 'How hate speech leads readily to violence, *New York Times*, 2 May 1995.
192. Ibid.

6 THE PRICE OF SILENCE

1. J. Franklin, 'First they kill your dog', *Muckraker*, Fall 1992.
2. *San Francisco Examiner*, 'Earth First! renounces tree spiking', 13 April 1990.
3. J. Bari, 'TV mystery: who bought KQED and Steve Talbot?', *SF Weekly*, Vol. X, No. 14, 5 June 1991.
4. E. Diringer, 'Environmental group says it won't spike trees', *SF Chronicle*, 11 April 1990.
5. A. Cockburn, 'Redwood murder plot', *San Francisco Examiner*, 6 June 1990, pp.A–17.
6. Committee for the Death of Earth First, letter to Betty Ball, no date.
7. W. Churchill, 'The FBI targets Judi Bari', *CovertAction Quarterly*, Winter 1993–4, No. 47.
8. R. Johnson, 'Activists bombed, busted', *The Mendocino Country Environmentalist*, 29 May–15 June 1990, p.1.
9. A. Furillo and J. Kay, 'Victim held for questioning on car bombing', *San Francisco Examiner*, 25 May 1990.
10. W. Churchill, 'The FBI targets Judi Bari', *CovertAction Quarterly*, Winter 1993–4, No. 47.
11. Ibid.
12. A. Furillo and J. Kay, 'Victim held for questioning on car bombing', *San Francisco Examiner*, 25 May 1990.
13. M. Geniella, 'Logging protesters claim pattern of violence', *Press Democrat*, 28 March 1990; A. Cockburn, 'Redwood murder plot', *San Francisco Examiner*, 6 June 1990, pp.A–17.
14. J. Bari, 'How to create a climate of violence', *The Press Democrat*, 20 November 1992.
15. D. W. Galitz, Letter to Kevin Eckery, Timber Association of California, 27 April 1990.
16. J. S. Zer, Judi Bari, *High Times*, June 1991, p.14.
17. R. Johnson, 'Activists bombed, busted', *The Mendocino Country Environmentalist*, 29 May–15 June 1990, p.1.
18. J. Bari, 'The bombing story, Part 2: FBI lies', *Earth First!*, 1994, Beltane, Vol. XIV, No. 5, p.14.
19. M. Geniella, 'FBI bomb drills preceded Bari blast', *Press Democrat*, 30 September 1994; S. Simac, 'FBI's nasty war against Earth First revealed', *Coastal Post*, Marin County's News Monthly, Vol. 20, No. 4, 1 April 1995, p.1.
20. W. Churchill, 'The FBI targets Judi Bari', *CovertAction Quarterly*, Winter 1993–4, No. 47; Affidavit for Search Warrant by Sergeant Chenault, Oakland Police Department.
21. M. Sitterud, *Follow-Up Investigation Report*, Oakland Police Department, No. 90–57171.
22. Ibid.
23. W. Churchill, 'The FBI targets Judi Bari', *CovertAction Quarterly*, Winter 1993–4, No. 47.

24. Deposition of Oakland Police Sergeant Sitterud, questioned by Cunningham.
25. W. Churchill, 'The FBI targets Judi Bari', *CovertAction Quarterly*, Winter 1993–4, No. 47.
26. Photographs released by the FBI show that the epicentre of the damage to Bari's car was under her seat and not behind it.
27. FBI Files; M. Taylor, 'Bomb materials linked to victim', *San Francisco Chronicle*, 6 July 1990; J. Bari, 'The bombing story, Part 2: FBI lies', *Earth First!*, 1994, Beltane, Vol. XIV, No. 5, p.15.
28. S. Simac, 'FBI's nasty war against Earth First revealed', *Coastal Post*, Marin County's News Monthly, Vol. 20, No. 4, 1 April 1995, p.1.
29. J. Bari, 'The bombing story, Part 2: FBI lies', *Earth First!*, 1994, Beltane, Vol. XIV, No. 5, p.14.
30. W. Churchill, 'The FBI targets Judi Bari', *CovertAction Quarterly*, Winter 1993–4, No. 47.
31. J. Bari, personal communication with author, March 1996.
32. J. Bari, notes on FBI documents released.
33. FBI documents.
34. Report of the FBI Laboratory, FBI, 14 June 1990; M. Taylor, 'Bomb materials linked to victim', *San Francisco Chronicle*, 6 July 1990.
35. Affidavit for Search Warrant by Sergeant Chenault, Oakland Police Department.
36. J. Bari, 'The bombing story, Part 2: FBI lies', *Earth First!*, 1994, Beltane, Vol. XIV, No. 5, p.15.
37. J.Bari, *Analysis of FBI Files*, no date.
38. FBI documents.
39. W. Churchill, 'The FBI targets Judi Bari', *CovertAction Quarterly*, Winter 1993–4, No. 47.
40. W. Churchill, 'The FBI targets Judi Bari', *CovertAction Quarterly*, Winter 1993–4, No. 47; S. Simac, 'FBI's nasty war against Earth First revealed', *Coastal Post*, Marin County's News Monthly, Vol. 20, No. 4, 1 April, p.12.
41. C. Berlet, 'The hunt for Red Menace', *CovertAction*, 1989, No. 31, p.4.
42. *San Francisco Bay Guardian*, 'Earth First!, terrorism and the FBI', 30 May 1990; W. Churchill, 'The FBI targets Judi Bari', *CovertAction Quarterly*, Winter 1993–4, No. 47; P. Rothberg, COINTELPRO, *Lies of Our Times*, September 1993, p.5; C. Berlet, 'Hunting the "Green Menace"', *The Humanist*, July/August 1991, p.28.
43. J. Ridgeway and B. Clifford, *Village Voice*, 25 July 1989; D. Helvarg, *The War Against the Greens: The 'Wise-Use' Movement, the New Right, and Anti-Environmental Violence*, Sierra Club Books, San Francisco, 1994, pp.392–3.
44. E. Volante, 'FBI tracked Abbey for 20-year span', *Arizona Daily Star*, 25 June 1989.
45. S. Burkholder, 'Red squads on the prowl', *The Progressive*, October 1988, pp.18–22.
46. D. Russell, 'Earth last!', *The Nation*, 17 July 1989.
47. J. Carlisle, 'Bombs, lies and body wires', *CovertAction*, 1991, No. 38, Fall, p.30.
48. Transcript of conversation of undercover agent Fain with other agents, 13 May 1989.
49. *The Animals Agenda*, 'Earth First! founder busted in possible set-up', September 1989.
50. S. Lawrence, 'Explosion', *Associated Press*, Sacramento, 25 April 1995.

51. BCTV News Hour, *Transcript*, 27 April 1995.
52. G. Hamilton, 'Fake bomb found in MacBlo office,' *The Vancouver Sun*, 29 April 1994; Western Canada, Wilderness Committee, Greenpeace Canada, Sierra Club, Friends of Clayoquot Sound, Valhalla Society, Sierra Legal Defense Fund, *Environmentalists Denounce Attempted Bombing of MacMillan Bloedel*, 2 May 1994.
53. FBI, *Document From FBI San Diego to Director FBI /Routine*, June 1990.
54. J. Bari, 'TV mystery: who bought KQED and Steve Talbot?', *SF Weekly*, Vol. X, No. 14, 5 June 1991.
55. Sahara Club, *Newsletter*, No. 8, 1991; Sahara Club, *Newsletter*, No. 5.
56. C. Berlet, 'Hunting the "Green Menace"', *The Humanist*, July/August, 1991, p.29.
57. E. Pell, 'Stop the Greens', *E Magazine*, November/December 1991.
58. J. Franklin, 'First they kill your dog', *Muckraker*, Fall 1992, p.7.
59. J. Franklin, 'Green blood', *San Francisco Bay Guardian*, 20 April 1994, p.21.
60. D. Helvarg, *The War Against the Greens: The 'Wise-Use' Movement, the New Right, and Anti-Environmental Violence*, Sierra Club Books, San Francisco, 1994, p.326.
61. Ibid., pp.324–91.
62. S. O'Donnell, interview with author, 9 November 1994.
63. Ibid.
64. Ibid.
65. Ibid.
66. C. Berlet, interview with author, 4 November 1994.
67. T. Ramos, interview with author, 22 December 1994.
68. P. de Armond, interview with author, 24 November 1994.
69. S. O'Donnell, interview with author, 9 November 1994.
70. S. O'Donnell, *Report of Investigation*, 11 May 1995.
71. T. Ramos, interview with author, 22 December 1994.
72. D. Helvarg, *The War Against the Greens: The 'Wise-Use' Movement, the New Right, and Anti-Environmental Violence*, Sierra Club Books, San Francisco, 1994, p.8.
73. CDFE, *The Reno Declaration of Non-Violence*, 1995.
74. D. Barry, interview with author, 7 November 1994.
75. T. Ramos, interview with author, 22 December 1994.
76. D. Barry, interview with author, 7 November 1994.
77. Ibid.
78. R. Sieman Interview on clean air, clean water, dirty fight, *60 Minutes*, CBS, 20 September 1992.
79. J. Bari, 'How to create a climate of violence', *The Press Democrat*, 20 November 1992.
80. D. Helvarg, *The War Against the Greens: The 'Wise-Use' Movement, the New Right, and Anti-Environmental Violence*, Sierra Club Books, San Francisco, 1994, p.300.
81. C. Berlet, interview with author, 4 November 1994.
82. L. Regenstein, *America the Poisoned: How Deadly Chemicals are Destroying Our Environment, Our Wildlife, Ourselves and – How We can Survive*, Acropolis, 1982, p.28.
83. D. Postrel, 'Will there ever be an end to it?', *Statesman Journal*, 23 March 1983.
84. *Public Eye*, 'How many more? Death At Duck Valley', 1979, Vol. 11, Issues 1 and 2.

85. *Waste Not*, Lynn 'Bear' Hill, 21 March 1991; S. O'Donnell, *Report of Investigation*, Ace Investigations, 24 April 1991.
86. *Waste Not*, Lynn 'Bear' Hill, 21 March 1991; S. O'Donnell, *Report of Investigation*, Ace Investigations, 24 April 1991; D. Russell, 'The mysterious death of Lynn Ray Hill', *In These Times*, Vol. 15, No. 28, 10–23 July 1991.
87. B. Selcaig, 'Inquiry into activist's death continues', *High County News*, 1 November 1993; D. Helvarg, *The War Against the Greens: The 'Wise-Use' Movement, the New Right, and Anti-Environmental Violence*, Sierra Club Books, San Francisco, 1994, p.386.
88. Ibid., pp.388–9.
89. K. Schill, 'Missing: another tribal environmentalist', *High Country News*, 17 October 1994.
90. B. Angel, *Indian Lands Action Update*, 11 August 1994; S. O'Donnell, *Report of Investigation*, 15 April 1995; S. Mydans, 'Tribe smells sludge and bureaucrats', *New York Times*, 20 October 1994, p.A8.
91. B. Angel, *Indian Lands Action Update*, 11 August 1994; S. O'Donnell, *Report of Investigation*, 15 April 1995; S. Mydans, 'Tribe smells sludge and bureaucrats', *New York Times*, 20 October 1994, p.A8.
92. D. Day, *The Eco Wars: A Layman's Guide to the Ecology Movement*, Harrap, 1989, p.213; *Anderson Valley Advertiser*, 'Judi Bari: misery loves company', 22 May 1991.
93. J. Franklin, 'First they kill your dog', *Muckracker*, Fall 1992, p.7.
94. D. Helvarg, *The War Against the Greens: The 'Wise-Use' Movement, the New Right, and Anti-Environmental Violence*, Sierra Club Books, San Francisco, 1994, pp.368–70, 380, 382–4.
95. Ibid., pp.368–70, 380, 382–4.
96. For further details read D. Helvarg, *The War Against the Greens: The 'Wise-Use' Movement, the New Right, and Anti-Environmental Violence*, Sierra Club Books, San Francisco, 1994, pp.195–218, 380–1; J. Franklin, 'Green blood', *San Francisco Bay Guardian*, 20 April 1994, p.21; J. Franklin, 'First they kill your dog', *Muckracker*, Fall 1992, pp.7–9.
97. J. Franklin, 'Green blood', *San Francisco Bay Guardian*, 20 April 1994, p.21; J. Franklin, 'First they kill your dog', *Muckracker*, Fall 1992, pp.7–9; D. Helvarg, *The War Against the Greens: The 'Wise-Use' Movement, the New Right, and Anti-Environmental Violence*, Sierra Club Books, San Francisco, 1994, pp.340–6.
98. S. O'Donnell, *Report of Investigation*, 6 May 1991.
99. J. Franklin, 'Green blood', *San Francisco Bay Guardian*, 20 April 1994, p.21.
100. W. B. Stone, *Statement to the New York Police Department*, 6 October 1990; S. O'Donnell, *Report of Investigation*, 6 May 1991.
101. S. O'Donnell, 'Targeting environmentalists', *CovertAction*, Summer 1992, No. 41, p.42.
102. Ibid.; S. O'Donnell, *Report of Investigation*, 9 April 1992.
103. For more information see S. O'Donnell, 'Targeting environmentalists', *CovertAction*, Summer 1992, No. 41, p.42; D. Helvarg, *The War Against the Greens: The 'Wise-Use' Movement, the New Right, and Anti-Environmental Violence*, Sierra Club Books, San Francisco, 1994, pp.371–6.
104. D. Helvarg, *The War Against the Greens: The 'Wise-Use' Movement, the New Right, and*

Anti-Environmental Violence, Sierra Club Books, San Francisco, 1994, pp.360–4; J. Franklin, 'First they kill your dog', *Muckracker*, Fall 1992, p.6.

105. J. Franklin, 'First they kill your dog', *Muckracker*, Fall, 1992 pp.7–9.
106. S. O'Donnell, *Report of Investigation*, 15 April 1995.
107. C. McCoy, 'Rafts of ire, U.S. Forest Service finds itself bedeviled by Hells Canyon plan', *The Wall Street Journal*, 18 August 1994; D. Helvarg, *The War Against the Greens: The 'Wise-Use' Movement, the New Right, and Anti-Environmental Violence*, Sierra Club Books, San Francisco, 1994, p.359; S. O'Donnell, *Report of Investigation*, 14 October 1995.
108. S. Allen, '"Wise Use" groups move to counter environmentalists', *The Boston Globe*, 20 October 1992.
109. D. Helvarg, *The War Against the Greens: The 'Wise-Use' Movement, the New Right, and Anti-Environmental Violence*, Sierra Club Books, San Francisco, 1994, pp.358–64.
110. S. O'Donnell, *Report of Investigation*, Ace Investigations, 15 April 1995.
111. S. O'Donnell, *Report of Investigation*, 11 May 1995.
112. G. Miller, Letter to Don Young, Chairman of the Committee on Resources, 8 May 1995.
113. D. Junas, 'The rise of the militias', *CovertAction*, Spring 1995, pp.20–21.
114. D. Helvarg, 'The anti-enviro connection', *The Nation*, 22 May 1995, p.722.
115. B. Clark, 'John Birch meets John Wayne', *The StranGer*, 17 May 1995.
116. D. Helvarg, 'The anti-enviro connection', *The Nation*, 22 May 1995, p.724.
117. B. Clark, 'John Birch meets John Wayne', *The StranGer*, 17 May 1995.
118. Fenton Communications, *Militia Linked to 'Property Rights' Movement: Federal Employees Leader Asks for Hearings*, news release, 2 May 1995.
119. G. Miller, Letter to Don Young, Chairman of the Committee on Resources, 8 May 1995; Fenton Communications, *Militia Linked to 'Property Rights' Movement: Federal Employees Leader Asks for Hearings*, news release, 2 May 1995.
120. Ibid.
121. G. Miller, Letter to Don Young, Chairman of the Committee on Resources, 8 May 1995.
122. R. Larson, 'GOP encourages acts of violence, Rep. Miller says', *The Washington Times*, 10 May 1995.
123. M. Janifsky, 'Accounts of violence by paramilitary groups', *The New York Times*, 12 July 1995.
124. *A CLEAR View*, 'Bomb blasts Forest Service Office', CLEAR, Vol. 3, No. 2, 29 January 1996; *A CLEAR View*, Vol. 3, Number 1, January 1996.
125. C. Dodd, 'SLAPP back!', *Buzzworm*, July/August 1992, Vol. 1V, No. 4.
126. G. W. Pring, and P. Canan, '"Strategic Lawsuits Against Public Participation" (SLAPPs): an introduction for Bench, Bar, and bystanders', *Bridgeport Law Review*, September 1992.
127. *CBS Magazines*, 24 September 1991.
128. Ibid.
129. P. Canan, M. Kretzmann, M. Hennessy, and G. Pring, 'Using law ideologically: the conflict between economic and political liberty', *The Journal of Law and Politics*, Spring 1992, Vol. VIII, No. 3, p.540.

130. *CBS Magazines*, 24 September 1991.
131. E. Pell, 'Corporate anti-environmentalism', *E magazine*, 7 November 1991.
132. G. W. Pring and P. Canan, '"Strategic Lawsuits Against Public Participation", (SLAPPs): an introduction for Bench, Bar, and bystanders', *Bridgeport Law Review*, September 1992.
133. T. Ramos, interview with author, 22 December 1994.
134. D. Helvarg, *The War Against the Greens: The 'Wise-Use' Movement, the New Right, and Anti-Environmental Violence*, Sierra Club Books, San Francisco, 1994, p.305.
135. *Technology Review*, 'Uncivil suits', Massachusetts Institute of Technology Alumni Association, April 1991, Vol. 94, p.14.
136. Ibid.
137. J. Bari, 'The Palco papers', *Anderson Valley Advertiser*, 27 March 1991.
138. D. Helvarg, *The War Against the Greens: The 'Wise-Use' Movement, the New Right, and Anti-Environmental Violence*, Sierra Club Books, San Francisco, 1994, p.305.
139. *CBS Magazines*, 24 September 1991.
140. E. Pell, 'Corporate anti-environmentalism', *E magazine*, 7 November 1991.
141. *Waste Not*, 'Ogden Martin threatens to sue the doctors of Orilla, Ontario, Canada. The doctors produced a report which led to the rejection of a $500 Million, 3,000 TPD solid waste incinerator', 13 September 1990.
142. *The Vancouver Sun*, 'MacBlo drops lawsuit SLAPP at Island opposition', 12 April 1993, p.D3.

7 TO CUT OR NOT TO CLEAR-CUT

1. P. Moore, B.C. Forest Alliance TV Ad, 1994, 12 September 1994.
2. *The Vancouver Sun*, 'Foresters "must fight ecologists"', 23 February 1980.
3. R. Arnold, *The Politics of Environmentalism*, Ontario Agricultural Conference, 8 January 1981.
4. T. McKegney, *'Only A Movement Can Combat A Movement', Environmental Campaigners Say*, Report of the Atlantic Vegetation Management Association 'Education Seminar' quoting Ron Arnold, Natural Resources – Forest Extension Service, 25 October 1984; L. LaRouche, 'The tragic state of USA counterintelligence', *EIR News Service*, Boston, 4 December 1987.
5. T. McKegney, *'Only A Movement Can Combat A Movement', Environmental Campaigners Say*, Report of the Atlantic Vegetation Management Association 'Education Seminar' quoting Ron Arnold, Natural Resources – Forest Extension Service, 25 October 1984.
6. R. Arnold, *The Environmental Movement and Industrial Responses*, Proceedings of Public Affairs and Forest Management: Pesticides in Forestry, Toronto, Ontario, 25–27 March 1985.
7. CBC Radio, 6 October 1986.

8. R. Arnold, 'Loggerheads over landuse', *Logging and Sawmilling Journal*, April 1988, reprinting paper that was presented to the Ontario Forest Industries Association in Toronto in February.
9. H. Goldenthal, 'Polarizing the public debate to subvert ecology activism', *Now Magazine*, 13–19 July 1989.
10. C. Emery, *Share Groups In British Colombia*, Library of Parliament Research Division, 10 December 1991; J. Danylchuk, '2 public servants run groups to battle environmentalists', *The Edmonton Journal*, 26 June 1989; *The Northern Miner*, 27 February 1989.
11. H. Goldenthal, 'Polarizing the public debate to subvert ecology activism', *Now Magazine*, 13–19 July 1989.
12. M. Hume, 'Battle of the forests: environmentalists tarred by a campaign of hate', *The Vancouver Sun*, 7 March 1990, p.A9
13. Ibid.
14. *British Colombia Environmental Information Institute*, Briefs, no date.
15. N. Parton, 'Canfor boss floating plan to counter "anti-everything"', *The Vancouver Sun*, 11 August 1989.
16. NorthCare, promotional leaflet, no date.
17. NorthCare, *Sharing our Resources . . . for Enjoyment and Employment*, no date; Municipality Membership list, June 1992.
18. K-A. Mullin, 'Northcare wants more women involved in group', *Northern Life*, 1 March 1989.
19. *The Evening Patriot*, 'Environmentalists in factional fight', 11 June 1991, p.22.
20. C. Emery, *Share Groups In British Colombia*, Library of Parliament Research Division, 10 December 1991, pp.1, 5.
21. Ibid., p.7
22. J. Nelson, 'Pulp and propaganda', *Canadian Forum*, July/August 1994, pp.15–19.
23. S. Hume, 'Anti-clearcut logger says he represents the silent majority', *The Vancouver Sun*, 16 August 1993.
24. M. Clayton and M. Trumbull, 'Canadians clash over future of forests', *The Christian Science Monitor*, 16 September 1993, p.7.
25. I. Gill, 'Moresby Park costs kept under wraps', *The Vancouver Sun*, 7 July 1987, p.A1.
26. Valhalla Society, personal communication with author, 15 December 1995; *Red Neck News*, Vol. 10, 12 July 1982; *Red Neck News*, Vol. 58, 11 June 1983; *Red Neck News* Vol. 7, 7 August 1985; letter to Colleen McCrory, 12 June 1985.
27. P. Armstrong, 'The Beban factor', *Logging and Sawmill Journal*, March 1988, pp.24–5.
28. K. Baldrey and G. Bohn, 'B.C. imposes moratorium on S. Moresby logging', *The Vancouver Sun*, 20 March 1987, p.D5.
29. K. Watt, 'Woodsman spare that tree!', *Report on Business Magazine*, March 1990, p.51.
30. M. Mason, *The Politics of Wilderness Preservation: Environmental Activism and Natural Areas Policy in British Columbia, Canada*, PhD Dissertation, Cambridge, 1992, p.182.
31. Forests Forever Advert, 10 September 1987.
32. M. Mason, *The Politics of Wilderness Preservation: Environmental Activism and Natural Areas*

Policy in British Columbia, Canada, PhD Dissertation, Cambridge, 1992, p.182.
33. Ibid., p.188.
34. B. Parfitt, 'Both sides dig in as verbal war intensifies in Stein', *The Vancouver Sun*, 19 May 1988, p.F1.
35. Ibid.
36. Ibid.
37. K. Goldberg, 'Share's right-wing links', *The Tribune*, 6 July 1992.
38. G. Bohn, 'Parliamentary study rekindles forests fight', *The Vancouver Sun*, 3 March 1992, p.B5.
39. North Island Citizens for Shared Resources, *Unions, Workers and Businesses Economic Defense Strategy*, 21 January 1991; M. Morton, *A History of Share: the Clayoquot Society*, October 1990; *Pennywise*, Kootenay West Share Society, 20 March 1991; L. Forman, 'Understanding the Share groups', *Forest Planning Canada*, Vol. 5, No. 1, January/February 1989, p.5; *Ad hoc* leaflets from Share Our Resources and Share our Forests and other share organisations.
40. A.M Gottlieb (ed.), *The Wise Use Agenda: The Citizen's Policy Guide to Environmental Resource Issues – A Task Force to the Bush Administration by the Wise Use Movement*, The Free Enterprise Press, 1989, pp.157–66.
41. M. Hume, 'Resource-use conference had links to Moonie Cult', *The Vancouver Sun*, 8 July 1989, p.A6.
42. C. Emery, *Share Groups In British Colombia*, Library of Parliament Research Division, 10 December 1991, Executive Summary.
43. Ibid., p.9
44. G. Bohn, 'Parliamentary study rekindles forests fight: "Share" camp counters claim its roots feed from U.S. movement', *The Vancouver Sun*, 3 March 1992; M. Morton, Letter to Carl Deal, 18 September 1992.
45. W. P. Pendley, *It Takes A Hero: The Grassroots Battle Against Environmental Oppression*, A Project of the Mountain States Legal Foundation, Free Enterprise Press, Bellevue, Washington, 1994, p.276.
46. *MABC Newsletter*, Share BC – Community Stability and Land Use in the 90's Conference and Workshop, November 1989, p.19.
47. G. Bohn, 'Parliamentary study rekindles forests fight', *The Vancouver Sun*, 3 March 1992, p.B5.
48. *Our Land*, List of Board Of Directors of Our Land Society, Vol. 1, No. 1, February 1989, p.5; W. Wilbur, 'Old-growth forests: the last stand', *The Nation*, Vol. 251, No. 2, 9 July 1990, p.37.
49. D. Harris, 'Wise use environmentalism', *Our Land*, Vol. 1, No. 1, February 1989.
50. Canadian Women in Timber, *Document Submitted to the Provincial Forest Resources Commission*, 16 March 1990; Canadian Women in Timber, *AGM*, 9 January 1990.
51. M. Mason, *The Politics of Wilderness Preservation: Environmental Activism and Natural Areas Policy in British Columbia, Canada*, PhD Dissertation, Cambridge, 1992, p.189.
52. G. Bohn, 'Parliamentary study rekindles forests fight: "Share" camp counters claim its roots feed from U.S. movement', *The Vancouver Sun*, 3 March 1992.

53. W. Sheridan, *The Origins and Objectives of Share Groups in British Colombia*, Library of Parliament, 10 July 1991, pp.1,5.
54. G. Bohn, 'Parliamentary study rekindles forests fight: "Share" camp counters claim its roots feed from U.S. movement', *The Vancouver Sun*, 3 March 1992.
55. *Business Information Wire*, 'ATTN: Environment Canada', 3 March 1992; *The Vancouver Sun*, 'Mohawk's move makes environmental waves', 11 January 1993, p.B3.
56. *Alberni Valley Times*, 'Share our resources disappointed by the apathy', 21 October 1992.
57. B. Parfitt, 'Fletcher challenge to abandon unmarked mailings, officials says', *The Vancouver Sun*, 26 October 1989.
58. *British Columbia Environmental Report*, Conference Reports, December 1992, p.19.
59. A. Edmondson, 'Militant group in US using wrong tactics says speaker', *Daily Townsman*, 13 February 1990.
60. *The B.C. Environmental Report*, Montana Forest Industry Advocate Tours B.C., Vol. 4, No. 2, May 1993, p.42.
61. *Western Horizons*, 'Around the West (and the world)', September 1993, p.13.
62. H. Goldenthal, 'Polarizing the public debate to subvert ecology activism', *Now Magazine*, 13–19 July 1989.
63. D. Wilson, 'Tension rises in timber county, *Globe and Mail*, 12 November 1990, p.A1.
64. K. Goldberg, 'Share's right-wing links', *The Tribune*, 6 July 1992, p.1.
65. C. Emery, *Share Groups In British Colombia*, Library of Parliament Research Division, 10 December 1991, pp.40,41.
66. H. Williams, *The Unfinished Agenda*, speech to Vancouver's Rotary Club, 2 August 1988.
67. P. Wilson, 'Losing ground', *The Truck Logger*, December/January 1988, p.25; R. Brunet, 'Changing the political landscape', *The Truck Logger*, December/January 1989/1990, Vol. 13, No. 1, p.20.
68. J. Van Allen, 'Loggers' rally', *The Essence*, 11 July 1989, p.7.
69. K. Williams, Letter to supplier, MacMillan Bloedel, 31 August 1989.
70. J. Mitchell, 'No deals', *Report on Business Magazine*, October 1989, pp.72, 77.
71. T. Corcoran, 'Noranda chief takes aim at so-called "Environmental Terrorists"', *Globe and Mail*, 22 November 1989, p.B2.
72. *The Vancouver Sun*, 'Premier, Parker at odds over environment, NDP says', 15 August 1989, p.A1.
73. Ibid.
74. *British Colombia Report*, 'Take the blame, Share the pain', November 1990, p.35.
75. B. Devitt, Letter to suppliers, Canadian Pacific Forest Products Limited, 18 April 1990.
76. T. Buell, Letter to employees, Weldwood of Canada Limited, 10 April 1991.
77. J. Lindsay, 'Boycotts are terrorism', *The Vancouver Sun*, 13 April 1991.
78. F. Shalom, 'Logging is a fact of life: "Moderate Environmentalist"', *The Gazette*, 27 March 1991.
79. A. Gibbon, 'Quebec project delays linked to "Eco-fascists"', *The Globe and Mail*, 4 September 1991.

80. S. Hume, 'Losers main share groups message', *The Vancouver Sun*, 25 November, 1992, p.A19.
81. Ibid.
82. S. Hume, 'We have met the enviro-terrorists, and they are us', *The Vancouver Sun*, 22 April 1991; S. Hume, 'Just what is MacMillan Bloedel up to?' *The Vancouver Sun*, 2 February 1990; S. Hume, 'Rage and resentment over our disputed forests', *The Vancouver Sun*, 17 August 1990.
83. K. Goldberg, 'More wise use abuse; logging industry in British Columbia and environmentalists', *Canadian Dimension*, May 1994, Vol. 28, p.27.
84. T. Stark, interview with author, 6 June 1995.
85. B. Parfitt, 'Supporters of timber boycott guilty of treason Munro says', *The Vancouver Sun*, 11 April 1991.
86. Ibid.; D. Suzuki, 'It's time for Jack Munro and Frank Oberle to chill out', *The Vancouver Sun*, 18 May 1991, p.B6.
87. A. Fletcher, 'PR link for forest firms and unions', *The Financial Post*, 11 April 1991; B. Parfitt, 'PR giants, President's men, and B.C. trees', *The Georgia Strait*, 21–28 February 1992.
88. B. Parfitt, 'PR giant in forestry drive linked to world's hotspots', *The Vancouver Sun*, 8 July 1991.
89. S. Hume, 'Forest "code" is an exercise in hypocrisy', *The Vancouver Sun*, 18 March 1992.
90. K. Goldberg, 'For the record', *Nanaimo Times*, 12 January 1993, p.A9.
91. B. Parfitt, 'PR giants, President's men, and B.C. trees', *The Georgia Strait*, 21–28 February 1992.
92. *PR Newswire*, 'Eleven join Alliance advisory board', Vancouver, 10 May 1991.
93. T. Stark, interview with author, 6 June 1995.
94. B.C. Forest Alliance, *Forests For All*, no date.
95. B. Parfitt, 'PR giant in forestry drive linked to world's hotspots', *The Vancouver Sun*, 8 July 1991.
96. J. Nelson, 'Pulp and propaganda', *Canadian Forum*, July/August 1994, p.16.
97. Canadia Newswire, *Alliance Opposes One-sided Information*, 23 December 1991; B.C. Forest Alliance, *Forests For All*, no date.
98. B. Parfitt, 'PR giants, President's men, and B.C. trees', *The Georgia Strait*, 21–28 February 1992.
99. *The Province*, 'Forest firms barking up wrong tree', 3 October 1990.
100. P. Marquis, *Canadian Resource Industries and Non-Governmental Organisations*, Political and Social Affairs Division, Library of Parliament, 5 January 1993, pp.10–11.
101. J. Schreiner, 'Giving forestry a good name', *The Financial Post*, 11 February 1992.
102. B. Parfitt, 'PR giants, President's men, and B.C. trees', *The Georgia Strait*, 21–28 February 1992; P. Marquis, *Canadian Resource Industries and Non-Governmental Organisations*, Political and Social Affairs Division, Library of Parliament, 5 January 1993, p.13.
103. K. Goldberg, 'All the news that's fun', *British Columbia Environmental Report*, March 1993, p.4.
104. Ibid.

105. Canada Newswire, *B.C. Forest Alliance Announces Economic Impact Study*, 22 April 1991.
106. The Forest Alliance of B.C., *The Forest and the People*, October 1991, Vol. 1, No. 3, p.1.
107. S. Hume, 'Forest "code" is an exercise in hypocrisy', *The Vancouver Sun*, 18 March 1992.
108. P. Marquis, *Canadian Resource Industries and Non-Governmental Organisations*, Political and Social Affairs Division, Library of Parliament, 5 January 1993, pp.10–11.
109. L. Manchester, *Letter to Jack Munro, Re: Forest Alliance Application to BC Environmental Network*, British Columbia Environmental Network, 14 December 1992.
110. T. Stark, interview with author, 6 June 1995.
111. K. Mahon, *Clayoquot Sound*, Greenpeace Canada, July 1993.
112. S. Bell, 'Loggers, supporters confront protesters', *The Vancouver Sun*, 16 August 1993.
113. K. Mahon, *Clayoquot Sound*, Greenpeace Canada, July 1993.
114. M. Clayton and M. Trumbull, 'Canadians clash over future of forests', *The Christian Science Monitor*, 16 September 1993, p.7.
115. S. Hume, 'Saws the real buzz in industry message', *The Vancouver Sun*, 27 January 1993.
116. *Reuter*, 'Canadian MP goes to jail for logging protest', Vancouver, 26 July 1994.
117. M. Clayton and M. Trumbull, 'Canadians clash over future of forests', *The Christian Science Monitor*, 16 September 1993, p.7.
118. E. Lazarus, '"Good-hearted" environmentalist persecuted by Crown, painter says', *The Vancouver Sun*, 20 June 1994.
119. *The Vancouver Sun*, 'Scientific panel given extension', 5 August 1994.
120. P. Kuiten Rouwer, 'Canada's forests in dire danger, crusader warns', *Calgary Herald*, 3 May 1992.
121. J. Fulton, Letter to Jack Munro, 8 February 1991.
122. The Forest Alliance of B.C., 'B.C. is NOT the Brazil of the North – Munro', *The Forest and the People*, November/December 1993, p.1; P. Luke, 'B.C.'s no Brazil: comparisons of forestry methods faulty: Munro', *Vancouver Province*, 2 November 1993; G. Hamilton, 'Jack's back and loaded for bear: trip to Brazil has Alliance chair condemning environmental tag', *The Vancouver Sun*, 2 November 1993; P. Luke, 'Mission is "Silly": B.C. Alliance is rapped on Brazil forest visit', *Vancouver Province*, 5 October 1993, p.A38.
123. S. Ward, *Canadian Press Newswire*, 6 November 1992.
124. P. Luke, 'Ex-Greenpeacer sold out, say foes', *The Province*, 8 January 1993.
125. T. Stark, interview with author, 6 June 1995.
126. *PR Newswire*, '"Eco-Judas" to address Forestry Association Annual Meeting, Portland', 18 February 1994; Dr P. Moore, *Speech to the Canadian Pulp and Paper Association*, Wood Pulp Section Open Forum, Montreal, Quebec, 27 January 1993.
127. R. Arnold, *At the Eye of the Storm: James Watt and the Environmentalists*, Regnery Gateway, Chicago, 1982, pp.27,35–8.
128. P. Moore, 'As the world turns', *The Vancouver Sun*, 5 February 1994.
129. *Timber Trades Journal*, 'Turning from extremes', 26 February 1994.

130. J. Fulton, 'Unveiling the real zero-tolerance extremist', *The Vancouver Sun*, 11 February 1994.
131. The Forest Alliance of B.C., 'Alliance welcomes new directors', *The Forest and the People*, August 1994, p.3.
132. *BC Report*, Save Our Jobs Committee, 18 April 1994; K. Fraser and S. Hamilton, '"Hitler" slurs worry Sihota', *Vancouver Province*, 8 April 1994, p.A26; *The Vancouver Sun*, 'Then again, some don't: Arcand of the IWA explains how to break the back of CORE', 8 April 1994; S. Hamilton and K. Fraser, 'Greens, officials are enemy, loggers told', *Vancouver Province*, 7 April 1994, p.A7.
133. *The Vancouver Sun*, 'Frustration fuels Arcand's words', 9 April 1994, p.A22.
134. Canadian Broadcasting Corporation, *The Fifth Estate*, 12 October 1993.
135. B. Kieran, 'Friends of the Clayoquot want it all', *The Province*, 6 July 1993; B. Kieran, 'Harcourt can breathe easier', *The Province*, 12 December 1993; B. Kieran, 'Enviro-loonies damage cause', *The Province*, 19 March 1993.
136. G. Hamilton, 'Harcourt beaten to promotions punch by paper', *The Vancouver Sun*, 3 February 1993, p.D2.
137. *Reuter*, 'Canada launches war of words with Greenpeace', Vancouver, 23 March 1994.
138. K. MacQueen, 'Forest industry strives to polish its image: environmental crusaders have struck fear into nation's largest employer', *The Ottawa Citizen*, 11 August 1991, p.E6.
139. J. Nelson, 'Pulp and propaganda', *Canadian Forum*, July/August 1994, pp.15–19.
140. Ibid.
141. *Canadian Press Newswire*, 'Environmental groups, say they are shocked by the Provincial Government's plans to give funding to the forest industry's lobby group', 1 May 1994.
142. V. Husband, Letter to Andrew Petter the Minister of Forests, The Sierra Club of Western Canada, 2 May 1994.
143. The Forest Alliance of B.C., 'The Forest Alliance welcomes new senior advisor, and television and newspaper ads: huge success', *The Forest and the People*, May/June 1994, p.1.
144. V. Husband, Letter to Andrew Petter the Minister of Forests, The Sierra Club of Western Canada, 2 May 1994.
145. The Forest Alliance of B.C., 'The Forest Alliance welcomes new senior advisor, and television and newspaper ads: huge success', *The Forest and the People*, May/June 1994, p.1.
146. MacMillan Bloedel, 'The Clayoquot compromise: when Greenpeace threatens our customers it's time to take a stand', *The Vancouver Sun*, 7 March 1994.
147. MacMillan Bloedel, 'The Clayoquot compromise', *The Vancouver Sun*, 19 March 1994.
148. J. Hunter, 'It's blackmail, Harcourt charges', *The Vancouver Sun*, 18 March 1994, p.D1.
149. M. Drohan, 'B.C. forest group meets its match: environmentalists' victory clear-cut at U.K. meeting on logging practices', *Globe and Mail*, 30 March 1994; S. Ward,

'Munro raises alarm over European wood market', *Victoria Times Colonist*, 30 March 1994; *Reuter*, 'German group said to move against MacBlo', Vancouver, 25 July 1994.

150. D. Hauka, 'War of woods takes to the airwaves', *Vancouver Province*, April 1994.
151. G. Bohn, 'Tale of forestry-ad rejection doesn't add up, *Times* says', *The Vancouver Sun*, 8 April 1994
152. *Reuter*, 'Canadian loggers stage mass protest', Vancouver, 21 March 1994; *Vancouver Province*, 'Hands off jobs; you'll pay', 21 March 1994.
153. K. Goldberg, 'More Wise Use abuse: logging industry in British Columbia and environmentalists', *Canadian Dimension*, May 1994, Vol. 28, p.27.
154. *Vancouver Province*, 'NDP to unveil Island plan', 20 June 1994.
155. *Vancouver Province*, 'Hands off jobs; you'll pay', 21 March 1994.
156. T. Stark, interview with author, 6 June 1995.
157. S. Hume, 'Facts and factoids in the public relations war over B.C.'s forests', *The Vancouver Sun*, 10 August 1994.
158. *The Vancouver Sun*, 'Petter peddles new logging rules in Europe', 29 September 1994.
159. *Canadian Newswire*, 'McCrory tells Europe "Brazil of the North" still applies to Canadian forest practices', 24 March 1995; WCWC, *Canada's Clearcutting Undermines Efforts to Slow Global Warming: Climate Change Implications Warrant Ban on Clearcut Logging*, 24 March 1995.
160. G. Hamilton, 'Forest Alliance directors hit choppy waters over Hollywood campaign', *The Vancouver Sun*, 7 April 1995.
161. P. Luke, 'Stone's new role: director shoots down clear-cut', *Province*, 24 March 1995.
162. T. Berman, Personal communication, 19 April 1995.
163. J. Nelson, 'Pulp and propaganda', *Canadian Forum*, July/August 1994, p.16.
164. P. Moore, B.C. Forest Alliance TV Ad, 12 September 1994.
165. J-P. Jeanrenaud and N. Dudley, Letter to Patrick Anderson, 15 September 1994.
166. C. Osterman, 'Environmentalists' hopes rise for Canada rainforest', *Reuter*, Vancouver, 29 May 1995; D. Thomas, 'Forest firms told to rethink Clayoquot logging practices', *The Financial Post* 30 May 1995; Greenpeace Canada, 'Greenpeace applauds as scientists recommend the end to clearcutting in Clayoquot', 29 May 1995; T. Berman, Personal communication regarding Clayoquot Science Panel, 29 May 1995.
167. B. Simon and A. Maitland-Montreal, 'Publishers to join fight against "Paper" protesters', *The Financial Times*, 31 January 1995; *Canadian Press Newswire*, 'NYT reviews MacBlo contract', 7 June 1995; *Reuter*, 'MacMillan loses NY Times contract-Greenpeace', 10 November 1995.
168. Ministry of Forests and Ministry of Environment, Lands and Parks, *Government Adopts Clayoquot Scientific Report Moves to Implementation*, press release, 6 July 1995.
169. J. Zarocostas, '"Green" group slams Canadian guidelines on forest products', *Knight-Ridder*, 26 May 1995.
170. *Arborvitae*, 'Canadian-Australian environmental certification proposal dropped', The IUCN/WWF Forest Conservation Newsletter, September 1995, p.13.

171. *Reuter*, 'Environmentalists urge end to Clayoquot logging', Vancouver, 8 November 1995.
172. B. Yaffe, 'Why clearcuts are not just topographical nightmares', *The Vancouver Sun*, 13 June 1995.

8 THE FIGHT FOR THE FORESTS OF CENTRAL AND LATIN AMERICA

1. T. Gross (ed.) *Fight for the Forest: Chico Mendes in his Own Words*, Latin American Bureau, 1989, p.6.
2. A. Revkin, *The Burning Season: The Murder of Chico Mendes and The Fight for the Amazon Rain Forest*, Collins, 1990, p.178; S. Hecht and A. Cockburn, *The Fate of the Forest: Developers, Destroyers and Defenders of the Amazon*, Penguin, 1990, p.196; G. Monbiot, 'Dispossessed without trace', *Index on Censorship*, 1992, Issue 5, p.9; Human Rights Watch and Natural Resources Defense Council, *Defending the Earth: Abuses of Human Rights and the Environment*, 1992, pp.3–4; S. Branford and O. Glock, *The Last Frontier, Fighting Over Land in the Amazon*, Zed Books, 1985, p.29.
   The texts mentioned above are excellent sources of information on the land struggle in the Amazon.
3. A. Revkin, *The Burning Season: The Murder of Chico Mendes and The Fight for the Amazon Rain Forest*, Collins, 1990, p.104.
4. Ibid., pp. 104, 105, 110.
5. M. Colchester, *Salvaging Nature: Indigenous Peoples, Protected Areas and Biodiversity Conservation*, United Nations Research Institute for Social Development, World Rainforest Movement, and the World Wide Fund for Nature, September 1994, p.12.
6. A. Revkin, *The Burning Season: The Murder of Chico Mendes and The Fight for the Amazon Rain Forest*, Collins, 1990, p.273.
7. Ibid., p.154.
8. Ibid., pp.172, 210.
9. Ibid., pp.7, 14, 291.
10. S. Hecht and A. Cockburn, *The Fate of the Forest: Developers, Destroyers and Defenders of the Amazon*, Penguin, 1990, p.193.
11. A. Revkin, *The Burning Season: The Murder of Chico Mendes and The Fight for the Amazon Rain Forest*, Collins, 1990, pp.137, 206, 224.
12. Ibid., p.180.
13. Ibid., p.181.
14. T. Gross (ed.), *Fight for the Forest: Chico Mendes in his Own Words*, Latin American Bureau, 1989, p.66.
15. S. Hecht and A. Cockburn, *The Fate of the Forest: Developers, Destroyers and Defenders of the Amazon*, Penguin, 1990, pp.193, 215–16.
16. G. Monbiot, interview with author, 18 November 1994.

17. A. Revkin, *The Burning Season: The Murder of Chico Mendes and The Fight for the Amazon Rain Forest*, Collins, 1990, pp.212–14, 227.
18. Ibid., p.202.
19. T. Gross (ed.), *Fight for the Forest: Chico Mendes in his Own Words*, Latin American Bureau, 1989, p.33.
20. Ibid., p.33; A. Revkin, *The Burning Season: The Murder of Chico Mendes and The Fight for the Amazon Rain Forest*, Collins, 1990, pp.243–8.
21. T. Gross (ed.), *Fight for the Forest: Chico Mendes in his Own Words*, Latin American Bureau, 1989, p.6.
22. Human Rights Watch and Natural Resources Defense Council, *Defending the Earth: Abuses of Human Rights and the Environment*, 1992, p.2; America's Watch, *On Trial In Brazil: Rural Violence and the Murder of Chico Mendes*, Washington, 9 December 1990.
23. *Reuter*, 'Brazil urges hunt for killers of forest activist', Rio De Janeiro, 4 July 1995.
24. T. Gross (ed.), *Fight for the Forest: Chico Mendes in his Own Words*, Latin American Bureau, 1989, p.66; A. Revkin, *The Burning Season: The Murder of Chico Mendes and The Fight for the Amazon Rain Forest*, Collins, 1990, p.286; America's Watch, *On Trial In Brazil: Rural Violence and the Murder of Chico Mendes*, Washington, 9 December 1990, p.7; *Index on Censorship*, Brazil, 1991, Issue 2, p.35; *Index on Censorship*, Brazil, 1990, Issue 8, p.34.
25. M. Adriance, *Promised Land: Base Christian Communities and the Struggle for the Amazon*, State University of New York Press, 1995.
26. D. Fass, *Human Rights Abuses Against People Involved in Environmental Issues*, presented to A Healthy Environment is a Human Right Conference, Montreal, 26–28 October 1995; Amnesty International, Brazil: *Manoel Pereira da Silva*, Urgent Action, UA 449/90, 7 November 1990.
27. *Index on Censorship*, Brazil, 1991, Issue 4 and 5, p.52; *Index on Censorship*, Brazil, 1991, Issue 6, p.36.
28. Human Rights Watch and Natural Resources Defense Council, *Defending the Earth: Abuses of Human Rights and the Environment*, 1992, p.7.
29. A. Revkin, *The Burning Season: The Murder of Chico Mendes and The Fight for the Amazon Rain Forest*, Collins, 1990, p.292.
30. M. Adriance, *Promised Land: Base Christian Communities and the Struggle for the Amazon*, State University of New York Press, 1995.
31. Amnesty International, *Brazil: Possible Extrajudicial Execution/Fear of Extrajudicial Execution: Valdinar Pereira Barros, Trade Unionist, Francisco Geronimo da Silva 'Dequinha, Trade Unionist*, UA 389/92, 9 December 1992.
32. *Rio Maria Bulletin*, 'Death threats against Father Ricardo Rezende: letters urgently needed!', Vol. 1V, No. 2, September 1994.
33. H. Paul, Personal communication, 13 October 1995.
34. *Index on Censorship*, Brazil, 1989, Issue 5, p.37.
35. *Friends of the Earth*, Please will you stop paying to have my people murdered? advert in *The Guardian*, 27 May 1995, p.11.
36. P. Grunter, 'Stars pay tribute to a peasant crusader', *The Evening Standard*, 13 June 1995, p.20.

37. Greenpeace International, *Greenpeace Condemns Shooting of Environmentalist*, Amsterdam, 30 April 1993; Greenpeace Brazil, Personal communication, 17 May 1993.
38. Greenpeace International, *Another Brazilian Environmentalist Murdered*, 5 May 1993.
39. Para Society in Defense of Human Rights, *President of a Rural Workers Union is Murdered in Para*, May 1995.
40. M. Adriance, *Promised Land: Base Christian Communities and the Struggle for the Amazon*, State University of New York Press, 1995.
41. Unless otherwise stated the text and quotes are taken from an interview with Judy Kimerling, 21 June 1995.
42. J. Kimerling, *Amazon Crude*, produced with the Natural Resources Defense Council, Washington, 1991, pp.39–40
43. A. Illianes, *Ecological Debate on the Problems Caused by the Oil Industry*, Amazon for Life Campaign Report, 1993, Quito.
44. C. Grylls, *Environmental Hooliganism in Ecuador*, Framtiden I Vare Hender, Norway, 1992, p.53.
45. *Lloyds List*, Texaco, 9 June 1992, p.2.
46. J. Kimerling, *Amazon Crude*, produced with the Natural Resources Defense Council, Washington, 1991, p.103.
47. Ibid., pp.31, 34; K. Gold and E. Bravo, *The Value of Tropical Forests of the Amazon*, in *Ecological Debate on the Problems Caused by the Oil Industry*, Amazon for Life Campaign Report, 1993, Quito.
48. C. Grylls, *Environmental Hooliganism in Ecuador*, Framtiden I Vare Hender, Norway, 1992, p.6.
49. J. Kimerling, *Amazon Crude*, produced with the Natural Resources Defense Council, Washington, 1991, p.43; *Mother Jones*, Crude, March/April 1992, p.41.
50. J. Kimerling, *Amazon Crude*, produced with the Natural Resources Defense Council, Washington, 1991, pp.48, 63, 65, 69; J. Kimerling, *Texaco: Its Past and Its Responsibilities*, in *Ecological Debate on the Problems Caused by the Oil Industry*, Amazon for Life Campaign Report, 1993, Quito.
51. J. Kimerling, *Amazon Crude*, produced with Natural Resources Defense Council, Washington, 1991, pp.63, 69; C. Mackerron, *Business in the Rainforests: Corporations, Deforestation and Sustainability*, Investor Responsibility Research Center, Washington, 1993, p.101.
52. C. Mackerron, *Business in the Rainforests: Corporations, Deforestation and Sustainability*, Investor Responsibility Research Center, Washington, 1993, p.102; A. Parlow, 'Of oil and exploration in Ecuador', *Multinational Monitor*, January/February 1991, p.22; J. Karten, *Oil Development and Indian Survival In Ecuador's Oriente*, TRIP, no date; Amazon for Life Campaign Report, *Ecological Debate on the Problems Caused by the Oil Industry*, 1993, Quito; A. Illianes, *Ecological Debate on the Problems Caused by the Oil Industry*, Amazon for Life Campaign Report, 1993, Quito.
53. J. Kimerling, interview with author, 7 September 1995.
54. *Latoil*, 'Environmental audit begins in February', January 1993, p.12.
55. Rainforest Action Network, *Texaco, Clean up Your Mess*, Action Alert, San Francisco, Number 86, July 1993.

56. Statement of International Delegates, Ecuador, 5–9 July 1993, signed by K. Baird, Senior Conservation Officer, Department of Conservation, New Zealand; H. Erake, Campaign Leader, Future in Our Hands, Norway; J. Kimerling, Environmental Attorney, Coalition in Support of Peoples of the Amazon and their Environment, USA; J. Lynard, Naturfolkenes Verden, Dinamarca; Terre Des Peuples Indigenes, Luxembourg; G. Marris, Rainforest Action Group, New Zealand; A. Pillen, Medical Student, Ku Leuven Belgium; C. Ross, Oxfam America, Coalition in Support of Peoples of the Amazon and their Environment, USA; M. Spencer, Researcher, Friends of the Earth, UK.
57. M. Spencer, Personal communication with author, 5 October 1993.
58. Indigenous Coordinating Body of the Amazon Basin (COICA), Confederation of Indigenous Nationalities of the Ecuadorian Amazon (CONFENIAE), Acción Ecológica, USA Coalition in Support of Amazonian Peoples and Their Environment, *Indigenous and Environmental Organisations Decry Lack of Participation by Affected Populations in 'Environmental Audit' of Texaco*, press release, Quito, 9 November 1993.
59. J. Kane, 'Huaorani goes to Washington', *The New Yorker*, 2 May 1994, pp.74–81.
60. Ibid.
61. T. Connor, 'Amazon Indians speak out against destruction', *United Press International*, New York, 3 November 1993; S. Maull, 'Texaco-Amazon', *Associated Press*, New York, 3 November 1993.
62. J. C. Kohn, M. H. Malman, M. J. D'Urso, D. Liberto, C. Bonifaz, J. Bonifaz, S. R. Donzinger and A. Damen, Civil action V Texaco in the United States District Court for the Southern District of New York, Kohn, Nast and Graf, Sullivan and Damen, 3 November 1993.
63. *Reuter*, 'Peruvian Indians sue Texaco, allege pollution', New York, 28 December 1994.
64. Amnesty International, *Peru: Torture of Community Leaders*, South Andean Action 09/92, AMR 46/58/92, London, 1 December 1992; D. Fass, *Human Rights Abuses Against People Involved in Environmental Issues*, presented to A Healthy Environment is a Human Right Conference, Montreal, 26–28 October, 1995.
65. Greenpeace Argentina, *Report of Incident*, 5 February 1993.
66. Greenpeace Argentina, Personal communication with author, 21 June 1994.
67. Amnesty International, *Urgent Action, María Elena Foronda, Environmental Activist and Oscar Díaz Barboza, Environmental Activist*, UA 352/94, 23 September 1994.
68. Greenpeace International, *Greenpeace International Condemns Death Threats to Uruguayan Environmentalist*, Amsterdam, 9 December 1994.
69. Caquetá Rainforest Amazonia Campaign, *Dead Green*, 18 November 1995.
70. *Index on Censorship*, 'Nuclear debators fired in Mexico', July/August 1989, Nos 6 and 7, p.53.
71. Global Response, *Native Rights and Forest Protection: Sierra Fired: Mexico*, September, 1994.
72. Greenpeace Latin America, personal communication with author, 2 June 1995.
73. Ibid.

74. AECO, *Press Release*, San Jos, Costa Rica, 7 December 1994.
75. Greenpeace Latin America, personal communication with author, 2 June 1995.
76. Ibid.
77. Ibid.
78. Ibid.
79. Ibid.
80. Ibid.
81. *IPS*, 'Honduras: ecologists demand murder investigation', 10 February 1995; J. Gollin, Letter to Malcolm Campbell, Global Response, 1995, no date: J. Gollin, *Trouble in paradise: the assassination of Jeannette Kawas*, 1995, no date.
82. *IPS*, 'Honduras: ecologists demand murder investigation', 10 February 1995; J.Gollin, Letter to Malcolm Campbell, Global Response, 1995, no date: J. Gollin, *Trouble in paradise: the assassination of Jeannette Kawas*, 1995, no date.
83. *IPS*, 'Award applauds "Green" activists', San Francisco, 22 April 1996.

9 DIRTY TRICKS DOWN UNDER

1. M. King, *Death of the Rainbow Warrior*, Penguin, 1986, pp.1–48.
2. Ibid., p.48.
3. Ibid., pp.193–4.
4. Ibid., pp.119–88
5. For further details see M. King, *Death of the Rainbow Warrior*, Penguin, 1986.
6. M. King, *Death of the Rainbow Warrior*, Penguin, 1986, pp.189–228.
7. Ibid., p.189.
8. Ibid., p.202.
9. A. Duval Smith, 'Paris planned virus attack on activists', *The Guardian*, 12 September 1995; *Reuter*, 'France mulled using virus on Greenpeace – Report', 11 September 1995.
10. R. Deacon, *The French Secret Service*, Grafton, 1990, p.315.
11. Ibid., pp.315–16; M. King, *Death of the Rainbow Warrior*, Penguin, 1986, pp.217–18.
12. M. Thurston, 'Ten years on, Rainbow Warrior agent speaks out', *Agence France Presse*, 12 May 1995.
13. *Reuter*, 'France's Rainbow Warrior saboteurs retired', 13 May 1995; M. Thurston, 'Ten years on, Rainbow Warrior agent speaks out', *Agence France Presse*, 12 May 1995.
14. M. King, *Death of the Rainbow Warrior*, Penguin, 1986, pp.120–3.
15. P. Chapman, 'Anger grows over French attack on Rainbow Warrior', *The Daily Telegraph*, 11 July 1995.
16. J. Gray, 'France "Over the top" in storming ship, NZ says', *Reuter*, 10 July 1995; P. J. Spielmann, 'Rainbow Warrior', *Associated Press*, Sydney, Australia, 10 July 1995; *UPI*, 'Australian FM condemns France', Jerusalem, 10 July 1995.
17. R. Meares, 'Greenpeace man's death changed little – daughter', *Reuter*, 10 July 1995.

18. B. Burton, interview with author, 19 June 1995.
19. Ibid.
20. Ibid.
21. *Arborvitae*, 'Australia goes for 15 per cent', The IUCN/WWF Forest Conservation Newsletter, September 1995, p.2.
22. I. Penna and B. Hare, 'Forest industry's advertising campaign – links to government and unions', *Australian Conservation Foundation Newsletter*, April 1987, pp.4–5.
23. Ibid.
24. *The Advocate*, 'Government pulls out of campaign on forestry', 3 February 1987; K. Nylander, 'Minister orders public service to quit forestry campaign', *The Examiner*, 3 February 1987.
25. *The Mercury*, 'Forestry now has one voice', 16 March 1987; *The Examiner*, 'Forestry interests to have their say', 13 March 1987.
26. National Association of Forest Industries Ltd, *Annual Report 1992–93*, 1993, p.5.
27. Dr R. Bain, *National Association of Forest Industries Invitation*, 16 May 1990, p.1.
28. B. Prismall, 'Forest war twist: industry takes view from conservation book', *The Mercury*, 23 November 1987, p.1.
29. Forest Protection Society, *Who Are We?*, brochure, 1993, p.1; *Forest Protection Society News*, Issue 1, Vol. 1, January 1987, p.1.
30. Forest Products Association, *Woodenbong Launches a New Branch of Forest Protection Society*, 27 February 1990; Forest Protection Society Ltd, *Who Are We?*, brochure, undated.
31. Information Australia, *Directory of Australian Associations*, 1994.
32. *Forest Protection Society News*, Issue 4, Vol. 7, July–August 1990, pp.24–5.
33. Network Fax, 20 August 1988, p.2.
34. *The Examiner*, 'Infiltration of greenies admitted', 7 February 1989.
35. *The Saturday Mercury*, Forest Protection Society State Coordinator, 17 April 1993, p.78; *The Forest Protection Society News*, June 1992.
36. Forest Protection Society Ltd, *Annual Report 1991–92*, 1992, p.8; Forest Protection Society Ltd, *Annual Report 1992–93*, 1993, p.10.
37. Forest Industries Association of Tasmania, *The Fight Against the Big Lie: Lines and Notes*, 1992, unpublished.
38. Forest Protection Society News Tasmania, *Annual Report 1992*, 1992, p.12; *FPS News*, Tasmania, April/May 1992, pp.5–6.
39. J. Snyder, title unknown, *People for the West!*, Western States Public Lands Coalition Public Lands Report, April 1992.
40. Notes of attendee, *PPF Membership Conference*, Glorieta, 31 May–2 June 1994; *The Mercury*, Forest Group Officer for US Conference, 26 May 1994.
41. *People for the West!*, International Agreement in Works, Western States Public Lands Coalition Public Lands Report, Vol. 7, No. 7, August, 1994, p.7.
42. *Australian Logging Council News*, Australian Forest, October 1993, p.4.
43. Greenpeace Australia, personal communication, 18 July 1995.
44. NAFI, 'Greenpeace co-founder to bring eco realist message to Australia', *Media Release*, 20 February 1996.
45. *New Zealand Herald*, 'Green campaign just cover says "Hit Man"', 19 March 1986.

46. D. Helvarg, *The War Against the Greens: The 'Wise-Use' Movement, the New Right, and Anti-Environmental Violence*, Sierra Club Books, San Francisco, 1994, pp.406–8.
47. B. Burton, personal communication with author, 22 June 1994.
48. B. Burton, 'Right wing think tanks go environmental', *Chain Reaction*, No.73–4, May 1995, pp26–9.
49. B. Burton, personal communication with author, 22 June 1994.
50. *IPA Review*, 1992, Vol. 45, No. 1; B. Burton, personal communication with author, 22 June 1994.
51. IPA, *The Environment in Perspective*, 1991.
52. R. Lindzen, *Global Warming: The Origin and Nature of the Alleged Scientific Consensus*, Environmental Backgrounder, IPA, 10–18 June 1992.
53. K. Hamilton, interview with Josselien Janssens, 24 April 1995.
54. J. Sinclair, 'Outlook: Changeable', *Listener*, 25 February 1995, p.19.
55. P. Shannon, 'Business assaults greenhouse', *Green Left Weekly*, 4 July 1994.
56. C. W. Baird, 'What garbage crisis? A market approach to solid waste management', *Policy*, CIS, Autumn 1992, p.23.
57. J. Byth, 'Green hysteria: scientists and the media join the stampede', *IPA Review*, Winter 1990, Vol. 43, No. 4; R. Brunton, *Environmentalism and Sorcery*, IPA, 31 January 1992; IPA, *The Environment in Perspective*, 1991.
58. D. Greason, 'Lyndon Larouche: a bad investment', *Australia/Israel Review*, 9–22 May 1994.
59. D. Greason, 'The LaRouchites: desperate and dateless?', *Australia/Israel Review*, 10–23 August 1993; S. MacLean, 'Seeds of unrest', *The Age*, 23 March 1991.
60. R. West, 'Peeling back the rhetoric of LaRouche's simple solutions', *The Age*, 29 May 1993.
61. Public Land Users Alliance, *The Truth about Wilderness*, 1994.
62. NSWPLUA, *Newsletter*, February 1996.
63. Forest Protection Society, '*Raglan Range Road: Our Cultural Heritage*, 1993.
64. B. Burton, 'Mining in National Parks', *The Examiner*, 31 July 1993; B. Burton and S. Cubit, *Ric Patterson ABC Radio 7ZR*, January 1993; *The Advocate*, 'Cradle turnaround: three areas may be reclassified for recreation', 12 April 1993, pp.1–2.
65. M. Stevenson, '"Dangerous fanatics" blasted at land rally', *The Examiner*, 17 January 1993.
66. *Northern District Times*, 'Pro-M2 group hits out at "Greenies"', 10 August 1994.
67. The Australian Federation for the Welfare of Animals, membership form, no date.
68. The Alliance for Beverage Cartons and the Environment, *Dioxins: 'No Need to Worry' Say Representatives of the Beverage Carton Industry*, 7 August 1990; D. Vincent, *Wrapped in PR*, Friends of the Earth, unpublished.
69. G. Van Rijswijk, *Letter to Queensland Conservation Council*, Association of Liquid Paper Board Carton Manufacturers, 27 October 1993; Association of Liquid Paper Board Carton Manufacturers, *Letter to Queensland Conservation Council*, 22 October 1993.
70. Mothers Opposing Pollution, *Mothers Environmental Group Seeks New Members*, undated.
71. *Sunshine Coast Daily*, 'Mums seek switch to milk cartons', 22 May 1993.
72. *Northwest News*, 'Mums in tree project', 29 September 1993.

73. *City Farm Association News*, 'Bogus green group warning', November 1993, p.3; I. Khastani, 'Who is Alana Maloney?', *City Farm Association News*, February 1994, p.1.
74. I. Khastani, 'Who is Alana Maloney?', *City Farm Association News*, February 1994, p.1.
75. Ibid.
76. *Food Week*, 'Move against plastic bottles is bogus', 26 October 1993, p.7.
77. B. Williams, 'Question over business links: greenie in carton war', *The Courier Mail*, 10 February 1995, p.1; B. Williams, 'Milk industry slams cancer scare tactics', *The Courier Mail*, 11 February 1995, p.8.
78. Mothers Opposing Pollution, *Letter to South Australia MPs*, 7 August 1995.
79. *The Mercury*, 'Publish despite "Threat"', 22 August 1972; B. Balfe, '"David Goliath" HEC case not likely', *The Mercury*, 22 August 1972.
80. B. Burton, personal communication with author, 26 June 1994.
81. Dobson, Mitchell and Allport, Letter to the Director of the Wilderness Society, 14 January 1993.
82. Total Environment Centre, *Helensburgh a Turning Point for Democracy in Planning*, 10 May 1994; *Illawarra Mercury*, 20 September 1993, p.2.; Lady Carrington Estates Pty Ltd vs. James Donohoe, Jennifer Donohoe, Timothy Tapsell, *Statement of Claim*, No. 18215, 1993.
83. A. De Blas, *The Environmental Effects of Mt Lyell Operations on Macquarie Harbour and Strahan*, Australian Centre for Independent Journalism, May 1994, p.119, citing K. Faulkner, General Manager, Mt Lyell Mining to Professor Jamie Kirkpatrick, 5 February; W. Bacon, Preface in A. De Blas, *The Environmental Effects of Mt Lyell Operations on Macquarie Harbour and Strahan*, Australian Centre for Independent Journalism, May 1994, p.ii; B. Montgomery, 'Thesis claims defame: mining company threatens legal action over pollution findings', *The Australian*, 1 June 1994; M. Fyfe, 'Acid water fears loom', *The Sunday Tasmanian*, 29 May 1994, pp.1, 6, 7; M. Fyfe, 'Concern at Uni stand on Mt Lyell thesis', *The Mercury*, 27 May 1993, p.6; 'Demand for full study on harbour fish woes', *The Mercury*, 30 May 1994; *The Sunday Tasmanian*, 'Making harbour fit state image', 29 May 1994.
84. Clean Seas Coalition, *Press Release*, 1 April 1993; *The Northern Star*, 'Sewage will still flow into sea, group says', 2 April 1993; *The Northern Star*, 'No sewage in outfall', 10 April 1993; The Council of the Shire of Ballina vs. W. Ringland, notice of motion in the Supreme Court of New South Wales, *Defamation List No. 11565*, 1993; P. Totaro and K. Gosman, 'Councils lose defamation right', *The Sydney Morning Herald*, 28 May 1994; P. Totaro, 'Court ruling hailed as victory for democracy', *The Sydney Morning Herald*, 28 May 1994; *The Northern Star*, 'Appeal court's judgement "Victory for Democracy"', 27 May 1994; M. Russel, 'Ballina weight critics rights', *The Sydney Morning Herald*, 30 May 1994.
85. Byron Environment Centre, *Club Med Threatens Legal Action*, press release, 11 March 1993; Byron Environment Centre, *Club Med – It's Not too Late*, 1993, p.3.
86. E. Rush, 'Protests halt bridge work', *The Adelaide Advertiser*, 29 October 1993; C. James, 'Bridge protesters to be sued', *The Advertiser*, 20 April 1994; C. James, *The Advertiser*, 27 April 1994.

87. B. Burton, *Bombs and Bloody Noses: Dirty Tricks and Violent Harassment*, paper given to Defending the Environment Conference, Adelaide, 21 May 1995.
88. B. Burton, personal communication with author, 3 June 1994.
89. J. McManus, 'Coromandel gold miners bite watchdog', *The (NZ) Independent*, 15 July 1994.
90. B. Burton, *Bombs and Bloody Noses: Dirty Tricks and Violent Harassment*, paper given to Defending the Environment Conference, Adelaide, 21 May 1995.
91. B. Burton, interview with author, 19 June 1995.
92. Ibid.
93. B. Burton, *The Corporate Counter Attack on the Environmental Movement*, presentation to the Ecopolitics Conference, Lincoln University, New Zealand, 8–10 July 1994.
94. B. Burton, personal communication with author, 26 June 1994.
95. P. Collenette, 'Police briefed on forest terror', *The Examiner*, 19 January 1993; B. Burton, personal communication with author, 26 June 1994.
96. R. Groom, *Media Release*, 11 March 1993.
97. S. Diwell, 'Explosives: brown points to pro-loggers', *The Mercury*, 13 March 1993.
98. Tasmania Police, *Progress Report Explosive Incident: Black River*, 11 March 1993; B. Burton, personal communication with author, 26 June 1994.
99. Evan Rolley Fan Club, *Large Scale Tree Spiking Campaign in Tasmania's Southern Forest was Announced Today*, press release, 21 March 1994.
100. B. Burton, interview with author, 19 June 1995.
101. B. Burton, 'Public Relations flunkies and eco-terrorism', *Chain Reaction*, December 1994, No. 72, pp.12–15.
102. H. Gilmore, 'Forests of blood', *Sunday Telegraph*, 1 January 1995, p.3.
103. B. Burton, *Bombs and Bloody Noses: Dirty Tricks and Violent Harassment*, paper given to Defending the Environment Conference, Adelaide, 21 May 1995, p.10.
104. B. Tobin, 'Oxy-gear used to cut pipes: police', *The Age*, 18 October 1991.
105. *Toxic Flash*, 'Sinister plots and overactive imaginations: sabotage at Coode Island', Hazardous Materials Action Group, Yarraville, 1992, p.1.
106. B. Tobin, 'Coode Fire was an accident police find', *The Age*, 11 June 1992; B. Burton, *Bombs and Bloody Noses: Dirty Tricks and Violent Harassment*, paper given to Defending the Environment Conference, Adelaide, 21 May 1995.
107. B. West, *Crisis Management Presentation to Members of the Australian Marketing Institute*, Marriot Hotel, 13 September 1995.
108. *National Business Review*, 'Greenpeace: a bunch of banana terrorists', 13 September 1991, p.8.
109. B. Burton, 'Public Relations flunkies and eco-terrorism', *Chain Reaction*, December 1994, No. 72, p.13.
110. Fake press release claiming to be from Northern Californian Earth First!, although it does not exist.
111. B. Burton, personal communication with author, 6 December 1995.
112. B. Burton, 'Public Relations flunkies and eco-terrorism', *Chain Reaction*, December 1994, No. 72, p.15.
113. W. Crawford, 'Sinister turn in lost plane drama', *The Mercury*, 13 September 1972;

The Mercury, 'Police probe about hangar', 14 September 1972; *The Mercury*, 'Plane search scaled down', 12 September 1972; *The Mercury*, 'Sabotage inquiry: Reece denies claim', 15 September 1972.

114. *The Examiner*, 'Conservationists begin 750km ride for rivers', 3 January 1981.
115. *The Advocate*, 'Meeting of peace at Tullah', 14 January 1981; *The Examiner*, 'No action against Tullah riot men', 17 January 1981; *The Examiner*, 'West Coast riot! SW dam critics ride into trouble at Tullah', 16 January 1981, p.1.
116. *The Mercury*, 'Brown bash at Strathan, says TWS', 14 January 1983; *The Mercury*, 'Four guilty of Brown assault', 15 January 1983, p.2; *The Advocate*, 'Brown bashed by youths', 14 January 1983.
117. *The Advocate*, 'TWS shop vandalised', 22 January 1983.
118. B. Burton, personal communication with author, 3 July 1994; *The Mercury*, '"No Dams" cars damaged', 21 June 1983.
119. *The Mercury*, 'Protestors claim maltreatment', 25 January 1983; *The Examiner*, 'Protestors suffered exposure', 29 January 1983.
120. *The Mercury*, 'Lemonthyme action fails to halt dozer', 6 March 1986; *The Examiner*, 'Protestors fail to stop work', 6 March 1986; L. Lester, 'Assault claims: greenies accuse police', *The Examiner*, 7 March 1986; *The Mercury*, 'Lemonthyme logging protestors allege violence: police "Turned backs on attacks"', 7 March 1986.
121. R. Kelley, 'Police stand back as workers attack', *The Mercury*, 8 March 1986, p.1; M. Binks, 'Brown blames Farmhouse Creek violence on govt', *The Advocate*, 8 March 1986.
122. R. Gray, media release, 8 March 1986, p.1.
123. *The Mercury*, 'Court told men petrified woman in Farmhouse Creek incident', 25 July 1986; *The Examiner*, 'Woman alleges assault during Farmhouse Creek protest', 25 July 1986, p.9; *The Advocate*, 'Farmhouse Creek appeal lodged', 23 September 1986; J Cox, *Reasons for Judgement Judith Ann Richter v Anthony Risby, Raymond Underwood, Kim Stanway and Michael Smith*, Serial No. 18/1987, List A, File no. LCA 107 and 108/1986, 8 April 1987; S. Diwell, 'Forest boss's action unlawful, says judge: no penalties for Risby and his three men', *The Mercury*, 9 April 1987;
124. B. Brown, media release, 9 April 1987; *The Examiner*, 'Government accused of mass forest injustice', 10 April 1987; *The Mercury*, 'Brown shot at', 10 March 1986, p.1.
125. *The Mercury*, 'Brown shot at', 10 March 1986, p.1.
126. *The Advocate*, 'Groom "Out of line in greenie competition"', 24 March 1987.
127. B. Burton, *Bombs and Bloody Noses: Dirty Tricks and Violent Harassment*, paper given to Defending the Environment Conference, Adelaide, 21 May 1995.
128. The Wilderness Society, *Blockade Continues Despite Intimidation*, media release, 22 February 1992; The Wilderness Society, *Details of Threats and Attacks Against People and Property in East Picton Forests*, letter to police, 24 February 1992; Tasmanian Police, *Burnt Out Cars Being Investigated*, media release, 24 February 1992; The Wilderness Society, *Police Slammed for Down-Playing Firebombing of East Picton Cars: Wrong Signals Being Sent to Violent Extremists in Community*, 24 February 1992; N. Clark and S. Diwell, 'Forest fear as cars torched', *The Mercury*, 25 February 1992, p.1; *The Examiner*, 'Protest dilemma in the forest', 28 February 1992.

127. *The Mercury*, 'Violence shocking, but not surprising', 26 February 1992.
129. The Western Tiers, *Police Report*, 21 October 1993, p.16; B. Burton, personal communication with author, 3 July 1994.
130. B. Burton, *Bombs and Bloody Noses: Dirty Tricks and Violent Harassment*, paper given to Defending the Environment Conference, Adelaide, 21 May 1995, p.4.
132. G. Lean, 'New Woodland Director destroyed virgin forest', *The Independent on Sunday*, 2 December 1995, p.5.
133. *The Advocate*, 'Groom "Out of line in greenie competition"', 24 March 1987.
134. B. Burton, *Bombs and Bloody Noses: Dirty Tricks and Violent Harassment*, paper given to Defending the Environment Conference, Adelaide, 21 May 1994, p.4.
135. M. Devine, *Telegraph Mirror*, 16 February 1995.
136. *The Advocate*, 'Death threats force move', 11 November 1992; B. Fuller, personal comment to Bob Burton, 15 June 1994; B. Burton, personal communication with author, 3 July 1994; *Manning River Times*, 'Sheed death threats', 8 February 1994; *Port Stephens Examiner*, 'They want to kill my dad', 8 February 1994; *The Canberra Times*, 'Protestors claim assault', 29 September 1993, p.2; *The Bombala Times*, 'Protestors say they were assaulted by loggers', 29 September 1993.
137. B. Burton, personal communication with author, 23 June 1994.
138. B. Burton, personal communication with author, 3 July 1994.
139. Ibid.
140. B. Burton, *Bombs and Bloody Noses: Dirty Tricks and Violent Harassment*, paper given to Defending the Environment Conference, Adelaide, 21 May 1995, p.6.
141. Ibid., pp.6–7.
142. B. Burton, interview with author, 19 June 1995.

10 SOUTH ASIA AND THE PACIFIC

1. R. Gillepsie, *Ecocide: Industrial Chemical Contamination and the Corporate Profit Imperative: The Case of Bougainville*, November 1994.
2. Amnesty International, *Malaysia: 'Operation Lallang': Detention Without Trial Under the Internal Security Act*, 20 December 1988, pp.1–4, 8.
3. Ibid., pp.15–23.
4. Ibid., p.27.
5. Friends of the Earth, *Malaysian Political Detentions: Protests Mount over Police Crackdown*, 9 November 1987.
6. Friends of the Earth, *Malaysia: Internal Security Act Detentions: Information Update*, 1 February 1988; Friends of the Earth, Survival International, International Union of Nature Conservation, Malaysia, *Sarawak Government Defends Destruction of Tribes and Rainforest International Mission Rebuffed*, 3 February 1988.
7. Friends of the Earth, *Information on Malaysian Detentions*, October–November 1987.

8. Friends of the Earth, *Malaysia: Internal Security Act Detentions – Information Update*, 1 February 1988; Friends of the Earth, Survival International, International Union of Nature Conservation, Malaysia, *Sarawak Government Defends Destruction of Tribes and Rainforest International Mission Rebuffed*, 3 February 1988.
9. B. Singh, *Unjust Detention of EPSM Vice-President*, Environmental Protection Society, Malaysia, 3 September 1987.
10. *Reuter*, 'Malaysia Bakun Dam to be finished ahead of time', Kuala Lumpur, 26 October 1994.
11. B. Singh, *Unjust Detention of EPSM Vice-President*, Environmental Protection Society, Malaysia, 3 September 1987; Internal Security Act, Name of Detainee: Tan Ka Kheng.
12. *BNA International Environment Daily*, 'Environmentalists call for review of Bakun hydroelectric project in Borneo', Bangkok, 17 July 1995.
13. *Reuter Textline Business Times* (Malaysia), 'Bakun project benefits far outweigh drawbacks', 20 July 1995.
14. *Reuter*, 'Sarawak tribes angry over lack of dam consultation', Kuala Lumpur, 22 August 1995; B. Tarrant, 'Malaysia Bakun Dam gets green light from key body', *Reuter*, 12 December 1995. B. Tarrant, 'Writs fly in Malaysia's Bakun Dam case', Reuters 26 July 1996.
15. Global Response, *Toxic Waste: Malaysia*, 1993, GR4; Consumers Association of Penang, *Wasted Lives: Radioactive Poisoning in Bukit Merah*, 1993.
16. *British Medical Journal*, Vol. 305, 29 August 1992, p.494; Consumers Association of Penang, *Wasted Lives: Radioactive Poisoning in Bukit Merah*, 1993; 'Philip', interview with author, 1994.
17. Global Response, *Action Status*, 3 February 1994; Global Response, *Toxic Waste: Malaysia*, 1993, GR4.
18. 'Philip', interview with author, 1994.
19. Ibid.
20. Ibid.
21. Ibid.
22. Ibid.
23. Ibid.
24. Human Rights Watch and Natural Resources Defense Council, *Defending the Earth: Abuses of Human Rights and the Environment*, 1992, p.46.
25. T. Selva, 'Taib: I only want to help the Penans', *The Star*, 16 September 1987; Human Rights Watch and Natural Resources Defense Council, *Defending the Earth: Abuses of Human Rights and the Environment*, 1992, p.51.
26. D. Dumanoski, 'Groups to campaign for release of Malaysian environmentalists', *Boston Sunday Globe*, 15 November 1987.
27. *New Strait Times*, 'All due to shifting cultivation: PM', 17 November 1987; Dr S. Chin, 'Shifting cultivation no threat to forests', *The Star*, 14 September 1989.
28. Human Rights Watch and Natural Resources Defense Council, *Defending the Earth: Abuses of Human Rights and the Environment*, 1992, p.59; J. Kendell, 'The children of the empty huts', *Index on Censorship*, July/August 1989, Vol. 18, No. 6 and 7, p.24.

29. Human Rights Watch and Natural Resources Defense Council, *Defending the Earth: Abuses of Human Rights and the Environment*, 1992, p.47.
30. *New Strait Times*, 'Environmentalists treating Penan like clowns, says Jabu', 18 March 1992, p.13.
31. *PR Week*, 'Growing strength of the Third World', date unknown.
32. C. Clover, 'Malaysia fury over Western calls for a timber boycott', *The Daily Telegraph*, 19 April 1988.
33. Ibid.
34. Ibid.; Friends of the Earth, Survival International, International Union of Nature Conservation, Malaysia, *Sarawak Government Defends Destruction of Tribes and Rainforest International Mission Rebuffed*, 3 February 1988.
35. 'Philip', interview with author, 1994.
36. Human Rights Watch and Natural Resources Defense Council, *Defending the Earth: Abuses of Human Rights and the Environment*, 1992, pp.47–8.
37. *Sabah Times*, 'UK office to fight anti-tropical timber lobby', 6 August 1993.
38. *PR Week*, 'Growing strength of the Third World', date unknown.
39. Global Response, *Rainforests and Indigenous People*, 12 February 1992; Global Response, *Action Status*, 14 March 1992; Survival International, *Crisis Deepens for Dayaks in Sarawak*, April 1992.
40. 'Philip', interview with author, 1994.
41. *Index on Censorship*, Malaysia, 1993, Issues 8 and 9, p.37.
42. *Index on Censorship*, Malaysia, 1991, Issue 10, p.55.
43. Human Rights Watch and Natural Resources Defense Council, *Defending the Earth: Abuses of Human Rights and the Environment*, 1992, pp.62–3.
44. Ibid., p.59.
45. Global Response, *Rainforests and Indigenous Peoples/Malaysia*, 1991, GR7.
46. Malaysian Timber Industry Development Council, 'Malaysia evergreen managing for perpetuity', *International Herald Tribune*, 22 March 1994.
47. S. Roberts, 'Account wins boost B-M to the tune of £300,000', *PR Week*, 7 October 1993.
48. Z. Tahir, 'Delegate to clarify situation', *Business Times*, 19 June 1995.
49. The Rainforest Information Centre, *Sarawak: The Struggle Continues*, 25 April 1995; Global Response, *Action Status*, 12 May 1993.
50. *AFP*, 'Five Malaysian states found to be overlogging', Kuala Lumpur, 3 October 1994.
51. The Rainforest Information Centre, *Sarawak: The Struggle Continues*, 25 April 1995.
52. *Setiakawan*, Campaign Against Western Environmentalists, No. 1, July–August 1989.
53. *New York Times*, advert 'Indonesia – Tropical Forests Forever', 18 August 1989.
54. *IPS*, 'Greens sue on reforestation', Jakarta, Indonesia, 24 August 1994.
55. N. Dudley, S. Stolton and J-P. Jeanrenaud, *Pulp Facts*, WWF International, in press.
56. *Index on Censorship*, Indonesia, 1989, Issue 8, p.38.
57. *TAPOL Bulletin*, 'Bob Hasan's ad ordered off the streets', No. 125, October 1994, p.23.
58. *Down to Earth*, 'Indigenous News', August 1994, No. 24, p.14; World Rainforest

Movement, Indonesian Government-Sponsored 'Development' and Logging Destroys Indigenous Peoples' Sustainable AgroForestry System, Bentian Case, Urgent Action, 31 March 1995.

59. *TAPOL Bulletin*, 'NGOs under threat from new decree', No. 125, October 1994, p.10.
60. G. Monbiot, 'Who will speak up for Irian Jaya?', *Index on Censorship*, July/August 1989, Vol. 18, No. 6 and 7, p.23.
61. Ibid.
62. *TAPOL Bulletin*, 'More killings in the Freeport drama', April 1995, No. 128, pp.22–3; J. Roberts, 'UK cash props up terror mine', *The Independent on Sunday*, 26 November 1995, p.13.
63. Australian Council for Overseas Aid, *Trouble at Freeport: Eyewitness Accounts of West Papuan Resistance to the Freeport-McMoRan Mine in Irian Jaya, Indonesia and Indonesian Military Repression: June 1994 – February 1995*, 1995, p.1.
64. Ibid., p. 3.
65. P. Jones, 'Mining protests met with massacre', *Peace News*, May 1995, p.7; Australian Council for Overseas Aid, *Trouble at Freeport: Eyewitness Accounts of West Papuan Resistance to the Freeport-McMoRan Mine in Irian Jaya, Indonesia and Indonesian Military Repression: June 1994 – February 1995*, 1995, p.1.
66. Ibid.
67. *Down to Earth*, 'Freeport in West Papua', 1995, No. 25, pp.2–3.
68. J. Roberts, 'UK cash props up terror mine', *The Independent on Sunday*, 26 November 1995, p.13.
69. R. Bryce, 'Struck by a golden spear', *The Guardian*, Society Section, 17 January 1996.
70. J. Roberts, 'UK cash props up terror mine', *The Independent on Sunday*, 26 November 1995, p.13.
71. Australian Council for Overseas Aid, *Trouble at Freeport: Eyewitness Accounts of West Papuan Resistance to the Freeport-McMoRan Mine in Irian Jaya, Indonesia and Indonesian Military Repression: June 1994 – February 1995*, 1995, pp.3–10.
72. Ibid., pp.3–10; *TAPOL Bulletin*, 'Freeport killings confirmed', No. 129, June 1995, pp.1–3.
73. J. Roberts, 'UK cash props up terror mine', *The Independent on Sunday*, 26 November 1995, p.13.
74. *TAPOL Bulletin*, 'Freeport killings confirmed', No. 129, June 1995, pp.1–3.
75. M. Miriori, *Statement To The Peoples' Right to Social Development International Conference*, Havana, Cuba, 18–20 November 1994.
76. R. Gillepsie, *Ecocide: Industrial Chemical Contamination and the Corporate Profit Imperative: The Case of Bougainville*, November 1994.
77. Ibid.; M. Miriori, *Statement To The Peoples' Right to Social Development International Conference*, Havana, Cuba, 18–20 November 1994.
78. R. Gillepsie, *Ecocide: Industrial Chemical Contamination and the Corporate Profit Imperative: The Case of Bougainville*, November 1994.
79. M. Miriori, *Bougainville: A Real Sad and Silent Human Tragedy in the South Pacific*, undated.
80. R. Gillepsie, *Ecocide: Industrial Chemical Contamination and the Corporate Profit Imperative:*

The Case of Bougainville, November 1994; M. Miriori, *Bougainville: A Real Sad and Silent Human Tragedy in the South Pacific*, undated; M. T. Havini, *A Compilation of Human Rights Abuses Against the People of Bougainville 1989–1995*, Bougainville Freedom Movement, April 1995.

81. M. Miriori, *Statement To The Peoples' Right to Social Development International Conference*, Havana, Cuba, 18–20 November 1994; M. T. Havini, *A Compilation of Human Rights Abuses Against the People of Bougainville 1989–1995*, Bougainville Freedom Movement, April 1995.
82. S. Heath, 'Destructive logging spreads in Solomons', *Greenleft*, 30 October 1994.
83. Greenpeace, *Greenpeace Calls on Solomon Island Government to Withdraw Army From Russell Islands*, Honiara, 18 April 1995; BBC, *Prime Minister Accuses NGOs of Stirring up Dispute Over Pavuvu Logging*, Summary of World Broadcasts, 23 May 1995; Source: Radio Australia External Service, Melbourne, 21 May 1995.
84. *AP Worldstream*, 'Solomon Islands blames activists for logging companies' bad image', 29 June 1995.
85. Greenpeace Solomon Islands, press release, 10 December 1995.
86. Ibid.
87. D. Callister, *Illegal Tropical Timber Trade: Asia Pacific*, Traffic Network, Cambridge, 1992.
88. Human Rights Watch and Natural Resources Defense Council, *Defending the Earth: Abuses of Human Rights and the Environment*, 1992, p.77.
89. Ibid., pp. 77–8.
90. M. Cohen, 'Dangerous pastures', *Index on Censorship*, July/August 1989, Vol. 18, No. 6 and 7, p.29.
91. Ibid.
92. Human Rights Watch and Natural Resources Defense Council, *Defending the Earth: Abuses of Human Rights and the Environment*, 1992, pp.77–80.
93. Ibid., p.81.
94. M. Cohen, 'Dangerous pastures', *Index on Censorship*, July/August 1989, Vol. 18, No. 6 and 7, p.29.
95. Human Rights Watch and Natural Resources Defense Council, *Defending the Earth: Abuses of Human Rights and the Environment*, 1992, pp.82–3; Global Response, *Action Status*, 30 March 1991.
96. Human Rights Watch and Natural Resources Defense Council, *Defending the Earth: Abuses of Human Rights and the Environment*, 1992, pp.84–5.
97. *Index on Censorship*, 'Philippines – attacks and threats on the press', 1990, No. 2, p.7; Human Rights Watch and Natural Resources Defense Council, *Defending the Earth: Abuses of Human Rights and the Environment*, 1992, p.86–7.
98. *Global Response*, Action Status, 26 December 1994.
99. M. Dodd, 'Political storm erupts over Cambodian logging deal', *Reuter*, 28 June 1994.
100. *AFP*, 'King calls for strict logging controls to avoid ecological catastrophe', 19 October 1994.
101. R. MacFarlane, *Citizens Pesticides Hoechst: The Story of Endosulfan and Triphenyltin*, Pesticide Action Network Asia and the Pacific, May 1994, p.iii.

102. Ibid.; *Global Pesticide Campaigner*, 1993, 'Hoechst sues Philippine journalists and activists', Vol. 3, No. 3, p.15.
103. *Pesticide Monitor*, 'Withdraw endosulfan and triphenlyin world-wide!', September 1994, Vol. 3, No. 3.
104. *Pesticide News*, 'Pesticide defense? Philippines groups question corporate strategies', June 1994, No. 24, p.11.
105. *Philippine Daily Inquirer*, 1 June 1994.
106. E. Hickey, 'International citizens' campaign targets Hoechst pesticides', *Global Pesticide Campaigner*, September 1994, p.14.
107. Ibid.
108. Ibid.
109. *Global Pesticide Campaigner*, 1993, 'Hoescht sues Philippine journalists and activists', Vol. 3, No. 3, p.15.
110. Ibid.
111. *Pesticide News*, 'Hoechst Philippines case thrown out', September 1994, No. 24, p.15.
112. R. MacFarlane, *Citizens Pesticides Hoechst: The Story of Endosulfan and Triphenyltin*, Pesticide Action Network Asia and the Pacific, May 1994, pp.30–1, 37.
113. M. Colchester, 'Unaccountable aid: secrecy in the World Bank', *Index on Censorship*, July/August 1989, Vol. 18, No. 6 and 7, p.11.
114. Ibid.
115. Amnesty International, *Indonesia: Four Shot Dead By Security Forces During Peaceful Demonstration in Madura, East Java*, 8 October 1993.
116. Ibid.
117. Amnesty International, *Amnesty International Report 1994*, 1994, p.287.
118. S. George and F. Sabelli, *Faith and Credit: The World Bank's Secular Empire*, Penguin, 1994, p.175.
119. *Lloyds List*, 'Damage to Dam, River Narmada, India', 6 January 1995.
120. P. Bidwai, 'India: Narmada Dam critics debate alternatives', *IPS*, 19 August 1994.
121. Ibid.; *Economic Times*, 'Sardar Sarovar project: exercising the other option', 8 October 1994.
122. Y. Brown, *Damming the Roads, Not the Rivers*, posted at gn:peacenewswri.news, 21 June 1994.
123. S. George and F. Sabelli, *Faith and Credit: The World Bank's Secular Empire*, Penguin, 1994, p.177.
124. Ibid., p.178.
125. P. McCully, 'World Bank president says SSP "Not to be ashamed of"', *Narmada Update*, International Rivers Network, 25 July 1994.
126. M. Colchester, *Salvaging Nature: Indigenous Peoples, Protected Areas and Biodiversity Conservation*, United Nations Research Institute for Social Development, World Rainforest Movement, and the World Wide Fund for Nature, September 1994, p.14, quoting World Bank, 'The relocation component in connection with the Sardar Sarovar (Narmada Project)', World Bank, 1982.
127. S. Dharmadhikary, General letter, Narmada Bachao Andolan, 20 November 1993.

128. Human Rights Watch and Natural Resources Defense Council, *Defending the Earth: Abuses of Human Rights and the Environment*, 1992, p.24.
129. Ibid., pp.26–33.
130. *The Ecologist*, 'Narmada Newsletter', May 1994; Human Rights Watch and Natural Resources Defense Council, *Defending the Earth: Abuses of Human Rights and the Environment*, 1992, pp.34,37; B. Desai, 'Cops rape tribal mother of eight', *Indian Express*, 14 April 1993; B. Desai, 'Andolan sticks to rape theory', *Indian Express*, 15 April 1993.
131. Survival International, *Narmada Dam: Key Activists Intimidated and Abused*, press release, 23 March 1994.
132. Y. Brown, *Damming the Roads, Not the Rivers*, posted at gn:peacenewswri.news, 21 June 1994.
133. *The Ecologist*, 'Narmada Newsletter', May 1994.
134. *Agence France Presse*, 'Indian authorities hospitalise fasting anti-dam crusader', 9 December 1994.
135. P. McCully, '"Constrained" review report released', *Narmada Update*, International Rivers Network, 14 December 1994; *United Press International*, 'India may review controversial dam', 17 December 1994; *IPS*, 'Anti-dam activists end hunger protest', 17 December 1994.
136. P. McCully, 'NBA ends fast', *Narmada Update*, International Rivers Network, 16 December 1994.
137. L. Udall, *Sardar Sarovar: Uncertain Future*, International Rivers Network, 15 May 1995.
138. Ibid.
139. V. Shiva, interview with author, 21 July 1995.
140. V. Shiva and H-B. Radha, 'The rise of the farmers' seed movement', *Third World Resurgence*, No. 39, November 1993, pp.24–7.
141. *IPS*, 'Greens unravel Dupont nylon plant', New Delhi, 31 January 1995; *AFP*, 'Anti-Du Pont protestors cremate slain comrade', Panaji, India, 25 January 1995; *Global Response*, Report, 1995, Issue 3.
142. *Reuter*, 'U.S. Du Pont says shifting India project site', New Delhi, 8 June 1995.
143. V. Shiva, interview with author, 21 July 1995.

11 'A SHELL-SHOCKED LAND'

1. *Index on Censorship*, 'Ogoni! Ogoni!', poem by Ken Saro-Wiwa, written in prison, 1994, No. 4/5, p.219.
2. K. Saro-Wiwa, interview with author and Andrea Goodall, 12 October 1992.
3. Shell Petroleum Development Company of Nigeria Limited, *The Ogoni Issue*, 25 January 1995.
4. Ibid., p.3.

5. P. Adams, 'Oil: the regime's Achilles' heel', *The Financial Times*, 11/12 November 1995.
6. Ogoni activist, interview with author, 1995.
7. C. Ake, interview with Catma Films, January 1994.
8. A. Rowell, *Shell-Shocked: The Environmental and Social Costs of Living with Shell in Nigeria*, Greenpeace International, July 1994, pp.14–15; R. Boele, *Ogoni: Report of the UNPO Mission to Investigate the Situation of the Ogoni of Nigeria, 17–26 February*, Unrepresented Nations and Peoples Organisation, 1 May 1995, p.8; Trocaire, *Struggling for Survival: the Ogoni People of Nigeria*, no date.
9. Hon O. Justice Inko-Tariah, Chief J. Ahaiakwo, B. Alamina, Chief G. Amadi, *Commission of Inquiry into the Causes and Circumstances of the Disturbances that Occurred at Umuechem in the Etche Local Government Area of Rivers State in the Federal Republic of Nigeria*, 1990, month unknown.
10. R. Boele, *Ogoni: Report of the UNPO Mission to Investigate the Situation of the Ogoni of Nigeria, 17–26 February*, Unrepresented Nations and Peoples Organisation, 1 May 1995, p.17.
11. Chief Dr H. Dappa-Biriye, Chief R. Briggs, Chief Dr B. Idoniboye-Obu, Professor D. Fubara, *The Endangered Environment of the Niger Delta: Constraints and Strategies*, an NGO memorandum of the Rivers Chiefs and Peoples Conference, for the World Conference of Indigenous Peoples on Environment and Development and the United Nations Conference on Environment and Development, Rio de Janeiro, 1992, pp.6, 11, 16; K. Saro-Wiwa, *Genocide in Nigeria: The Ogoni Tragedy*, Saros International, 1992, pp.11–13; A. Ilenre, 'The ethnic minority question', *Nigerian Tribune*, 1 November 1993, p.8.
12. D. Moffat and O. Lindén, 'Perception and reality: assessing priorities for sustainable development in the Niger River Delta', *Ambio*, December 1995, Vol. 24, No. 7–8, p.531.
13. Chief Dr H. Dappa-Biriye, Chief R. Briggs, Chief Dr B. Idoniboye-Obu, Professor D. Fubara, *The Endangered Environment of the Niger Delta: Constraints and Strategies*, an NGO memorandum of the Rivers Chiefs and Peoples Conference, for the World Conference of Indigenous Peoples on Environment and Development and the United Nations Conference on Environment and Development, Rio de Janeiro, 1992, pp.59–60.
14. C. Ake, interview with author, 1 December 1995.
15. A. Rowell, *Shell-Shocked: The Environmental and Social Costs of Living with Shell in Nigeria*, Greenpeace International, July 1994, pp.9–13.
16. J. Vidal, 'Born of oil, buried in oil', *The Guardian*, 4 January 1995.
17. D. Penman, 'Bellamy urged to drop Shell study', *The Independent*, 11 December 1995, p.2.
18. C. Ake, *Shelling Nigeria*, press statement, 15 January 1996.
19. D. Moffat and O. Lindén, 'Perception and reality: assessing priorities for sustainable development in the Niger River Delta', *Ambio*, December 1995, Vol. 24, No. 7–8, p.529.
20. Chief Dr H. Dappa-Birige, Chief R. Briggs, Chief Dr B. Idoniboye-Obu, Professor D. Fubara, *The Endangered Environment of the Niger Delta: Constraints and Strategies*, an

NGO memorandum of the River Chiefs and Peoples Conference, for the World Conference of Indigenous Peoples on Environment and Development and the United Nations Conference on Environment and Development, Rio de Janeiro, 1992, p.2.

21. Ogoni activist, interview with author, 1995.
22. Sister M. McCarron, interview with author, 16 November 1994.
23. C. Ake, interview with author, 1 December 1995.
24. D. Moffat and O. Lindén, 'Perception and reality: assessing priorities for sustainable development in the Niger River Delta', *Ambio*, December 1995, Vol. 24, No. 7–8, p.532.
25. Oil Spill Intelligence Report, *Custom Oil Spill Data: Shell's Ten Year Spill Record*, Cutter Information Corporation, 1992, Arlington; Cutter Information is an independent research company and bears no responsibility for the way in which its data is used.
26. Shell Petroleum Development Company of Nigeria, *Environmental Programme*, 5 April 1995.
27. A. Rowell, *Shell-Shocked: The Environmental and Social Costs of Living with Shell in Nigeria*, Greenpeace International, July 1994, pp.9–13.
28. D. Moffat and O. Lindén, 'Perception and reality: assessing priorities for sustainable development in the Niger River Delta', *Ambio*, December 1995, Vol. 24, No. 7–8, p.532; C. Ake, *Shelling Nigeria*, press statement, 15 January 1996; The Ecological Steering Group on the Oil Spill in Shetland, *The Environmental Impact of the Wreck of the Braer*, Scottish Office, Edinburgh, 1994, p.62.
29. Ogoni activist, interview with author, 1995.
30. O. Douglas, interview with author, 10 November 1994.
31. K. Saro-Wiwa, *Genocide in Nigeria: The Ogoni Tragedy*, Saros International, 1992, p.82.
32. Catma Films, *Delta Force*, shown on Channel Four, 4 May 1995.
33. P. Lewis, 'Blood and oil: a special report: after Nigeria represses, Shell defends its record', *The New York Times*, 13 February 1996.
34. P. Clothier and E. O'Connor, 'Pollution warnings "ignored by Shell"', *The Guardian*, 13 May 1996.
35. R. Boele, *Ogoni: Report of the UNPO Mission to Investigate the Situation of the Ogoni of Nigeria, 17–26 February*, Unrepresented Nations and Peoples Organisation, 1 May 1995, p.9.
36. C. Eke-Ejelam, 'Ordeal of oil communities', *Daily Sunray*, 17 March 1994.
37. L. Donegan and J. Vidal, 'Shell haunted by close ties to military regime', *The Guardian*, 13 November 1995.
38. D. Moffat and O. Lindén, 'Perception and reality: assessing priorities for sustainable development in the Niger River Delta', *Ambio*, December 1995, Vol. 24, No.7–8, p.536.
39. *African Concord*, 'Oloibiri: in limbo', 3 December 1990, p.29.
40. Sister M. McCarron, interview with author, 16 November 1994.
41. D. Abimboye, 'Massacre at dawn', *African Concord*, 3 December 1990, p.27.
42. Environmental Rights Action, *Shell in Iko: The Story of Double Standards*, 10 July 1995.

43. Ibid.
44. Catma Films, *The Heat of the Moment*, shown on Channel Four, 8 October 1992.
45. J. R. Udofia, Threat of Disruption of our Operations at Umuechem by Members of the Umuechem Community, Letter to Commissioner of Police, 29 October 1990.
46. Catma Films, *The Heat of the Moment*, shown on Channel Four, 8 October 1992.
47. Hon O. Justice Inko-Tariah, Chief J. Ahaiakwo, B. Alamina, Chief G. Amadi, *Commission of Inquiry into the Causes and Circumstances of the Disturbances that Occurred at Umuechem in the Etche Local Government Area of Rivers State in the Federal Republic of Nigeria*, 1990.
48. R. Tookey, Letter to Mrs Farmer concerning Shell's operations in Nigeria, 11 June 1993.
49. E. Bello *et al.*, 'On the war path', *African Concord*, August 1992, pp.20–21.
50. Ibid., p.18.
51. Ibid., p.19.
52. Human Rights Watch/Africa, *Nigeria: The Ogoni Crisis: A Case Study of Military Repression in Southeastern Nigeria*, July 1995, Vol. 7, No. 5, p.33; A. Rowell, *Shell-Shocked: The Environmental and Social Costs of Living with Shell in Nigeria*, Greenpeace International, July 1994, p.16.
53. A. Rowell, *Shell-Shocked: The Environmental and Social Costs of Living with Shell in Nigeria*, Greenpeace International, July 1994, p.16.
54. Human Rights Watch/Africa, *Nigeria: The Ogoni Crisis: A Case Study of Military Repression in Southeastern Nigeria*, July 1995, Vol. 7, No. 5, pp.36–7.
55. Ibid., pp.37–8.
56. Ibid., pp.34–5.
57. Ibid., pp.35–6, 33.
58. C. Bakwuye, 'Ogonis protest over oil revenue', *Daily Sunray*, 6 January 1993, pp.1, 20.
59. K. Saro-Wiwa, 'Message from prison', *The News*, 8 August 1994, p.10.
60. R. Boele, *Ogoni: Report of the UNPO Mission to Investigate the Situation of the Ogoni of Nigeria, 17–26 February*, Unrepresented Nations and Peoples Organisation, 1 May 1995, p.11.
61. Ibid., p.11.
62. Ibid., pp.9–10.
63. Sister M. McCarron, interview with author, 16 November 1994.
64. Amnesty International, *Nigeria: Military Government Clampdown on Opposition*, 11 November 1994, p.6.
65. Chief W. Nzidee, F. Yowika, N. Ndegwe, E. Kobani, O. Nalelo, Chief A. Ngei and O. Ngofa, Humble Petition of Complaint on Shell-BP Operations in Ogoni Division, letter to His Excellency the Military Governor, 25 April 1970.
66. A. Nedom and C. Kpakol, Damages Done to Our Life-Line by the Continued Presence of Shell-BP Company of Nigeria. Her Installations and Exploration of Crude Oil on Our Soil and Adequate Compensations There-of, letter to the Manager Shell-BP Company of Nigeria, 27 July 1970.
67. P. Badom, A Protest Presented to Representatives of the Shell-BP Dev. Co of Nig.

Ltd. by the Dere Youths Association. Against the Company's Lack of Interest in the Sufferings of Dere People which Sufferings are Caused as a Result of the Company's Operations, no date.

68. K. Saro-Wiwa, *Genocide in Nigeria: The Ogoni Tragedy*, Saros International, 1992, p.80.
69. L. Loolo, Letter to the Editor Overseas Newspapers Limited, 1970, exact date unknown.
70. Catma Films, *The Drilling Fields*, shown on Channel Four TV, 23 May 1994.
71. Amnesty International, *Possible Extrajudicial Execution/Legal Concern*, 19 May 1993; Unrepresented Nations and Peoples Organisation, *Developments in Ogoni, January–July 1993, Nigeria*, Office of the General Secretary, The Hague, 26 July 1993.
72. R. Boele, *Ogoni: Report of the UNPO Mission to Investigate the Situation of the Ogoni of Nigeria, 17–26 February*, Unrepresented Nations and Peoples Organisation, 1 May 1995, p.18.
73. *The Observer*, 'An unlikely warrior for justice', 11 July 1993.
74. Human Rights Watch/Africa, *Nigeria: The Ogoni Crisis: A Case Study of Military Repression in Southeastern Nigeria*, July 1995, Vol. 7, No. 5, p.10; Unrepresented Nations and Peoples Organisation, *Developments in Ogoni, January–July 1993, Nigeria*, Office of the General Secretary, The Hague, 26 July 1993.
75. S. Kiley, 'Nigeria accused of attempted oilfield genocide', *The Times*, 19 July 1993; UNPO, 1993, 'Saro-Wiwa falls unconscious after interrogation', 19 July 1993.
76. Greenpeace, *Drop Charges Against Nigerian Human Rights Activist, Says Greenpeace*, press release, 29 July 1993.
77. T. Owolabi, 'Genocide in Ogoniland', *Sunday Tribune*, October 1993, p.12.
78. K. Saro-Wiwa, *Report to Ogoni Leaders Meeting at Bori*, 3 October 1993.
79. Amnesty International, *Extrajudicial Executions, At Least 35 Members of the Ogoni Ethnic Group from the Town of Kaa in Rivers State, Including Mr Nwiku and Three Young Children*, Urgent Action, 10 August 1993; R. Boele, *Ogoni: Report of the UNPO Mission to Investigate the Situation of the Ogoni of Nigeria, 17–26 February*, Unrepresented Nations and Peoples Organisation, 1 May 1995, p.24.
80. Ogoni interviewed on *Delta Force*, Catma Films, shown on Channel Four, 4 May 1995.
81. C. Ake, interview with Catma Films, January 1994.
82. Human Rights Watch/Africa, *Nigeria: The Ogoni Crisis: A Case Study of Military Repression in Southeastern Nigeria*, July 1995, Vol. 7, No. 5, p.12.
83. Ibid.
84. Catma Films, *Delta Force*, shown on Channel Four, 4 May 1995; Catma Films, *The Drilling Fields*, shown on Channel Four TV, 23 May 1994.
85. Human Rights Watch/Africa, *Nigeria: The Ogoni Crisis: A Case Study of Military Repression in Southeastern Nigeria*, July 1995, Vol. 7, No. 5, pp.12–13.
86. Catma Films, *Delta Force*, shown on Channel Four, 4 May 1995; R. Boele, *Ogoni: Report of the UNPO Mission to Investigate the Situation of the Ogoni of Nigeria, 17–26 February*, Unrepresented Nations and Peoples Organisation, 1 May 1995, p.24.
87. Catma Films, *Delta Force*, shown on Channel Four, 4 May 1995.
88. R. Boele, *Ogoni: Report of the UNPO Mission to Investigate the Situation of the Ogoni of*

Nigeria, 17–26 February, Unrepresented Nations and Peoples Organisation, 1 May 1995, pp.26–7.

89. Human Rights Watch/Africa, *Nigeria: The Ogoni Crisis: A Case Study of Military Repression in Southeastern Nigeria*, July 1995, Vol. 7, No. 5, p.13.
90. Ogoni interviews in *Delta Force*, Catma Films, shown on Channel Four, 4 May 1995.
91. Catma Films, *Delta Force*, shown on Channel Four, 4 May 1995.
92. S. Olukoya, 'The Ogoni agony', *Newswatch*, 26 September 1994, p.26; M. Birnbaum, *Nigeria: Fundamental Rights Denied: Report of the Trial of Ken Saro-Wiwa and Others*, ARTICLE 19 in Association with the Bar Human Rights Committee of England and Wales and the Law Society of England and Wales, June 1995, Appendix 10: Summary of Affidavits alleging bribery; Ogoni activist, interview with author, 1995.
93. No author, *Crisis in Ogoniland. How Saro-Wiwa Turned Mosop into a Gestapo*, no publisher, handed out by the Nigerian Minister for Information, at a meeting at Oxford, 1 May 1995.
94. Ibid.
95. Ogoni activist, interview with author, 1995.
96. The Commissioner of Police, *Restoration of Law and Order in Ogoni Land*, Operation Order No. 4/94, 21 April 1994, p.1.
97. Sister M. McCarron, interview with author, 16 November 1994.
98. Major P. Okuntimo, *RSIS Operations: Law and Order in Ogoni, Etc*, memo from the Chairman of Rivers State Internal Security (RSIS) to His Excellency, the Military Administrator, Restricted, 12 May 1994.
99. Ibid.
100. Ogoni activist, interview with author, 1995.
101. R. Boele, *Ogoni: Report of the UNPO Mission to Investigate the Situation of the Ogoni of Nigeria, 17–26 February*, Unrepresented Nations and Peoples Organisation, 1 May 1995, p.28.
102. Catma Films, *Delta Force*, shown on Channel Four, 4 May 1995.
103. Amnesty International, *Nigeria: Military Government Clampdown on Opposition*, 11 November 1994, pp.6–7.
104. Human Rights Watch/Africa, *Nigeria: The Ogoni Crisis: A Case Study of Military Repression in Southeastern Nigeria*, July 1995, Vol. 7, No. 5, p.14.
105. Unrepresented Nations and Peoples Organisation, *Arrested Ogoni Leader Rejects Nigerian Government Accusations*, press release, 25 May 1994.
106. R. Boele, *Ogoni: Report of the UNPO Mission to Investigate the Situation of the Ogoni of Nigeria, 17–26 February*, Unrepresented Nations and Peoples Organisation, 1 May 1995, p.15.
107. M. Birnbaum, *Nigeria: Fundamental Rights Denied: Report of the Trial of Ken Saro-Wiwa and Others*, ARTICLE 19 in Association with the Bar Human Rights Committee of England and Wales and the Law Society of England and Wales, June 1995, p.27.
108. Amnesty International, *Ken Saro-Wiwa, Writer and President of the Movement for the Survival of the Ogoni People (MOSOP)*, Urgent Action, 24 May 1994.
109. Ogoni activist, interview with author, 1995.
110. O. Douglas, 'Ogoni: four days of brutality and torture', *Liberty*, May-August 1994,

p.22; J. Vidal, 'Born of oil, buried in oil', *The Guardian*, 4 January 1995; W. Soyinka, 'Nigeria's long steep, bloody slide', *The New York Times*, 22 August 1994.

111. Human Rights Watch/Africa, *Nigeria: The Ogoni Crisis: A Case Study of Military Repression in Southeastern Nigeria*, July 1995, Vol. 7, No. 5, p.16.
112. S. Olukoya, 'The Ogoni agony', *Newswatch*, 26 September 1994, p.23.
113. Catma Films, *Delta Force*, shown on Channel Four, 4 May 1995.
114. Amnesty International, *Nigeria: Security Forces Attack Ogoni Villages*, 29 June 1994; Human Rights Watch/Africa, *Nigeria: The Ogoni Crisis: A Case Study of Military Repression in Southeastern Nigeria*, July 1995, Vol. 7, No. 5, p.17.
115. Human Rights Watch/Africa, *Nigeria: The Ogoni Crisis: A Case Study of Military Repression in Southeastern Nigeria*, July 1995, Vol. 7, No. 5, p.23.
116. Catma Films, *Delta Force*, shown on Channel Four, 4 May 1995.
117. T. Ajayeoba, 'The killing field', *TELL*, 25 July 1994, p.30; G. Brooks, 'Slick alliance: Shell's Nigerian fields produce few benefits for region's villagers', *The Wall Street Journal*, 6 May 1994.
118. Human Rights Watch/Africa, *Nigeria: The Ogoni Crisis: A Case Study of Military Repression in Southeastern Nigeria*, July 1995, Vol. 7, No. 5, p.24.
119. S. Olukoya, 'The Ogoni agony', *Newswatch*, 26 September 1994, p.23.
120. U. Maduemesi, 'This is conquest', *TELL*, 18 July 1994; Amnesty International, *Nigeria: Military Clampdown on Opposition*, 11 November 1994, p.9.
121. *Amnesty International Report*, Nigeria – 1995, 1995; Human Rights Watch/Africa, *Nigeria: The Ogoni Crisis: A Case Study of Military Repression in Southeastern Nigeria*, July 1995, Vol. 7, No. 5, pp.18–22.
122. O. Douglas, 'Ogoni: four days of brutality and torture', *Liberty*, May-August 1994, p.22.
123. Catma Films, *Interview With Lt. Col. Komo*, 23 May 1994.
124. G. Brooks, 'Slick alliance: Shell's Nigerian fields produce few benefits for region's villagers', *The Wall Street Journal*, 6 May 1994.
125. G. Brooks, 'Questions annoy military in Nigeria and a reporter is questioned herself', *The Wall Street Journal*, 6 May 1994.
126. M. Adekeye, 'The Rwanda here', *The News*, 8 August 1994, p.11.
127. MOSOP, *Reaction of the Rivers State Military Government to the 1994 Right Livelihood Award Won By Saro-Wiwa and MOSOP*, press release, 10 November 1994.
128. The Right Livelihood Award Foundation, *1994 Right Livelihood Awards Stress Importance of Children, Spiritual Values and Indigenous Cultures*, press release, 12 October 1994.
129. The Goldman Environmental Prize, *The World's Largest Environmental Prize*, 17 April 1995; M. Birnbaum, *Nigeria: Fundamental Rights Denied: Report of the Trial of Ken Saro-Wiwa and Others*, ARTICLE 19 in Association with the Bar Human Rights Committee of England and Wales and the Law Society of England and Wales, June 1995, p.13.
130. M. Birnbaum, *Nigeria: Fundamental Rights Denied: Report of the Trial of Ken Saro-Wiwa and Others*, ARTICLE 19 in Association with the Bar Human Rights Committee of England and Wales and the Law Society of England and Wales, June 1995, pp.1–2.
131. Ibid., p.19.

132. Ibid., p.2.
133. Ibid., p.2.
134. Ibid., pp.8–9.
135. Ibid., pp. 33, 45, 57, 58; Human Rights Watch/Africa, *Nigeria: The Ogoni Crisis: A Case Study of Military Repression in Southeastern Nigeria*, July 1995, Vol. 7, No. 5, p.28.
136. M. Birnbaum, *Nigeria: Fundamental Rights Denied: Report of the Trial of Ken Saro-Wiwa and Others*, ARTICLE 19 in Association with the Bar Human Rights Committee of England and Wales and the Law Society of England and Wales, June 1995, p.9.
137. K. Saro-Wiwa, 'Statement from prison', *The Nation*, 10 May 1995.
138. S. Kiley, 'Nigeria accused of attempted oilfield genocide', *The Times*, 19 July, 1993.
139. M. Birnbaum, *Nigeria: Fundamental Rights Denied: Report of the Trial of Ken Saro-Wiwa and Others*, ARTICLE 19 in Association with the Bar Human Rights Committee of England and Wales and the Law Society of England and Wales, June 1995, p.36.
140. Ibid., p.36.
141. Ibid., Appendix 10: Summary of Affidavits Alleging Bribery.
142. Ibid.
143. B. Anderson, *Statement*, Managing Director of the Shell Petroleum Development Company of Nigeria, 8 November 1995.
144. K. Saro-Wiwa, *Closing Statement To The Military Appointed Tribunal*, October 1995.
145. C. Ake, interview with author, 1 December 1995.
146. N. Ashton-Jones, *Shell Oil In Nigeria*, August 1994.
147. SPDC, *Meeting at Central Offices on Community Relations and the Environment* (15/16 February in London, 18 February in the Hague), draft minutes, 1993.
148. Ibid.
149. C. Ake, interview with Catma Films, January 1994.
150. Shell Petroleum Development Company of Nigeria Limited, *The Ogoni Issue*, 25 January 1995, p.2.
151. C. McGreal, 'Spilt oil brews up a political storm', *The Guardian*, 11 August 1993, p.8.
152. Sister M. McCarron, interview with author, 16 November 1994.
153. Shell Petroleum Development Company of Nigeria Limited, *The Ogoni Issue*, 25 January 1995, p.6.
154. C. Ake, interview with author, 1 December 1995.
155. S. Braithwaite, interview with author, 16 September 1995.
156. Shell Petroleum Development Company of Nigeria, *The Environment: Nigeria Brief*, May 1995.
157. C. Ake, interview with author, 1 December 1995.
158. R. Tookey, Letter to Shelley Braithwaite, 9 December 1992.
159. Shell Petroleum Development Company of Nigeria Limited, *The Ogoni Issue*, 25 January 1995, p.7.
160. C. Ake, interview with Catma Films, January 1994.
161. C. Ake, interview with author, 1 December 1995.
162. Sister M. McCarron, interview with author, 16 November 1994.
163. Oil Spill Intelligence Report, *Custom Oil Spill Data: Shell's Ten Year Spill Record*, 1992, Cutter Information Corporation, Arlington.

164. R. Dowden, '"Green" Shell shares sold in protest at spills', *The Independent*, 24 October 1994.
165. Shell International Petroleum Company, *Shell Nigeria Launches Major Environmental Survey*, 2 February 1995.
166. *Reuter*, 'Shell says Nigerian eco-study not pressure-driven', 3 February 1995.
167. Minutes by a potential contractor of a meeting held with Shell Environmental Division, 1995.
168. Ibid.
169. Niger Delta Environmental Survey, *Niger Delta Environmental Survey Underway*, 24 May 1995.
170. Unrepresented National and Peoples Organisation, *Ogoni: Urgent Update*, relaying information received from MOSOP, 13 February 1995.
171. O. Douglas, interview with author, 10 November 1994.
172. C. Ake, Letter to Mr Onosonde, 15 November 1995.
173. C. Ake, interview with author, 1 December 1995.
174. Nigeria High Commission, *Record of the Meeting Held Between The High Commissioner Alhaji Abubakar Alhaji and Four Senior Officials of Shell International Petroleum Company Ltd (SIPC)*, Shell House, 16 March 1995.
175. C. Ake, 'War and terror', *The News*, 22 August 1994, p.9.
176. O. Isralson, 'Washington lawmakers: under the influence; US lobbyists have the world's interests at heart', *World Paper*, February 1990, p.10; B. Vora, 'Beyond hiring a lobbying form, the ethnic newswatch', *News India*, 4 March 1994, Vol. 24, No. 9, p.54.
177. C. McGreal, 'Nigeria accuses Ogonis of acting for foreigners', *The Guardian*, 7 February 1996, p.10.
178. C. Ake, interview with author, 1 December 1995.
179. Ogoni activist, interview with author, 1995.
180. Human Rights Watch/Africa, *Nigeria: The Ogoni Crisis: A Case Study of Military Repression in Southeastern Nigeria*, 1995, July, Vol. 7, No. 5, pp.38, 41.
181. Sister M. McCarron, interview with author, 16 November 1994.
182. N. Ashton-Jones, *Shell Oil In Nigeria*, August 1994.
183. Major P. Okuntimo, *RSIS Operations: Law and Order in Ogoni, Etc*, memo from the Chairman of Rivers State Internal Security (RSIS) to His Excellency, the Military Administrator, restricted, 12 May 1994.
184. O. Douglas, 'Ogoni: four days of brutality and torture', *Liberty*, May-August 1994, p.22.
185. O. Douglas, interview with author, 10 November 1994.
186. N. Ashton Jones, letter to author, 9 August 1995; N. Ashton Jones, *Detention Notes*, 2 July 1994.
187. K. Saro-Wiwa, 'Message from prison', *The News*, 8 August 1994, p.9.
188. O. Wiwa, *Testimony*, 1 December 1995.
189. *The Sunday Times*, 'Shell axes "Corrupt" Nigeria staff', 17 December 1995, p.26.
190. Human Rights Watch/Africa, *Nigeria: The Ogoni Crisis: A Case Study of Military Repression in Southeastern Nigeria*, July 1995, Vol. 7, No. 5, p.38.
191. R. Boele, *Ogoni: Report of the UNPO Mission to Investigate the Situation of the Ogoni of*

Nigeria, 17–26 February, Unrepresented Nations and Peoples Organisation, 1 May 1995, p.34.

192. SPDC, *Background Brief: SPDC Answers Allegations of Bribery*, 8 November 1995.
193. *ThisDay*, 'Still Shell-Shocked', Vol. 2, No. 293, 9 February 1996, p.5.
194. Human Rights Watch/Africa, *Nigeria: The Ogoni Crisis: A Case Study of Military Repression in Southeastern Nigeria*, July 1995, Vol. 7, No. 5, p.39.
195. C. Duodo, 'Shell admits importing guns for Nigerian police', *The Observer*, 28 January 1996; C. Ake, *Shelling Nigeria*, press statement, 15 January 1996; Sister M. McCarron, interview with author, 16 November 1994.
196. P. Ghazi and C. Duodu, 'How Shell tried to buy Berettas for Nigerians', *The Observer*, 11 February 1996.
197. C. Ake, interview with author, 1 December 1995.
198. Shell, *Shell and Nigeria*, May 1995.
199. C. Ake, interview with author, 1 December 1995.
200. P. Lewis, 'Blood and oil: a special report: after Nigeria represses, Shell defends its record', *The New York Times*, 13 February 1996.
201. *The Sunday Times*, 'Shell Axes "Corrupt" Nigeria staff', 17 December 1995, p.26.
202. N. Ashton-Jones, letter to author, 9 August 1995.
203. B. Anderson, *Statement*, Managing Director of the Shell Petroleum Development Company of Nigeria, 8 November 1995.
204. O. Wiwa, *Testimony*, 1 December 1995; P. Ghazi, 'Shell refused to help Saro-Wiwa unless protest called off', *The Observer*,19 November 1995, p.1.
205. J. Jukwey, 'Shell stays in Nigeria despite pressure on hangings', *Reuter*, 15 November 1995.
206. C. Ake, interview with author, 1 December 1995.
207. Shell International Limited, Shell Nigeria offers plan for Ogoni, News Release, 8 May 1996
208. C. Ake, interview with author, 1 December 1995.
209. B. Okri, 'Listen to my friend', *The Guardian*, 1 November 1995, p.19.

12 THE ROAD TO NOWHERE

1. J. Torrance, interview with author, 16 October 1994.
2. N. Schoon, 'Spate of sceptical books provokes green backlash', *The Independent*, 10 March 1995, p.12.
3. BBC Radio Four, *Special Assignment*, 22 April 1995.
4. R. North, *Life on a Modern Planet: A Manifesto for Progress*, Manchester University Press, 1995, p.ix.
5. H. Simonian, 'Economy v. ecology', *The Financial Times*, 15 February 1995, p.13.
6. Euro Chlor Federation, *Chlorine in Perspective*, September 1992.
7. *European Chemical News*, 'French defend $C1_2$ on economic grounds', 10 October 1994.

8. F. Engelbeen, Letter to *European Chemical News*, 28 September 1994.
9. *European Chemical News*, 'French defend Cl_2 on economic grounds', 10 October 1994.
10. B. Lyons, personal communication, 3 February 1994.
11. R. North, *Life on a Modern Planet: A Manifesto for Progress*, Manchester University Press, 1995, p.8.
12. *Kyodo News Service Japan Economic Newswire*, 'Sri Lankan NGOs protest "piracy" of natural resources', 22 August 1995.
13. *The Economist*, 'Flowing uphill – water: wars of the next century will be over water', 12 August 1995.
14. W. Beckerman, *Small is Stupid: Blowing the Whistle on the Greens*, Duckworth, 1995, pp.viii, 22, 50, 103.
15. Ibid., pp.125, 171.
16. G. Mulgan, 'A bitter pill for tougher greens', *The Independent*, 23 March 1995.
17. M. Ridley, *Down to Earth: A Contrarian View of Environmental Problems*, Institute of Economic Affairs, in association with the *Sunday Telegraph*, 1995, pp.25, 40.
18. Ibid., pp.19–28; C. Clover, 'So, the end isn't nigh', *The Daily Telegraph*, 14 March 1995.
19. M. Ridley, 'Down to earth: Denmark, the great hypocrite of greenery', *Sunday Telegraph*, 25 June 1995.
20. *The Daily Telegraph*, 'Green for danger', 21 June 1995, p.28.
21. B. Wynne and C. Waterton, *Why the ecobacklash is half-right – and wholly wrong*, The Centre for Study of Environmental Change, Lancaster University, 8 August 1995.
22. P. Routledge, 'Eggar hits out at Greenpeace "Terrorism"', *The Independent on Sunday*, 3 September 1995, p.2.
23. D. Lawson, 'Giants fight and the nation loses', *The Daily Telegraph*, 24 June 1995; H. Gurdon, 'Silence that cost Shell the battle', *The Daily Telegraph*, 24 June 1995.
24. J. Vidal, 'Eco-soundings', *The Guardian*, 9 August 1995, Society Section, p.4.
25. *The Daily Telegraph*, 'Green for danger', 21 June 1995, p.28.
26. D. Lawson, 'Giants fight and the nation loses', *The Daily Telegraph*, 24 June 1995; H. Gurdon, 'Silence that cost Shell the battle', *The Daily Telegraph*, 24 June 1995; C. Clover, 'How they see the world', *The Daily Telegraph*, 22 June 1995.
27. A. Culf, 'Greenpeace used us, TV editors say', *The Guardian*, 28 August 1995.
28. Greenpeace employee, personal communication, 1995.
29. D. Hughes and D. Norris, 'Red-faced greens admit: We got it wrong', *The Daily Mail*, 6 September 1995, p.11; W. Stewart, 'Dark side of Greenpeace do-gooders', *Daily Express*, 6 September 1995; *The Times*, 'Grow up, Greenpeace', 6 September 1995; *The Independent*, 'Better to blunder than to lie', 6 September 1995.
30. T. Eggar, Interview on *Radio Four One O'Clock News*, 5 September 1995.
31. Dr C. Fay, Interview on *Radio Four One O'Clock News*, 5 September 1995.
32. B. Wynne and C. Waterton, '*Why the ecobacklash is half-right – and wholly wrong*', The Centre for Study of Environmental Change, Lancaster University, 8 August 1995.
33. S. Baxter, 'The secret powers of the new persuaders', *The Sunday Times*, 18 September 1994, p.6.

34. M. Dowie, *Losing Ground: American Environmentalism at the Close of the Twentieth Century*, Massachusetts Institute of Technology, 1995, p.134.
35. J. Hunt, 'Attack on green policies', *The Financial Times*, 4 October 1989, p.10.
36. *The Times*, 'Environmental Protection Bill proposals criticized', 14 February 1990; J. Hunt, 'Call for free market pricing system on pollution', *The Financial Times*, 14 February 1990, p.11.
37. M. Ridley, *Down to Earth: A Contrarian View of Environmental Problems*, the Institute of Economic Affairs, 1995.
38. A. McHallam, *The New Authoritarians: Reflections on the Greens*, Institute for European Defence and Strategic Studies, 1991, Occasional Paper 51, pp.19, 45, 58.
39. J. Porrit, 'Eco-terrors and the illiberal tendency', *The Guardian*, 29 November 1991.
40. N. Nuttal, 'Of pixies, monsters . . . and environmentalists', *The Times*, 11 October 1993.
41. S. Watts, 'Environment "fads" attacked: right-wingers question value of cloth nappies and recycling', *The Independent*, 11 October 1993, p.4.
42. R. Bate and J. Morris, *Global Warming: Apocalypse or Hot Air?*, Institute of Economic Affairs, March 1994.
43. C. Clover, 'Carbon tax no use, say greenhouse sceptics', *The Daily Telegraph*, 17 March 1994, p.10.
44. *The Independent*, 'Will scepticism endanger greens?' 4 April 1994, p.13.
45. Institute of Economic Affairs, *Environmental Risk, Perception and Reality*, conference at St. Ermmin's Hotel, London, 20 October 1995.
46. J. Porrit, 'Eco-terrors and the illiberal tendency', *The Guardian*, 29 November 1991.
47. G. Lean, 'Hague assuages green anger over Redwood', *The Independent on Sunday*, 30 July 1995, p.2.
48. G. Lean, '"Batty" Redwood wants to privatise Snowdon', *The Independent on Sunday*, 22 January 1995, p.1.
49. *The Guardian*, 'Snowdonia privatisation plan angers environmentalists', 23 January 1995; G. Lean, '"Batty" Redwood wants to privatise Snowdon', *The Independent on Sunday*, 22 January 1995, p.1; C. Clover, 'Nature reserve "sell-off" plan splits cabinet', *The Daily Telegraph*, 23 January 1995; O. Tickell, 'Redwood in tooth and claw', *BBC Wildlife Magazine*, March 1995, p.62.
50. G. Lean, 'Hague assuages green anger over Redwood', *The Independent on Sunday*, 30 July 1995, p.2.
51. M. Fletcher, 'Redwood crusade wins cash from American Right', *The Times*, 14 September 1995, p.2; *London Evening Standard*, 'Moonie grabber', 6 September 1995.
52. S. Castle, 'Britain's New Right', *The Independent on Sunday*, 18 June 1995, p.19.
53. M. Hayes, *The New Right in Britain: An Introduction to Theory and Practice*, Pluto Press, 1994, p.12.
54. Ibid., p.48.
55. Ibid., p.48.
56. G. Murray, *Enemies of the State: A Sensational Exposé of the Security Services*, Pockett Books, 1993, p.86.
57. Ibid., p.104.

58. J. Cutler, 'Surveillance and the nuclear state', *Index on Censorship*, July/August 1989, Vol. 18, No. 6, 7.
59. G. Murray, *Enemies of the State: A Sensational Exposé of the Security Services*, Pockett Books, 1993, pp.147, 160, 186.
60. Ibid., pp.210, 214.
61. J. Cutler, 'Surveillance and the nuclear state', *Index on Censorship*, July/August 1989, Vol. 18, No. 6,7, p.44.
62. J. Torrance, interview with author, 16 October 1994.
63. B. Hanson, Letter to *New Civil Engineer*, 30 June 1994.
64. Department of Transport, *Roads to Prosperity*, Her Majesty's Stationery Office, 1989.
65. A. Rowell and M. Furgusson, *The National Road Traffic Forecasts: Fact, Fiction, or Fudge?*, 1989, unpublished briefing paper for WWF.
66. D. Nicholson-Lord, 'Tide of anger rises over roads that ruin ancient sites', *The Independent On Sunday*, 20 June 1993, p.7.
67. M. Tresidder, 'Maid Marian to the tribe in the treetops', *The Guardian,* 6 May 1995, p.27.
68. G. Stern and N. Tuckman, interview with author, 22 September 1995.
69. N. Tuckman, '*Notes of a meeting with Michael Howard*', 18 May 1993; G. Stern and N. Tuckman, interview with author, 22 September 1995.
70. D. Nicholson-Lord, 'Tide of anger rises over roads that ruin ancient sites', *The Independent On Sunday*, 20 June 1993, p.7.
71. J. Torrance, interview with author, 16 October 1994.
72. R. Lush, interview with author, 23 August 1995.
73. Ibid.
74. Ibid.
75. J. Vidal, 'The real earth movers', *The Guardian*, Society, 7 December 1995, pp.4, 5.
76. R. Lush, interview with author, 23 August 1995.
77. Ibid.
78. J. Torrance, interview with author, 16 October 1994.
79. J. Vidal, 'The fluffy and the bloody', *The Guardian*, 24 June 1994, Section 11, p.17.
80. J. Gallagher, 'Unguarded behaviour', *New Statesman and Society*, 31 March 1995, p.23.
81. *Press Association*, 'Government attacked on M-way protest court move', 2 December 1994.
82. R. Lush, interview with author, 23 August 1995.
83. Ibid.
84. *The Guardian*, 'Private detectives photographing M3 protestors', 11 November 1992, p.2.
85. Home Affairs Committee, *The Private Security Industry, Volume 11, Minutes of Evidence and Appendices*, Her Majesty's Stationery Office, 1995, Appendix 28, Memorandum by Liberty, p.195.
86. S. Fairlie, 'SLAPPs come to Britain', *The Ecologist*, September/October 1993, Vol. 23, No. 5, p.165.
87. P. Brown, 'Twyford M3 protestors win £50,000', *The Guardian*, 7 January 1995, p.6.

88. *Press Association*, 'Government attacked on M-way protest court move', 2 December 1994.
89. Ibid.
90. M. Swartz, interview with author, 15 August 1995.
91. Ibid.
92. P. Brown, 'Twyford M3 protestors win £50,000', *The Guardian*, 7 January 1995, p.6.
93. S. Fairlie, 'SLAPPs come to Britain', *The Ecologist*, September/October 1993, Vol. 23, No. 5, p.165.
94. J. Torrance, interview with author, 16 October 1994.
95. M. Swartz, interview with author, 15 August 1995.
96. Ibid.
97. P. Morotzo, interview with author, 25 October 1994.
98. R. Lush, interview with author, 23 August 1995.
99. 'Small world, unreasonable force', *Undercurrents*, no date.
100. M. Swartz, interview with author, 15 August 1995.
101. Ibid.
102. P. Morotzo, interview with author, 25 October 1994.
103. Ibid.
104. 'Small world, unreasonable force', *Undercurrents*, no date.
105. P. Morotzo, interview with author, 25 October 1994.
106. M. Swartz, interview with author, 15 August 1995.
107. R. Geffen, *The Forced Eviction of Leytonstonia*, No. M11 Link Campaign, 13 June 1994; J. Vidal, 'The fluffy and the bloody', *The Guardian*, 24 June 1994, Section 11, p.16 quoting statements from Oliver Nutkins and Zoe Chater.
108. P. Morotzo, interview with author, 25 October 1994.
109. P. Ghazi, 'Lollipop lady put under house arrest', *The Observer*, 30 January 1994, p.12.
110. *Do or Die*, News from the Autonomous Zones, 1994, Issue No. 4, p.23.
111. P. Morotzo, interview with author, 25 October 1994.
112. *The Guardian*, 'Two jailed for fire attack on M-way protestors', 6 September 1994; M. Whitfield, 'Gang was paid to force anti-road protestors from the tree house', *The Independent*, 1 September 1994; C. Wolmer, 'Two men paid to burn M11 protesters' camp are jailed', *The Independent*, 6 September 1994.
113. L. Jury, 'Road protestors allege rise in violence by security guards', *The Guardian*, 18 June 1994, p.5; J. Vidal, 'The fluffy and the bloody', *The Guardian*, 24 June 1994, Section 11, p.16.
114. J. Vidal, 'The fluffy and the bloody', *The Guardian*, 24 June 1994, Section 11, p.16.
115. P. Ghazi and R. Tredre, 'Anti-road demos hit by law on trespass', *The Observer*, 19 June 1994, p.3.
116. G. Monbiot, interview with author, 18 November 1994.
117. Ibid.
118. L. Jury, 'Road protestors allege rise in violence by security guards', *The Guardian*, 18 June 1994, p.5.
119. C. Moreton, '"Resistance culture" sets up news network, *The Independent on Sunday*, 5 February 1995.

120. J. Vidal, 'The fluffy and the bloody', *The Guardian*, 24 June 1994, Section 11, p.16.
121. Public Eye, *Out of Order*, 21 October 1994.
122. *Road Alert* . . . Stop Press, May 1995, p.7.
123. J. Gallagher, 'Unguarded behaviour', *New Statesman and Society*, 31 March 1995, p.23.
124. Public Eye, *Out of Order*, 21 October 1994.
125. Ibid.
126. 'Small world, unreasonable force', *Undercurrents*, no date.
127. H. Jones, 'M11 security guard escapes charges', *Red Pepper*, March 1995, p.5.
128. M. Swartz, interview with author, 15 August 1995.
129. R. Lush, interview with author, 23 August 1995.
130. D. Penman, 'Road rage', *The Independent*, 9 January 1996.
131. J. Vidal, 'The bypass of justice', *The Guardian*, 9 April 1996.
132. L. Johnson, interview with author, 28 February 1996.
133. J. Vidal, 'The parallel lines', *The Guardian*, Society, 7 February 1996, pp.4–5.
134. L. Johnson, interview with author, 28 February 1996.
135. L. Johnson, interview with author, 28 February 1996 and 25 April 1996.
136. J. Vidal, 'In the forest, in the dark', *The Guardian*, 25 January 1996, Section 2–3; *The Guardian*, Reliance Security Services Ltd, 23 February 1996.
137. J. Vidal, 'Eco-soundings', *The Guardian*, 28 February 1996; R. Lush, interview with author, 28 February 1996.
138. L. Johnson, interview with author, 28 February 1996.
139. J. Griffiths, 'Throwing down a mitt in the mud', *The Guardian*, 31 January 1996, Society, p.4; R. Lush, interview with author, 28 February 1996.
140. J. Vidal, 'Eco-soundings', *The Guardian*, 28 February 1996.
141. R. Lush, interview with author, 28 February 1996.
142. Ibid.
143. Home Affairs Committee, *The Private Security Industry, Volume I*, Her Majesty's Stationery Office, 1995, p.v.
144. Ibid., p.xiii.
145. Ibid., p.xv.
146. Home Affairs Committee, *The Private Security Industry, Volume II, Minutes of Evidence and Appendices*, Her Majesty's Stationery Office, 1995, Appendix 28, Memorandum by Liberty, p.191.
147. Ibid., p.195.
148. *The Guardian*, 'Protest greets the Criminal Justice Act', 4 November 1994.
149. Home Affairs Committee, *The Private Security Industry, Volume II, Minutes of Evidence and Appendices*, Her Majesty's Stationery Office, 1995, Appendix 28, Memorandum by Liberty, p.190.
150. Ibid., p.193.
151. L. Parrot, Speech to the Green Party Conference on the Criminal Justice Act, Friends Meeting House, Euston, 17 November 1994.
152. D. Taylor, Speech at Green Party Conference on the Criminal Justice Act, Friends Meeting House, Euston, 17 November 1994.

153. H. Mills and D. Penman, 'Bill ends its stormy passage into law', *The Independent*, 14 November 1994, p.8.
154. *The Guardian*, 'Protest greets the Criminal Justice Act', 4 November 1994.
155. M. Swartz, interview with author, 15 August 1995.
156. Green Party Conference on the Criminal Justice Act, Friends Meeting House, Euston, 17 November 1994.
157. R. Lush, interview with author, 23 August 1995.
158. *Construction News*, 'Highways agency to review its contract', 2 June 1994.
159. S. Purnell and T. Moore, 'Bill for security guards at M11 extension could top £26m', *The Daily Telegraph*, 1 May 1995, p.1.
160. R. Lush, interview with author, 23 August 1995.
161. *New Civil Engineer*, 'Growing doubts', 26 June 1994, p.12.
162. J. Harlow, 'Green guerrillas booby-trap sites', *The Sunday Times*, 3 July 1994.
163. Ibid.
164. *Construction News*, 'Barbaric protestors turn to trapping to halt roads', 30 June 1994.
165. M. Bunting, 'All creatures great and small', *The Guardian*, 6 July 1994.
166. Freedom Network *et al.*, Letter to the Editor of *The Guardian*, 1995, no date.
167. J. Harlow, 'Revealed: manual for eco-terrorists', *The Sunday Times*, 11 September 1995, p.5.
168. J. Torrance, interview with author, 16 October 1994.
169. L. O'Hara, *Turning Up the Heat: M15 After the Cold War*, Phoenix, 1994, p.88.
170. Ibid.
171. T. Magire and P. Gruner, '"Green Firebomb" group set to attack business', *The Evening Standard*, 20 July 1993, p.14.
172. L. O'Hara, 'At war with the greens', *Open Eye*, 1995, Issue 3, p.40.
173. Ibid.
174. R. Norton-Taylor, 'Police concern at new role for M15', *The Guardian*, 28 February 1995, p.11; S. Milne, 'Spies, lies sabotage', *Red Pepper*, pp.20–1.
175. J. Vidal and R. Norton-Taylor, 'Blurred lines of protest', *The Guardian*, 26 July 1995, Society, pp.4–5.
176. Ibid.
177. J. Vidal, 'On the bumpy trail of an unfathomable beast', *The Guardian*, 20 May 1995, p.25.
178. 'Small world, unreasonable force', *Undercurrents*, no date.
179. C. Elliott and D. Campbell, 'Police chiefs want anti-terror squad to spy on green activists', *The Guardian*, 27 March 1996, p.1.
180. R. Lush, interview with author, 23 August 1995.
181. N. Crossley, 'Watching us, watching you', *The Big Issue*, 24–30 April 1995, No. 127, p.14.
182. N. Dyle and H. Russell, 'Roadbuilders admit spying on protestors', *New Construction Engineer*, 26 May 1994.
183. R. Lush, interview with author, 23 August 1995.
184. 'Small world, unreasonable force', *Undercurrents*, no date.
185. *The Surveyor*, 'Cost of law and order on the road', 13 April 1995.

186. S. Purnell and T. Moore, 'Bill for security guards at M11 extension could top £26m', *The Daily Telegraph*, 1 May 1995, p.1.
187. M. Linton, 'Majority back ban on cars in city centres', *The Guardian*, 9 August 1995, p.1.
188. C. Clover and T. Moore, 'Is that enough roads?', *The Daily Telegraph*, 19 October 1994.
189. R. Smithers, 'Public transport "Key to Pollution Fight"', *The Guardian*, 27 October 1995, p.2.
190. S. Bosely, 'Bun fight', *The Guardian*, 11 July 1995, Section Two, p.5
191. J. Carey, 'Big Mac versus the little people', *The Guardian*, 15 April 1995, p.23.
192. 'Small world', *Undercurrents 2*, 1994.
193. Ibid.
194. M. Marqusee, 'The big beef bun fight', *The Guardian*, 17 January 1995, Section Two, pp.2/3.
195. McLibel Support Campaign, Weekly Bulletin – Weeks Nine and Ten, 10–21 October 1994; M. Marqusee, 'The big beef bun fight', *The Guardian*, 17 January 1995, Section Two, pp.2/3; D. Morris, *Interview on Undercurrents 2*, Small World, 1992.
196. McLibel Support Campaign, Weekly Bulletin – Weeks Nine and Ten, 10–21 October 1994; M. Marqusee, 'The big beef bun fight', *The Guardian*, 17 January 1995, Section Two, pp.2/3; A. Beale, 'Twelve months of McLibel lunacy', *Peace News*, July 1995, p.5.
197. S. Midgeley, 'Big Mac gets a mouthful of abuse', *The Independent*, 28 October 1994, p.1.
198. A. Beale, 'Twelve months of McLibel lunacy', *Peace News*, July 1995, p.5.
199. McLibel Support Campaign, *Legal Update*, August 1994.
200. J. Carey, 'Big Mac versus the little people', *The Guardian*, 15 April 1995, p.23; D. Morris, Interview on Undercurrents 2, Small World, 1992.
201. B. Appleyard, 'Big Mac vs small frys', *The Independent*, 4 July 1994, Section 11, p.1.

13 A FISHY TALE TO FINISH

1. M. Gudmundsson, press conference, organised by the High North Alliance, Puerto Vallarta, 25 May 1994.
2. *Sunnmoersposten*, 19 August 1986; quoting H. Asgrimsson, remarks made at a fishing conference in Akureyri, Iceland, 18 August 1986.
3. Arbitration Court, Verdict, 15 November 1993; the arbitration court was appointed by Swedish Chamber of Commerce.
4. The Icelandic Parliamentary, Althingistidindi, for 1988-89 Session show that Halldor Asgrimsson, the Minister of Fisheries confirmed that Gudmundsson had been paid 400,000 ISK; S. Aintila, Letter from the Nordic Council of Ministers to Janus

Hillgrad, Greenpeace Denmark, 24 May 1989; T. Arabo, Letter from the Project Leader of Vestnorden to S. Aintila, 19 May 1989.

5. Timinn, 3 March 1993.
6. Documentation from the National Press Club shows that *21st Century Science and Technology* paid for the room on 8 June 1989.
7. Greenpeace Norge V. Magnus Gudmundsson and Anor, 12 May 1992.
8. A. Heiberg, Letter to Duncan Currie, Goran Olenborg and Jeanne Herlofsson, 9 October 1992.
9. *Lofotposten*, 15 May 1992; Norwegian Ministry of Fisheries, press release, 21 February 1994.
10. Putting People First, Letter to Honourable Fortney Pete Stark, United States House of Representatives, 24 June 1993.
11. J. Veldheim, *Groenn Kontact*, 1993, Number 3.
12. *High North News*, 'US law defines Watson a pirate: US authorities obliged to intervene', 19 April 1993. The article was written by William Wewer.
13. K. Barthelmess, 'IWC and anti-whaling sentiments', *IWC47 Conservation Tribune*, 26 May 1995, No. 11, p.5.
14. NAMMCO, Opening Statement to the 47th Annual Meeting of the IWC, 29 May–2 June 1995.
15. *The International Harpoon*, Puerto Vallarta, 1994, No. 1.
16. Press release from the Norwegian Ministry of Fisheries, 21 February 1994; A. Johan Johansen, *Greenpeace Interview*, HNA's Information Secretary, Spring 1994.
17. *Eurofish Report*, 'Greenpeace accused of buying anti-whaling votes', No. 386, 13 August 1992.
18. F. Palacio, Greenpeace Article and Distortion of Statements Made to Reporter, Letter to *Forbes*, 5 November 1991.
19. M. Gudmundsson, press conference, organised by the High North Alliance, Puerto Vallarta, 25 May 1994.
20. Ibid.
21. A. Goodall, *Report on the Alliance for America's Fourth Annual Fly-in for Freedom*, Washington, 17–21 September.1994.
22. *Timinn*, 8 March 1990.
23. Notes from a conference participant, 22–24 March 1990; *Verdens Gang*, 'Fight together', 11 May 1994.
24. *Timinn*, 'A declaration of support for Icelandic whaling', 23 March 1991.
25. Conference proceedings.
26. *Folha de São Paulo*, 3 May 1994, pp.3–2; *O Estado de São Paulo*, 3 May 1994, p.A: 15.
27. G. Kennedy, 'Kiwi fishing industry tours Greenpeace critic', *The National Business Review*, 8 April 1993.
28. *Suisna Keizai Shinby*, 'Expansion from anti-whaling to anti-fishing: criticism of Greenpeace's activity', 21 February 1994; M. Gudmundsson, Speech to the New Zealand Fishing Industry Association, 1 April 1993; *2ZB Radio*, Auckland, NZ, 5 April 1993, Bill Ralston interviewing M. Gudmundsson; *The Independent* (New Zealand), 8 April 1994.

29. *2ZB Radio*, Auckland, New Zealand, 5 April 1993 – Bill Ralston interviewing Magnus Gudmundsson (text by Transcript Services Auckland); *Christchurch Press*, 26 April 1993; *The Independent*, (New Zealand), 8 April 1994.
30. M. Gudmundsson, paper presented at the Taiji Symposium on Utilisation of Living Marine Resources, March 1992.
31. M. Gudmundsson, lecture entitled Fisheries, Problems and Greenpeace actions, Greater Japan Fisheries Conference Hall, 15 February 1994.
32. *Fiskaren*, 'North American grass roots support whaling', 4 August 1993.
33. *Dagbladat*, 'Hval-stotte fra vapentorkjempere', 4 September 1993 and eye witness sitting at the conference in 1994.
34. *Verdens Gang*, 'Fight together', 11 May 1994.
35. K. Westrheim and F. Graff, 'Nordland's harpoon', *Dagbladat*, 9 July 1994.
36. M. Gudmundsson, lecture entitled Fisheries, Problems and Greenpeace actions, Greater Japan Fisheries Conference Hall, 15 February 1994.
37. *Verdens Gang*, 'Fight together', 11 May 1994.
38. *21st Century Science and Technology*, 'Greenpeace's financial misconduct and terrorism exposed in new documentary produced by Danish television', 15 November 1993.
39. T. Bode, statement made by the Executive Director of Greenpeace Germany, 20 July 1994.
40. P. Rasmussen, 'Greenpeace's financial misconduct exposed in Danish documentary', *21st Century Science and Technology*, Winter 1993, p.56.
41. R. Timm, 'How to manage looking bad', *Süddeutsche Zeitung*, 19 July 1994.
42. Application for pronouncement of interim injunction in the matter of F. Palacio Versus Norddeutscher Rundfunk, 5 July 1994.
43. Ruling of the case of Francisco J. Palacio PhD, 8730 S.W. 51 Street Miami, Florida, Versus Norddeutscher Rundfunk, Landgericht, Hamburg, 7 July 1994.
44. Application for pronouncement of interim injunction in the matter of David McTaggart Versus Norddeutscher Rundfunk, 6 July 1994; Ruling in the case of David McTaggart Versus Norddeutscher Rundfunk, 7 July 1994.
45. Ruling in case of Greenpeace Versus Norddeutscher Rundfunk, 12 July 1994.
46. A. Goodall, *Report on the Alliance for America's Fourth Annual Fly-in for Freedom*, Washington, 17–21 September.1994.
47. Ibid.
48. Ibid.
49. NAMMCO, Opening statement to the 47th Annual Meeting of the IWC, 29 May–2 June 1995.
50. *IWMC*, International Whaling Commission, 29 May – 2 June 1995.
51. IWMC, brochure, no date.
52. *O'Dwyer's Washington Report*, 'Parsons seeks nod for whale killing by "Coastal Peoples"', Vol. IV, No. 9, 25 April 1994; S. Warner, 'Know your friends: pro-outdoors lobbying organizations; directory', *Outdoor Life*, Vol. 189, No. 6, June 1992, p.60.
53. *IWMC*, International Whaling Commission, 29 May–2 June 1995, p.3.
54. *Conservation Tribune*, 'IWMC delegation experts', 9 November 1994, p.6.

55. M. Gudmundsson, Facts or fiction; human rights versus animals rights, press conference, 17 November 1994.
56. M. Gudmundsson, 'Inspiration for an environmental film-maker', *Conservation Tribune*, 16 November 1994, pp.4–5; G. Blichfeldt, 'Species survival network harpoons its own credibility', *Conservation Tribune*, 25 November 1994, p.1.
57. J. Speart, 'When Congress goes gunning: sportsmen take aim at endangered species; Congressional Sportsmen's Caucus efforts to weaken environmental laws and protection for endangered species', *National Audubon Society*, Vol. 96, No. 5, September 1994, p.16.
58. Ibid.
59. S. Boynton, 'US should clean up illegal policy on wildlife and marine resources, Washington Legal Foundation', *Legal Backgrounder*, 31 March 1995; R. Bellant, *The Coors Connection: How Coors Family Philanthropy Undermines Democratic Pluralism*, South End Press, Boston, 1991, p.11.
60. J. Speart, 'When Congress goes gunning: sportsmen take aim at endangered species; Congressional Sportsmen's Caucus efforts to weaken environmental laws and protection for endangered species', *National Audubon Society*, Vol. 96, No. 5, September 1994, p.16.
61. Ibid.
62. *Morgunbladid*, 'A US organisation on rational utilisation of natural resources want Iceland to rejoin the IWC', 20 May 1995.
63. *Morgunbladid*, 17 November 1995.
64. IWMC, *Conservation Tribune*, 11 November 1994.
65. Extract from *Cornerstone*, magazine on property rights, date unknown; Other members include R. J. Smith also from the Competitive Enterprise Institute; Wayne Hage; Myron Thomas Ebell; Margaret Ann Reigle; Mark Pollot; William Perry Pendley; Jane Shaw, Political Economy Research Center; Nancie Marzulla, Defenders of Property Rights; Jim Burling, Pacific Legal Foundation; Bruce Vincent, Community For A Great Northwest; Robert E. Gordon, National Wilderness Institute; James R. Streeter, National Wilderness Institute and David Howard, Land Rights Foundation.
66. *Nikkan Suisan Keizai*, 'Explicit stance of confrontation with extremist organisations', 29 September 1994.
67. *GCT*, 'Science as the bridge for sustainable use', Global Guardian Trust Newsletter, February 1994.
68. N. Akao, 'A double standard on sustainable development', *GCT*, Global Guardian Trust Newsletter, February 1994, No. 1.
69. *Nikkan Suisan Keizai*, 'Explicit stance of confrontation with extremist organisations', 29 September 1994.
70. Ibid.
71. Ibid.
72. Ibid.
73. Ibid.
74. A. Romero, Letter to J. Rodriguez, FBI, 27 February 1995.
75. Icelandic National Broadcasting Service, 14 May 1995.

76. Icelandic National Broadcasting Service, interview with M. Gudmundsson, 14 May 1995, news hour at 12:20.
77. L. Selvik, press release, Dublin, 29 May 1995; Global Guardian Trust, opening statement to the 47th Annual Meeting of the IWC, 29 May – 2 June 1995.
78. IWMC, 'Japanese Parliament member salutes wildlife managers for devotion to sustainable use of marine resources', *News Release*, 30 May 1995.
79. IWMC, Save The Whales: The Propaganda War, 31 May 1995.
80. Y. Kaneko, 'Lethal vs. non-lethal research', *IWC47 Conservation Tribune*, 2 June 1995, No. V, p.7; F. Herrera-Teran, 'Dolphins and tuna', *IWC47 Conservation Tribune*, 26 May 1995, No. 11, special edition; S. Boynton, 'U.S. should clean up illegal wildlife and marine resource policy', *IWC47 Conservation Tribune*, 26 May 1995, No. 11, p.4M; De Alessi, 'Some whales need saving more than others', *IWC47 Conservation Tribune*, 29 May 1995, No. 111.
81. D. Rephan, personal communication, 25 August 1995.
82. Ibid.
83. Joint declaration of non-governmental organizations interested in responsible aquatic resource utilization.
84. T. Platt and B. Vincent, letter to Bill Clinton, 8 February 1996.

14 CONCLUSION

1. B. Stall, 'Creating a national movement dedicated to the environment', *Los Angeles Times*, 16 April 1995, p.3.
2. R. D. Bullard, Introduction, in *Confronting Environmental Racism: Voices from the Grassroots*, ed. R. D. Bullard, South End Press, 1993, p.49.
3. P. Montague, 'A "Movement" in disarray', *Rachel's Environment & Health Weekly*, 19 January 1995, No. 425.
4. V. Shiva, interview with author, 21 July 1995.
5. Ibid.
6. M. Dowie, *Losing Ground: American Environmentalism at the Close of the Twentieth Century*, Massachusetts Institute of Technology, 1995, pp.262–3.
7. D. Smith, *In Search of Social Justice*, The New Economics Foundation, March 1995, p.2.
8. Ibid., p.3.

INDEX